Satellite Remote Sensing for Water Management

This textbook reflects the changing landscape of water management by combining the fields of satellite remote sensing and water management. Divided into three major sections, it begins by discussing the information that satellite remote sensing can provide about water, and then moves on to examine how it can address real-world management challenges, focusing on precipitation, surface water, irrigation management, reservoir monitoring and water temperature tracking. The final part analyses governance and social issues that have recently been given more attention as the world reckons with social justice and equity aspects of engineering solutions. This book uses case studies from around the globe to demonstrate how satellite remote sensing can improve traditional water practices and includes end-of-chapter exercises to facilitate student learning. It is intended for advanced undergraduate and graduate students in water resource management, and as a reference textbook for researchers and professionals.

Faisal Hossain is a John R. Kiely Endowed Professor in the Department of Civil and Environmental Engineering at the University of Washington, Seattle. He has been carrying out research and educational activities in the field of water management for over 20 years. His research focuses on remote sensing, water resources engineering, and water–food–energy security. He is the recipient of numerous awards, including a US Fulbright Faculty Award, the NASA New Investigator Award, the Charles Falkenberg Award and International Award from the American Geophysical Union, and the Walter Huber Award from the American Society of Civil Engineers. He is a Fellow of the American Meteorological Society and the Environmental and Water Resources Institute of the American Society of Civil Engineers. Hossain has published over 180 peer-reviewed journal articles. He is the author of the undergraduate textbook *Modern Concepts of Water Resources* (University Grants Commission of Bangladesh, 2007), and editor of many books, including *Resilience of Large Water Management Infrastructure: Solutions from Modern Atmospheric Science* (Springer, 2019); *Earth Science Satellite Applications: Current and Future Prospects* (Springer, 2016); and *Satellite Rainfall Applications for Surface Hydrology* (Springer, 2007). He has produced and directed several documentary and fiction films, such as *Bay of Hope* (2015) and *Cotton Fields from the Ivory Tower* (2017). Hossain is also the author of books for children on *The Secret Lives of Scientists, Engineers, and Doctors* (in two volumes), and *Robots and Other Amazing Gadgets Invented 800 Years Ago* (all with Mascot).

"Faisal Hossain's *Satellite Remote Sensing for Water Management* offers a clear and timely treatise on how to use remote sensing technologies to track freshwater resources around the world. It is an excellent resource for all students, teachers, and practitioners of remote sensing hydrology."

Laurence C. Smith, Professor of Earth and Planetary Science, Brown University, and author of *Rivers of Power: How a Natural Force Raised Kingdoms, Destroyed Civilizations, and Shapes Our World*

'*Satellite Remote Sensing for Water Management* is a comprehensive treatment of satellite applications to a variety of topics in hydrology, hydraulics, and irrigation which constitute the essential ingredients for planning, design, development, operation, and management of water resources projects. The real strength of the book lies in the description of real-world case studies and end-of-chapter exercises. This will be an excellent textbook for an undergraduate or a beginning graduate class in civil, agricultural, environmental, watershed, and Earth systems engineering. Faculty members teaching a course in remote sensing will find the book to be useful and informative."

Professor Vijay P. Singh, Texas A&M University

Satellite Remote Sensing for Water Management

Faisal Hossain

University of Washington

CAMBRIDGE
UNIVERSITY PRESS

Shaftesbury Road, Cambridge CB2 8EA, United Kingdom

One Liberty Plaza, 20th Floor, New York, NY 10006, USA

477 Williamstown Road, Port Melbourne, VIC 3207, Australia

314–321, 3rd Floor, Plot 3, Splendor Forum, Jasola District Centre, New Delhi – 110025, India

103 Penang Road, #05–06/07, Visioncrest Commercial, Singapore 238467

Cambridge University Press is part of Cambridge University Press & Assessment, a department of the University of Cambridge.

We share the University's mission to contribute to society through the pursuit of education, learning and research at the highest international levels of excellence.

www.cambridge.org
Information on this title: www.cambridge.org/highereducation/isbn/9781009453530

DOI: 10.1017/9781009453509

© Faisal Hossain 2026

This publication is in copyright. Subject to statutory exception and to the provisions of relevant collective licensing agreements, no reproduction of any part may take place without the written permission of Cambridge University Press & Assessment.

When citing this work, please include a reference to the DOI 10.1017/9781009453509

First published 2026

A catalogue record for this publication is available from the British Library

Library of Congress Cataloging-in-Publication Data
Names: Hossain, Faisal author
Title: Satellite remote sensing for water management / Faisal Hossain, University of Washington.
Description: Cambridge, United Kingdom ; New York, NY :
Cambridge University Press, 2026. | Includes bibliographical references and index.
Identifiers: LCCN 2025006313 | ISBN 9781009453493 hardback |
ISBN 9781009453530 paperback | ISBN 9781009453509 ebook
Subjects: LCSH: Remote-sensing images | Water-supply – Management
Classification: LCC G70.4 .H675 2026 | DDC 363.6/10284–dc23/eng/20250528
LC record available at https://lccn.loc.gov/2025006313

ISBN 978-1-009-45349-3 Hardback
ISBN 978-1-009-45353-0 Paperback

Cambridge University Press & Assessment has no responsibility for the persistence or accuracy of URLs for external or third-party internet websites referred to in this publication and does not guarantee that any content on such websites is, or will remain, accurate or appropriate.

For EU product safety concerns, contact us at Calle de José Abascal, 56, 1°, 28003 Madrid, Spain, or email eugpsr@cambridge.org

To the young generation of the world seeking freedom from water challenges through data and information technology.

To the youth of Bangladesh and their July 36, 2024.

To my parents, my sister Nini, Mizan bhai, my wife Sayma, my boys Adeeb and Nehan.

Contents

Preface

This book is the culmination of many years of teaching a course with a similar name. While teaching the course to a wide-ranging audience, I felt the need for a book written at the interface between "water management" and "satellite remote sensing." In this era of heightened awareness on social issues and how equitable current solutions are, I have found students from diverse backgrounds to be always keen to learn about water as a resource. My experience of interacting with a broad audience informed me that such a book should take into account the way the world of water management has been recently evolving, owing to the rapid changes in information technology and satellite remote sensing to solve current water problems. Therefore, this book is not a book on water management or on satellite remote sensing. I believe there are plenty of great books on these two topics. Another book on either of those topics rehashing the well-known theories is really not what the world needs now, in my opinion.

Today, climate change, water scarcity or excess, and competition for natural resources, many of which depend on water, are all around us. At the same time, we are now witnessing the fourth industrial revolution. There have been great strides made in technology for measuring the world's water using satellites in space. Data is also all around us. And there is certainly data on water and our surrounding environment in vast amounts, known as "Big Data" on water. The satellite remote sensing data on water continues to become more accessible with each passing year as information technology, comprising the internet, cloud computing, and distributed data storage facilities, becomes more affordable for more open and inclusive access. Such technological innovation has already spurred changes in many fields. Water management practice is gradually waking up to the possibilities afforded by information technology and satellite sensing to address newer challenges such as climate change, hydro-politics, and a decline in traditional measurements using ground infrastructure. For example, we can now track and share information on how much rainfall the world is experiencing at any given time in most places, and this data is now made freely available for the public good. The same can be said about how agricultural and irrigation systems are using water, or how dams are altering the flow of rivers downstream. Such data is not perfect, has uncertainty, and does not replace ground information. But we have come a long way over the past few decades to make remote sensing of water a practical tool rather than just an idea for research. The playing field for accessing water data continues to be leveled further each year for everyone to participate and contribute to a more robust solution-building process. As our awareness of social justice and equity aspects for water management continues to increase in the twenty-first century, we need to ask ourselves what this changing landscape of greater access to water data really means.

Currently, at the undergraduate level of education, we still lack the foundational and introductory book that sits at the interface between the two topics of satellite remote sensing and water management. Thus, those who are now going through a university education are not being exposed early on to the promise of information technology and how it can solve the emerging challenges of water through satellite remote sensing. While teaching my course titled "Satellite Remote Sensing for Water Management" and practicing the same around the world, I came to realize that if we are to train the next generation of water practitioners, managers, and scientists in the basics of publicly available satellite remote sensing data for innovating our water management practice, we first need an introductory textbook. Such a book should be designed for breadth and a wide audience, using real-world case studies and tutorials to help the next generation of students see what remote sensing data is and what it can do in water management. This book is the product of that realization.

This book has three major sections. The first section (Chapters 1 through 4) is on the what, where, why, and how. The second section (Chapters 5 through 10) is on specific domains of water, showing how real-world management challenges can be addressed with satellite remote sensing. This section covers precipitation, surface water, irrigation management, reservoir monitoring, and water temperature tracking. We provide real-world success stories in the form of case studies where the traditional water practice is improving with satellite remote sensing. The third section (Chapters 11 and 12) is on governance and social science issues that have recently shot to prominence as the world reckons with social justice and equity aspects of engineering solutions. Chapter 11 is on social justice in water management, where it challenges the reader to think of how remote-sensing-based water management solutions need to work for all. Chapter 12 is on citizen science in water management and shows the potential of an interconnected world where everyone can now be a participating scientist or water practitioner, using their phones to generate data. These chapters are intentionally short and lack the rigor and depth that a course on social justice or citizen science may demand. This is because these topics are not my active area of research, and I do not claim expertise. However, I recognize that readers need to start thinking about them in the context of increasing availability and access of satellite remote sensing for water management.

My hope is that the book will cater to a wide range of readers and not just those who are pursuing a STEM degree. I hope it will be appealing to those who are majoring in arts and various interdisciplinary studies. The book could also be used by those in mid-career, thinking of upgrading their skill and knowledge in this world of information technology where remote sensing and data systems will be in strong demand for jobs in the twenty-first century.

How to Use This Book

This book is intended for students from science and engineering backgrounds as well as from liberal arts and interdisciplinary studies. To allow for such a broad audience, the book is organized into 12 chapters. The entire contents of the book can be taught in a 14-week semester if there is an additional lab component to practice the tutorials and data-based exercises. In a 10-week quarter, Chapters 11 and 12 can be skipped. The book can be used in two different modes by the instructor, with an optional third mode. In the first mode, the entire book can be completed by those who want only a theoretical and text-based understanding rather than to build actual data literacy using cloud computing with real satellite data. This mode will be particularly appealing for those from backgrounds in liberal arts and interdisciplinary studies, or for mid-career professionals continuing their education. In this mode, students can skip the data tutorials and data exercises in Chapters 4 to 10 (except Chapter 7). The second mode is designed for those students looking for a bigger intellectual and data science challenge. This mode can be suitable for those majoring in science or engineering programs. In this mode, students can use the entire contents of each chapter, including the data-based tutorials and exercises that are provided in Chapters 4 to 10. In this mode, there is a Google Earth Engine (GEE) component and a traditional local computing option using Python codes with sample data downloaded from the publisher's cloud. These two options (cloud computing and local computing) are provided as not everyone may have the necessary cloud-computing literacy in GEE or have access to background training in one academic term. Lastly, the third (optional) mode is the nontechnical part of the book in Chapters 11 and 12 (social justice and citizen science), where there are intentionally no exercises. These two chapters can be easily skipped if students desire only a theoretical and text-based understanding. Alternately, the instructor and reader can just use these last two chapters for a short course, workshop, or webinar.

I will be the first to admit that we have embarked on an ambitious and near-impossible education agenda by trying to cater to a wide range of audience skill levels in a way that will remain relevant in the coming years. The book has been designed in a manner in which the components on tutorials/hands-on and case studies in each chapter could be used in a modular fashion without affecting the learning of the big picture and basic concepts. The tutorials and exercises in various computing and cloud environments have been tested by independent testers. However, the landscape of information technology is also changing rapidly. The software architecture, cloud paradigms, and programming languages that are in wide use today may become less relevant in a few years and require a future edition for updates. Wherever possible, I have therefore tried to set the tone by reminding readers of this "potential" change that is constant, and I have exhorted readers to focus more on the big picture concepts, basic principles, and logic that are less affected by information technology specifics. The same can be said of the case studies and success stories of satellite remote sensing for water management in this book. Many of these may be replaced with newer and better examples, or no longer be in effect for a variety of reasons that cannot yet be foreseen. While using this book, it is therefore important for students and instructors alike not to get too caught up in the specifics, syntaxes, or unique case studies of satellite remote sensing missions and the data software systems that were in use at the time of writing (2023–2024).

Acknowledgements

This book could not have been completed without the vast community of students, colleagues, research collaborators, and stakeholder water agencies who helped shape the contents for teaching water management using satellite remote sensing. First, I must acknowledge all my students since 2004 who collectively played a role in developing specific examples, case studies, and tutorials. In particular, I am grateful to Sanchit Minocha, Pritam Das, George Darkwah, Sarath Suresh, and Shahzaib Khan, who actively helped me develop exercises, tutorials, exercises based on real data, and case studies for this book. They did this on top of their graduate research responsibilities, and while their names are explicitly mentioned wherever their contributions appear, I do not think that is sufficient recognition of their hard work. I could not have made this book relevant to this generation of students interested in building data science skills without their contributions. I am also thankful to my friends and colleagues, Dr. Matthew Bonnema of NASA Jet Propulsion Laboratory and Dr. Michael Durand of the Ohio State University, for their specific contributions in a section on cloud computing and discharge estimation, respectively. My former students (Abebe, Abel, Wondie, Safat, Nishan, Shahryar, Hisham, Ling, Claire, Kensey, Xiaodong, Mehedi, Akbor, Caitlin), who pursued their research on improving the application of science and remote sensing for solving water problems, are deeply acknowledged. Together, they built the global network of problem cases and stakeholders that helped generate wisdom on how to make satellite remote sensing relevant for water management applications. I am also thankful to all my colleagues who have worked with me or taught me many things on remote sensing on water that I could have never known by reading a book. Special thanks to Tamlin, Hyongki, Ernesto, Soroosh, Brad, Dennis, Bart, Erkan, Jessica, Jeff, Ashutosh, Dan, Colin, Kostas, Gordon, Stefano, and many more whom I cannot fit in these pages. Special thanks to many programs and agencies that support our effort to build the collective wisdom on the use of satellite remote sensing for solving water problems of today, notably the John R. Kiely Endowment, NASA Applied Science program, National Science Foundation, USAID, Asian Development Bank, World Bank, Ivanhoe Foundation (thank you Cheryl for your generous endowment!), Bangladesh Water Development Board, Indian Institute of Technology-Kanpur and Bombay, NASA Jet Propulsion Laboratory, Pakistan Council of Research in Water Resources, Alexandria University, Asian Disaster Preparedness Center, Department of Agricultural Extension (Bangladesh), Columbia River Inter-tribal Fish Commission, and many more. The perspectives from the user world kept us humble about how to make science and technology work for real impact, which I hope will somehow be evident in this book.

I owe a lot to my two boys, Adeeb and Nehan, and my super-patient wife Sayma. My boys suddenly started taking an interest in my work when their science teachers were making them do hands-on experiments on water. And Sayma allowed me to spend long hours in the coffee shop "far from the madding crowd" at home to concentrate on writing the book. I want to thank my father, Delawer Hossain, who has always been by my side for such educational projects. He and I wrote our first textbook 20 years ago, titled *Modern Concepts of Water Resources*, which was a lot of fun. Now I hope he lives to see this book that owes so much to his mentoring and encouragement as I grew up.

Lastly, I want to thank Dr. Matt Lloyd, the editor of Cambridge University Press, who kept me sane throughout, and the entire community of staff (especially Emma, Maya, Lindsay and Charles). I'm especially thankful to Emma Collison for her guidance on navigating the complex landscape of permissions and copyrights. Thanks to Maria Tursi for her help with the indexing of the book.

1   Why Manage Water with Satellite Remote Sensing?

1.1   Chapter Overview

The goal of this first chapter is to introduce ourselves to the growing importance of using satellite remote sensing to manage our water. We will try to understand this in the context of the underlying challenges and new global forces developing this century that are expected to make traditional ways of managing water using in-situ data more challenging.

1.2   Water Management and the Global Forces

Water is precious, especially freshwater, which is often not available in the right amount and right quality where needed most. Any resource which we depend on a large scale for the health of the world population and the planet needs to be managed. Water management can be considered as the control and distribution of water resources to minimize damage to life and property and to maximize beneficial use. This is achieved through the process of planning, developing, and managing water resources. Some examples of water management are flood risk reduction with dams and levees, and efficient water use for food production with irrigation systems. Water has been managed for thousands of years, from ancient civilizations to modern-day nations. But there are global forces shaping up that have been challenging the way we have historically managed this water.

Current world population is a little over 8 billion in 2023, and a majority of the population lives in what we call "developing" economies. Add to this the fact that humanity has experienced a great march towards cities to pursue better livelihood opportunities. According to the United Nations, more than 70 percent of the world's population is expected to be living in highly urbanized and organized settlements by 2050 (UN, 2019; Liu et al., 2020; Figure 1.1). This percentage is higher in many countries, such as in the United States, where, based on the last census, more than 80 percent of the population now lives in cities. Through this march towards urbanization, towns have become cities, cities have become mega cities, and adjoining mega cities have merged to become megapolitan archipelagos (Lang et al., 2020). Such global trends mean that natural resources need to be "managed" in a concentrated manner, and responsibly, to meet the competing needs of humans and the natural environment they live in. At the core of these resources is water, which is needed for thriving cities, for human consumption (domestic, industrial), for growing food, producing energy, and ensuring healthy ecosystems and public health (Figure 1.2).

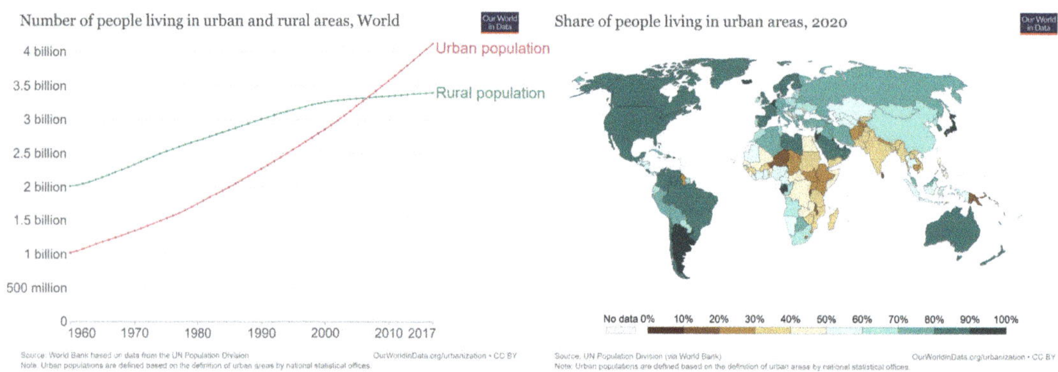

Figure 1.1 Left panel: rapid increase in people living in urban areas since 1960. Right panel: the geographic extent of urbanization for countries. Urban populations are based on the definition of urban areas by national statistical offices. [Source: UN Population Division (via World Bank); Our World in Data https://creativecommons.org/licenses/by/4.0/]

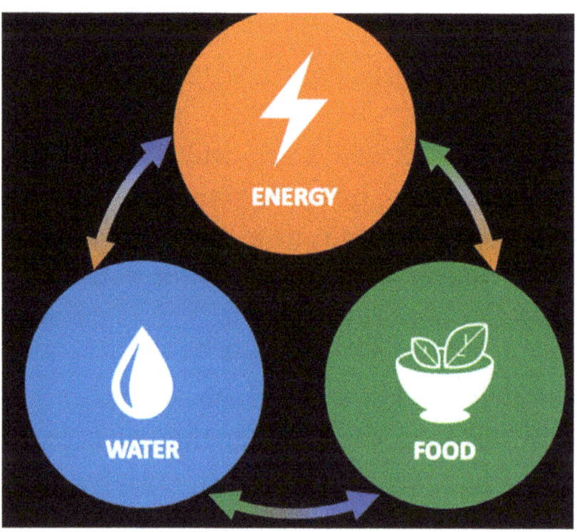

Figure 1.2 The interconnectedness of water with food and energy.

Let us just ponder for a moment the stark numbers – 8 billion and growing, with 70 percent of this population needing highly dense and managed water resources systems to ensure sustainable cities and healthy ecosystems. How are we going to manage the water under such circumstances, as demand keeps rising to support better livelihood opportunities around the world? According to some estimates, if the whole world had the same standard of living as the highly developed nations, the demand on Earth's resources would be equivalent to supporting 80 billion people today (Brockman, 2014). So where will the freshwater come from to sustain such anticipated demand due to economic development? Will there be enough water to maintain such a lifestyle? How challenging or costly will it become to maintain a healthy planet with sustainable livelihoods? Lastly, where will the data on water come from? Here the "we" refers to the broader community of water management practice – comprising scientists, practitioners, planners, policy makers, governance, and various other stakeholders who depend on freshwater resources.

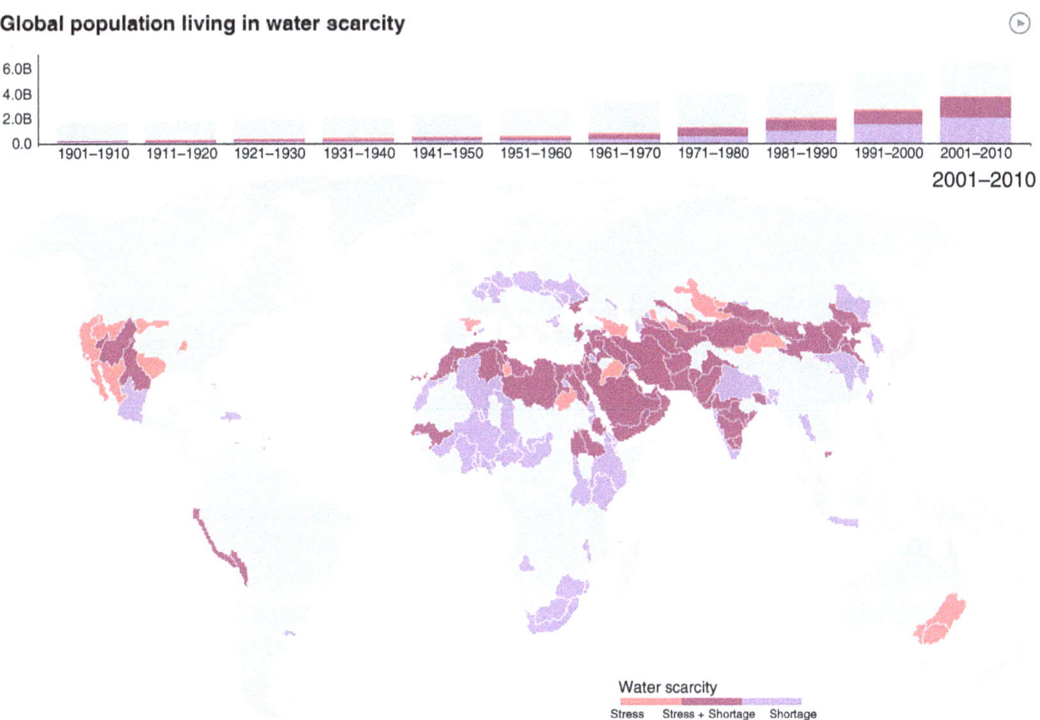

Figure 1.3 Atlas on global water scarcity. [Credit: Used with permission from https://waterscarcityatlas .org, based on Kummu et al. (2016), https://doi.org/10.1038/srep38495]

In addition to the above questions, we have witnessed a changing climate in the twenty-first century. Extremes in water availability are becoming exacerbated in ways that have challenged traditional solutions for water management (Figure 1.3). We have also made ourselves more vulnerable to natural hazards and extreme swings of nature by concentrating settlements along coasts, rivers, and lakes. An interesting trend that traditional water management theory did not factor in earlier in the twentieth century is that more than 50 percent of the world's population now lives very close to a shoreline (Ceola et al., 2015). Just like the great migration towards cities, there has also been a great migration towards the land–water boundary in many cases. Yet, most of today's large water management solutions and practices were developed during the twentieth century when climate change and the underlying trends of development, mass-migration, and urbanization were not recognized as necessary projections in their design.

1.3 How Can We Manage Our Water More Responsibly in the Twenty-First Century?

The traditional (twentieth-century) way of managing water resources has been mostly interventionist – where we have built physical structures to store, divert, and treat water. Most often, the solution has been to build storage structures such as dams to create artificial

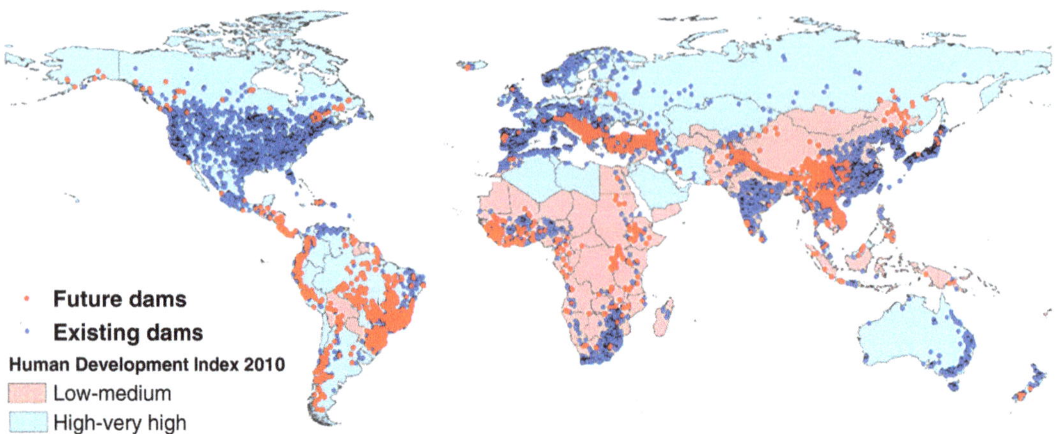

Figure 1.4 Location of large dams (planned and existing) around the world, overlaid on development status defined by the UN Human Development Index 2010. [Information on current and planned dams extracted from Zarfl et al. (2015).]

reservoirs to store freshwater (Figure 1.4). According to some estimates, the world today may have close to a million artificial reservoirs ranging in area from the smallest (a few acres) to largest size (several hundred square kilometers). We have also converted land for agricultural production around the world that requires irrigation systems using groundwater or surface water (Figure 1.5).

Although solutions for water management have worked well in the twentieth century, they have not come without a cost. Today, questions about sustainability are being asked. For example, reservoirs are silting up, with decreasing active storage in many places (Kondolf et al., 2014). At the same time, these reservoirs are expected to meet increasing demand for water, flood control, or hydropower under environmental conditions they were not designed for. Irrigation projects of the previous century have similar challenges. Many irrigation systems today are not able to meet the food production needs of a much larger population today than what they were designed for. A good example is the Indus Basin Irrigation System (IBIS) in Pakistan, where supplementary sources of water now need to be secured to keep the system functioning (Bose et al., 2021).

The question of sustainability of current water management solutions has rightly placed a focus on reuse and recycling of spent freshwater, where many innovative solutions have been developed. One example out of many is the solar-powered desalination of sea water for irrigation (Caldera and Breyer, 2019). However, in developing nations with limited economic opportunities, the traditional interventionist solutions of the twentieth century continue to be adopted (Figure 1.4). Dams continue to be built in those locations, new irrigation projects based on twentieth-century ideas continue to come up, rivers continue to be diverted, altered, or choked, affecting water availability in other countries. Today, our rivers are more fragmented than they have ever been since civilization began (Grill et al., 2019).

So, to manage our water resources responsibly into the twenty-first century, we first need to know what has happened to the state of our water resources over the past many decades, owing

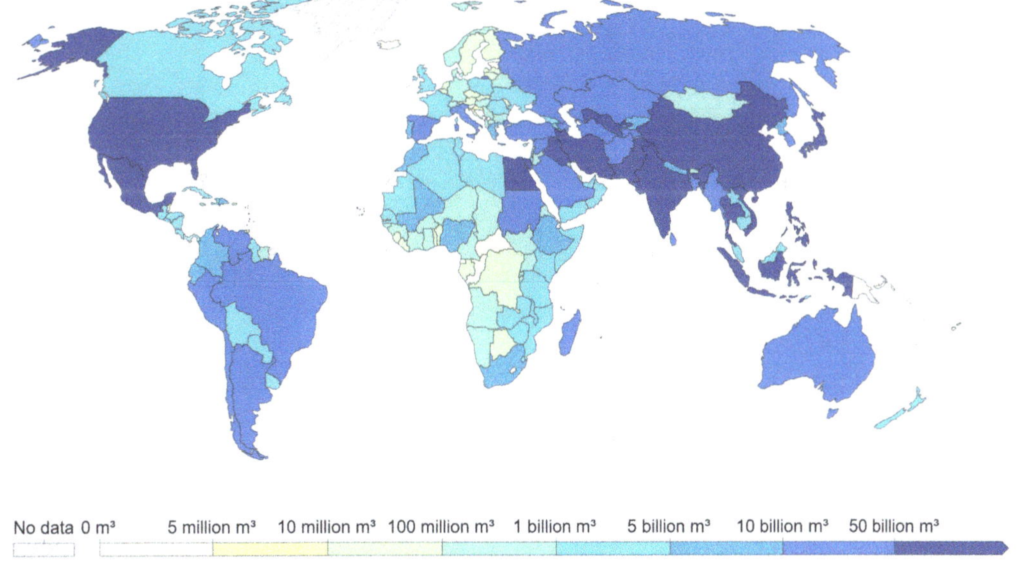

Figure 1.5 Global map of agricultural water withdrawals according to the Food and Agriculture Organization. [Source: Food and Agriculture Organization of the United Nations – AQUASTAT; OurWorldInData.org/water-use-stress https://creativecommons.org/licenses/by/4.0/]

to the traditional ways of water management. We need to understand what has worked and what has not. Where are the opportunities for innovative solutions to re-adjust our current traditional solutions and make water use more efficient? When we talk about what has worked or has not worked, the answers we seek should not be just in terms of meeting demands through sufficient quantity and quality of water, but also in terms of who the solution has or has not worked for.

Once we are able to explore how well traditional water management solutions have worked, we then need to ask ourselves whether traditional interventionist water management solutions of the twentieth century are a good fit in places that are developing rapidly. As mentioned earlier, many developing regions around the world are essentially replicating many of the old and myopic ideas of the twentieth-century developed world that are now known to have come at a very high social and environmental cost. Take, for example, the case of the Mekong region, where there is tremendous dam development in the foreseeable future for hydropower generation and its export. Such dam development will likely disrupt fisheries, among other ecosystem services, and have negative ripple effects on the livelihood of many nations in the region (Orr et al., 2012; Figure 1.6).

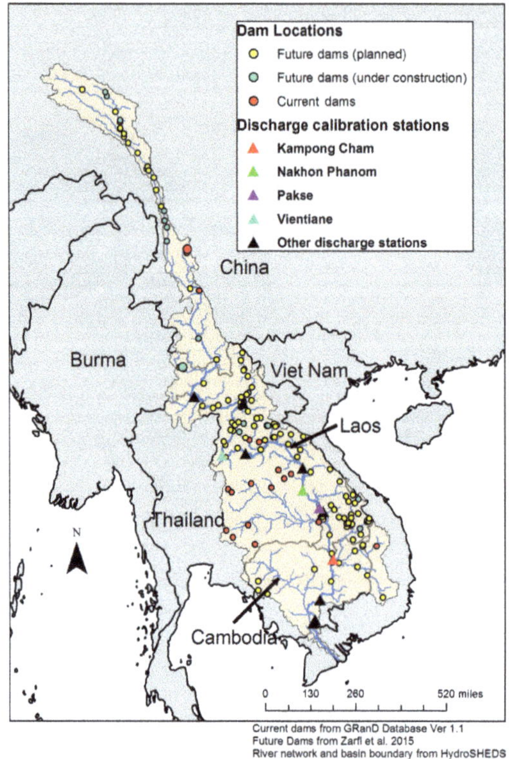

Figure 1.6 An example of extensive interventionist approaches to water management (i.e. dam building) that continue in the developing world. The example here is Lao PDR in Southeast Asia. [Data for current dams from GRanD Database Ver 1.1, data for future dams from Zarfi et al., 2015, and data for river network and basin boundary from HydroSHEDS.]

1.4 The Key Challenges of Water Management in the Twenty-First Century

The Hydropolitical Challenge

At the very core, any type of water management solution, whether it is at the conceptual, planning, or operational stage, needs water data. For water management, data on water are essential, such as precipitation data, streamflow data, and groundwater data. Other geophysical data such as geologic data, land use/cover data, and population data also go hand in hand in any development of water management solutions.

While developed regions have put comparatively more resources than developing regions into having a functional ground-based network for measurement of water, the world in general is faced with increasing scarcity of water data from traditional sources of on-ground measurement.

The first less-than-obvious fact is the hydropolitical nature of water data. Today, about 40 percent of the land mass is part of one of the international river basins (Figure 1.7)

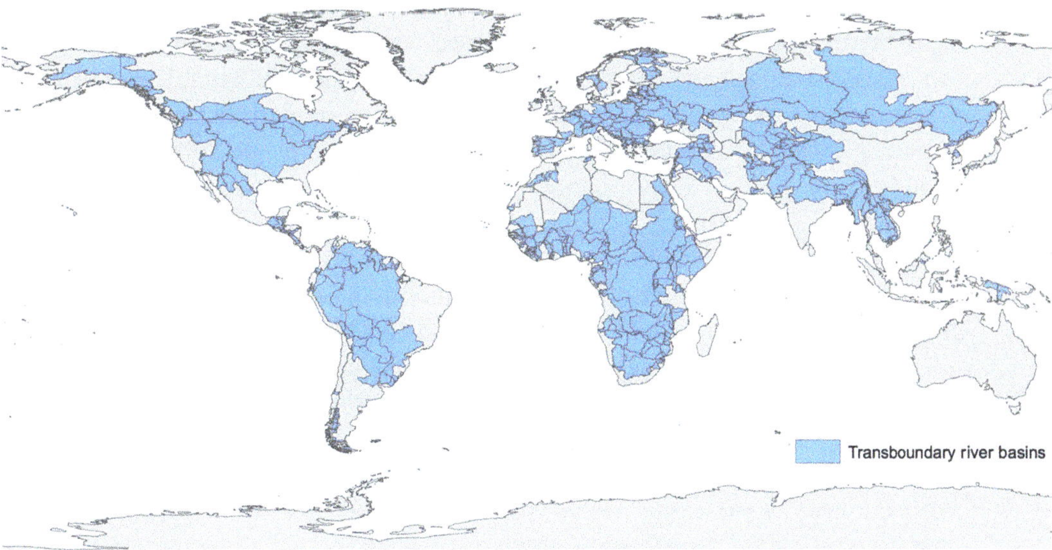

Figure 1.7 Map of transboundary river basins. [Source: Product of the Transboundary Freshwater Dispute Database, College of Earth, Ocean, and Atmospheric Sciences, Oregon State University. Additional information about the TFDD can be found at: http://transboundarywaters.science.oregonstate .edu and in Giordano et al., 2014]

that together account for more than 50 percent of the total water resources of the world (Giordano et al., 2014). This means that the water we use, or need to manage, probably has its source in another country or in a different jurisdiction from where it is consumed. It can also mean that the water management solution will affect another country or political jurisdiction downstream. This makes the creation of seamless and openly shared water data for exploring more holistic and transparent management solutions a challenge. Internationally, we can think of India–Pakistan for the Indus river, India–Bangladesh for the Ganges river, Laos, Cambodia, Vietnam, Thailand, and China for the Mekong river, Egypt–Sudan and Ethiopia for the Nile. These are countries that are often unable to adapt or build water management solutions fully independently, owing to lack of water data over the entire river basin. In the United States, the Colorado river water poses somewhat similar challenges, where there is sharing with Mexico, and also as far west as California via the states of Arizona and Nevada. In Europe, we can think of the Danube or the Rhine. In each of these cases, the open and public sharing of water data is often considered a national security risk (Hossain et al., 2023). In other words, the challenge of water management is compounded by the national interests or international politics of water data sharing where the lack of publicly accessible data can lead to difficulties in sound water management.

The Infrastructure and Financial Challenge

The next challenge is that of declining on-ground networks for water measurement. Around the world, streamflow gauges appear to be declining or less well maintained than they used to be (Shiklomanov et al., 2002). In the US, we have more than 8,000 streamflow gauges, according to the United States Geological Survey (see: www.usgs.gov/mission-areas/water-resources/science/usgs-streamgaging-network), and a similar number of rainfall gauges. These in-situ gauges measure water data at a given point and not the entire river reach or over a river basin. For lakes, we are currently collecting data on lakes such as height and water quality for a very small fraction. Natural lakes are rarely gauged using in-situ gauges, especially if they are not large and well known. In the developing regions of the world, there is an inadequate network of ground gauges (Shikhlomanov et al., 2022). Yet these are the regions that need water data for developing responsible water management solutions. The remoteness and difficulty in accessing a region can also make it harder to place and maintain in-situ gauges.

The economics of gauge networks is also a challenge. First, there is a cost in maintaining ground-based measurement networks. Maintaining, calibrating, preventing vandalism or theft, and training skilled practitioners for data collection and data quality control can be costly endeavors. While developed and economically more prosperous regions may be able to afford costly but necessary in-situ water measurement networks, most developing nations cannot do so at the level needed for effective water management.

When we combine this pervasive lack of adequate ground data on water with the equally pervasive issues of hydropolitics of water data and the economic challenges, one has to ask, then, *"How else can we measure and record water data to explore twenty-first-century water management challenges and explore practical solutions?"*

One issue that we should mention in the context of challenges is that of ensuring equity and social justice in the management of water. In an era of heightened awareness on how well management solutions are working for all, the challenge of ensuring equity and social justice is something that is often intertwined with these other more tangible challenges. In this book, we will explore how satellite remote sensing can play a positive role for social justice and equity.

1.5   Satellite Remote Sensing for Water Management

Since October 4, 1957, when the Russian satellite Sputnik was launched, hundreds of Earth-observing satellites have followed. According to the Union of Concerned Scientists, there have been about 1,000 satellites launched for Earth observation or Earth science out of more than 7,000 launched in total to date (UCS, 2021). The first satellite designed specifically for Earth observation was Vanguard 2, followed by TIROS-1 in 1960, which produced the first television footage of weather patterns from space. The success of TIROS-1 led to a stream of meteorological satellites and provided the basis for subsequent development of devices designed specifically for land observation.

For monitoring water bodies over land and ocean, we have had reliable satellite altimeter missions measuring water height since the 1980s and optical sensors to measure water extent and land cover since the 1970s. Both these efforts can be considered operational – that is, running continuously – as they have endured for many decades and have not been discontinued. For example, the Topex-Poseidon, Jason, and Sentinel satellite series combined with the Landsat satellite series provide more than a 30-year time history of change of Earth's water and land surfaces.

For tracking water quality such as temperature and color (sediment concentration, harmful algal bloom), we also have more than 30 years of global record, also from the Landsat mission, that can help improve our understanding of water management problems and solutions needed for the twenty-first century. Similarly, meteorological satellites have continued to be operational and provide round-the-clock global estimates of precipitation and weather at scales relevant for water management since the launch of the Tropical Rainfall Measuring Mission (TRMM) in 1997, and later the Global Precipitation Measurement (GPM) mission in 2014. Satellites for tracking surface water volume, such as the Surface Water and Ocean Topography (SWOT) mission, and for tracking soil moisture, such as Soil Moisture Active and Passive (SMAP) and Soil Moisture and Ocean Salinity (SMOS), have opened new horizons for quantitative water management in the field of operations, planning, and management of reservoir and irrigation systems (Durand et al., 2021). Many of these satellites can also be used to track the state of our cryosphere and the extent of snow, which is a critical parameter for water management. Since the early 2000s, we have also had a satellite tailored for tracking groundwater storage changes at large spatial scales, called the Gravity Recovery and Climate Experiment (GRACE) (Richey et al., 2015).

If we think of the entire cycle of water, we have actually been tracking most of the components using the vantage point of space (Figure 1.8) since the late twentieth century. Given that today we are able to track most of the water cycle and have a long historical record from a source other than ground networks, why would we *not* want to use such a convenient source of data to manage, day to day, our water resources from space? Such data is global with large spatial coverage, immune to data-sharing hurdles on the ground and not as affected by on-ground network maintenance and cost challenges. This is in no way to suggest that on-ground data

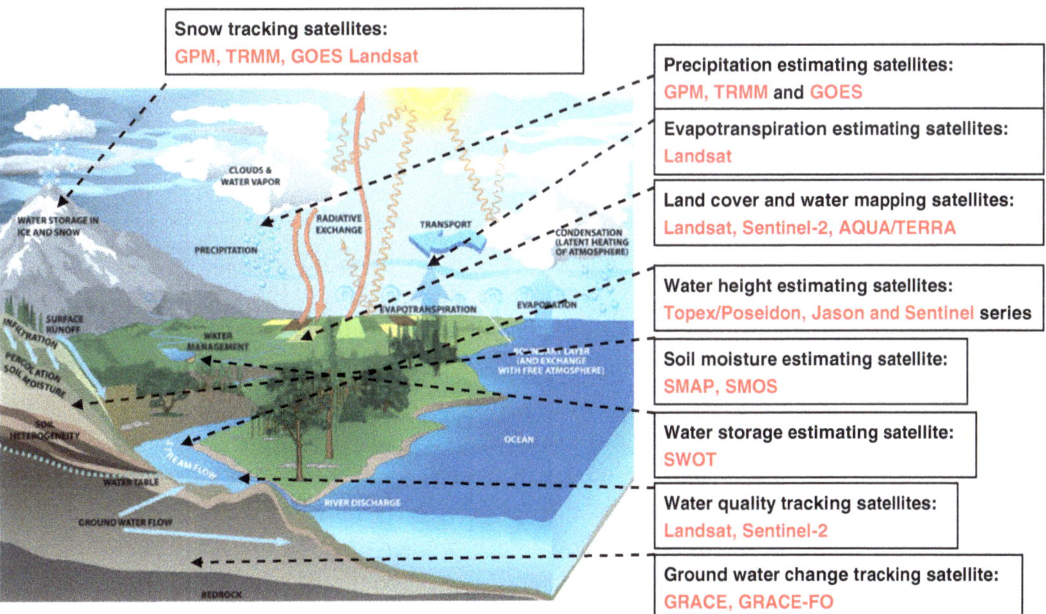

Figure 1.8 Water cycle components that are tracked by today's satellites. GOES, Geostationary Operational Environmental Satellite; others are defined in the text.

Table 1.1 **Overview of satellite missions and their primary purpose and relevance for water management**

Name of satellite mission	Primary purpose and data	Year launched	Mission weblink
Global Precipitation Measurement	To estimate precipitation data globally at spatial resolution of 10 km and 1-hour intervals	2014	https://gpm.nasa.gov/missions/GPM
Tropical Rainfall Measurement Mission	Same as above but for the Tropics, where most of the rainfall amount occurs today	1997	https://gpm.nasa.gov/missions/trmm
Topex/Poseidon	To track sea surface temperature and height globally	1992	https://sealevel.jpl.nasa.gov/missions/topex-poseidon/summary/
Jason series	Same as above	2001	www.eumetsat.int/our-satellites/jason-series
Sentinel-3A/B	Same as above	2016	https://sentinels.copernicus.eu/web/sentinel/missions/sentinel-3
Landsat	To provide imagery over land in visible and thermal wavelengths to track how Earth is changing	1972	https://landsat.gsfc.nasa.gov/
SMAP	To estimate upper layer soil moisture globally	2015	https://smap.jpl.nasa.gov/
SWOT	To estimate global surface water extent and elevation changes	2022	https://swot.jpl.nasa.gov
GRACE	To track gravity anomalies and estimate total water storage change at monthly frequency and around 500-km resolution	2002	https://grace.jpl.nasa.gov/mission/grace/
GRACE-FO	Similar (but different resolutions)	2018	https://gracefo.jpl.nasa.gov/

from gauges on water should be ignored or replaced in the twenty-first century. On the contrary, ground gauges are a critical source for quality control and validation of satellite data, and can be used synergistically to minimize the weakness of satellite estimates of water that are indirect measurements. In addition, if satellite data can be combined with ground-based data, then the limitations of each source can be creatively minimized to develop innovative water management in challenging places. Indeed, we have many success stories to showcase how water can be managed effectively from space (see the case studies).

Today, the data we receive on Earth's water cycle components from the vantage of space probably exceed tens of terabytes a day. Fortunately, many of such satellite data on water from space have been tested, experimented with, improved continuously, and become a lot easier to access, use and apply for water management thanks to advancements made in data informatics, open-source software, and cloud computing. For example, today a user of water data can be located anywhere on the planet and just need internet connectivity to access, visualize, and run analysis on very large satellite datasets in the cloud to generate actionable knowledge the size of a few kilobytes. Such actionable knowledge can be on historical patterns, nowcasts, forecasts, risks (probability), or a statistical summary of extremes that are often a design input for water management systems and operations. Today, there is in fact no need to download massive datasets as most water managers will have little need to store so much data.

There are many successful examples of application of satellite data on water for developing management solutions. Most of these started from the early 2000s. Data from GPM is now used in flood forecasting systems around the world; data on river height from satellite altimeters is being used to monitor reservoir storage changes, or to understand the effects of drought, or to drive adaptation plans downstream. Precipitation and temperature data is now frequently used to estimate crop water demand and estimate crop yield – both of which can drive the planning and operation of irrigation systems. Data on river height, river width, and land cover can be used in models to predict and forecast streamflow. Some satellites can track, at coarser scale, groundwater storage changes and are currently being used in operational drought monitoring in many countries. In other instances where snowpack is an important parameter for water supply forecasting, satellite data on snow-covered area is routinely used to plan seasonal operations of reservoirs and water utilities.

1.6 The Future of Satellite Remote Sensing for Water Management

Over the past 20 years, there has been a dramatic shift in the way we manage the world's water resources. With the explosion of Big Data on water sensing from satellites in space, innovations in affordable information technology, and cloud computing to democratize accessibility to timely information, today's water management will continue to innovate and evolve rapidly well into the middle of the twenty-first century. Satellite remote sensing for water management is a topic on which the community of practitioners and scientists needs to build literacy. This will have to start with a brief introduction to remote sensing in the context of water resources and hydrology. Water management professionals will take inspiration from successful examples of application of water data from space. They will need to apply basic theory to learn how the water remote sensing and management can be woven together to solve real-world water management challenges that will continue to spread into the rest of the twenty-first century.

As newer innovations are developed in water management with advancements in information technology, it is important that we get exposed to how the field is changing. The water management community needs to experience "hands on" how rapidly water management has changed and will continue to change even more as it uses Big Data and data informatics as essential tools for solving water problems. Water management is no longer the water management of the twentieth century using point-based and limited gauge data. We will continue to launch water-specific global satellite missions, such as SWOT (launched in December 2022), complemented by other space missions of the past and future. Citizen science (i.e. data harvesting by citizens) with satellite data is also becoming common to address the need to re-design and "re-manage" water to address environmental degradation, sustainability, social injustice, and inequity.

As mentioned earlier, key strengths of the vantage of space for water data are that such data is immune to the hydropolitics of water sharing and requires minimal cost for use, as the data is mostly public. For this reason, we will continue to see many more successful examples of application of satellite data to derive innovative water management solutions around the world as data access becomes easier over time. For example, we can use satellite water data to adapt the operation of High Aswan dam in Egypt, based on new upstream reservoirs built by Sudan and Ethiopia. In Pakistan, reduced water storage capacity in the dams of upper Indus poses a major problem to flood management, where many of the rivers originate from India and data is not shared in near real-time – a problem that can be addressed by satellite management. Thus, satellite data in near real-time on rivers and reservoirs in transboundary locations is proving to be effective in water management. This concept is not unique to Pakistan and

---

Case Study 1.1: **Monitoring Reservoir Operations from Space**

Today, by piecing together data on the water cycle and land from multiple current and historical satellites, we can operationally nowcast, forecast, and track historical behavior of most artificial reservoirs. One such example is the operational Reservoir Assessment Tool (RAT) that can now routinely monitor large reservoirs in developing regions where ground data on reservoirs are nonexistent or not shared to develop such a system (Figure 1.9). RAT can also help infer reservoir operating patterns based on historical operation, which can help downstream and upstream stakeholders to manage their water resources more effectively, because such operating data is usually not shared openly in most regions. Today, RAT is in operation by several water management agencies in South and Southeast Asia, Middle East, and North Africa (Minocha et al., 2024).

- **1,598** reservoirs routinely monitored from 1985
- Storage change, inflow, outflow, evaporative loss monitored
- **8+** satellite missions
- **2 TB/day** of daily data
- **Open-source** tool for dam monitoring
- Currently used for management in the Mekong, Tigris–Euphrates, Nile Indus, and Columbia rivers
- Any agency or user can set up such a system

Figure 1.9 The Reservoir Assessment Tool, which uses satellite data to track how reservoirs are being operated. © openstreetmap.org, www.openstreetmap.org/copyright

India, but also seen in Bangladesh, and Southeast Asian nations, where the vantage of space in tracking water flowing across political boundaries is driving innovative water management solutions responsive to twenty-first century challenges. In the US, many agencies are now routinely assimilating satellite data on water, such as precipitation or reservoir area, to forecast freshwater supply. In short, the practice of water management using satellite remote sensing has already begun.

Case Study 1.2: Drought Monitoring Using Satellites

For hydrological drought, observing all of the relevant hydrological variables (i.e. snow, surface water, soil moisture, and groundwater) necessary for characterization, across the appropriate temporal and spatial scales, remains challenging. NASA's GRACE satellite mission provides monthly, integrated information about water storage variations throughout all components of the surface and subsurface water balance, which was previously unobtainable (Figure 1.10).

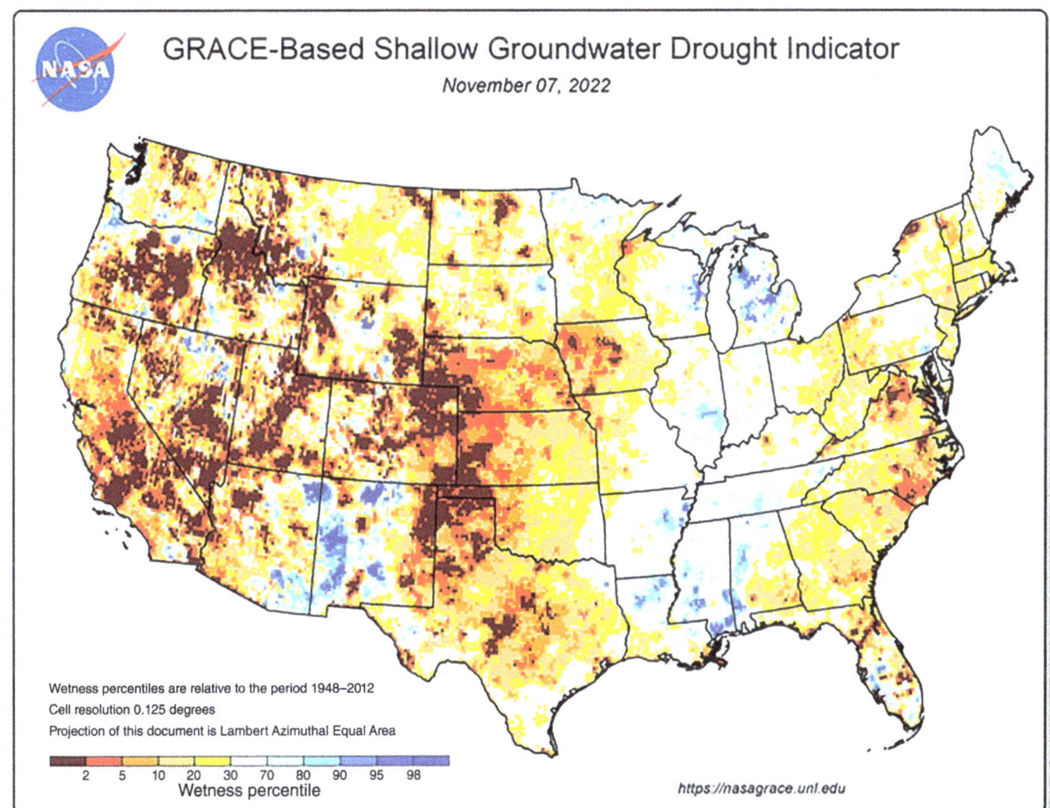

Figure 1.10 Monitoring groundwater drought using satellite data from a mission called GRACE (Gravity Recovery and Climate Experiment). The GRACE-Based Shallow Groundwater Drought Indicator is available at nasagrace.unl.edu through a partnership with the National Drought Mitigation Center.

Case Study 1.2 (cont.)

In the image in Figure 1.10, the shallow groundwater drought indicator is based on terrestrial water storage observations derived from GRACE satellite data and integrated with other observations, using a numerical model of land surface water and energy processes. The drought indicators describe current wet or dry conditions, expressed as a percentile showing the probability of occurrence within the period of record from 1948 to the present, with lower values (warm colors) meaning drier than normal, and higher values (blues) meaning wetter than normal. For more information, please visit https://nasagrace.unl.edu/.

Case Study 1.3: Minimizing Unnecessary Irrigation for Food Production

Food production accounts for more than 50 percent of the total human freshwater consumption. In many places, this is as high as 90 percent. So, any sustainable water management solution based on finite supply of water has to address food production and its water use efficiency. In many places, an overwhelming majority of farmers continue to waste water and over-irrigate owing to old irrigation practices, archaic irrigation financing, or just lack of awareness of how much water the plants need. Today, we can monitor crop water need using satellites. We can also track evapotranspiration of crops using satellites. By analyzing the two and also keeping an

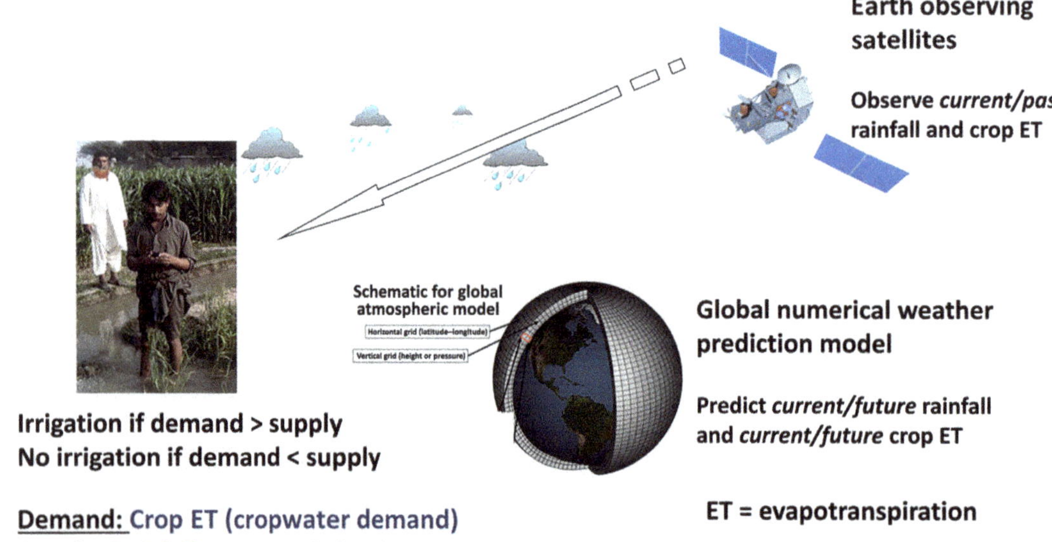

Figure 1.11 Synergistic use of various satellites and weather models to provide farmers in South Asia with advice on irrigation that is based on crop needs and can save water. [Credits: Faisal Hossain and Pakistan Council of Research in Water Resources www.pcrwr.gov.pk]

eye on weather forecast, it is possible to alert farmers on their cellphones of situations when irrigation can be avoided, or when they really need to keep the irrigation going. Such an approach can also guide surface water irrigation systems on when and where to direct water to a cropped area (called "command area"). Satellite-based irrigation management has witnessed much success since 2015 in areas of South Asia, where it is estimated that farmers can lower their water use footprint by 40 percent and still generate similar crop yields (Figure 1.11). In places where this water is harvested from rain for use during dry season or where the water needs expensive fuel to be pumped, think of the savings that satellites can enable and thus improve a farmer's livelihood (Hossain et al., 2017).

1.7 Conclusion

In this chapter, we introduced ourselves to the growing importance of managing our precious water resources using satellite remote sensing. This technique in space represents an alternative, growing, and increasingly accessible source of data on water that water managers must learn to harness. We discussed how traditional methods of water management using in-situ data are becoming more challenging because of underlying limitations such as hydropolitics and declining in-situ measurement infrastructure.

Without rewriting water management theory, the future chapters of this book will show how we can do better and more effective water management using a water data source that is generated in space using sensors looking down on Earth. As we head into the middle of the twenty-first century, the challenges from climate change, rising population, increasing demand, competing needs, and failures of traditional management based on limited in-situ data sources will remain. It is important that we learn the basic theories of remote sensing, their limitations and strengths for estimating water cycle variables, so that water management is applied appropriately in developed and developing regions. In the next two chapters, our goal will be to build this basic understanding of theory and fundamentals of remote sensing for water.

EXERCISES: CHAPTER 1

Q1.1 Name and describe one water management solution that you depend on for your lifestyle. Describe where the water comes from and how you think it is managed to ensure reliability of supply of water at your end.

Q1.2 What is an international river basin?

Q1.3 How many international river basins are present today? Do you think you currently live or have ever lived in an international river basin (name the river)?

Q1.4 There are many global forces that will affect the availability of water in the middle of the twenty-first century. Name the ones you think are important and explain briefly why. [Hints: Climate change; urbanization; population growth]

Q1.5   Use the internet to find out the total number of ground gauges that are installed for (a) streamflow measurement; (b) precipitation measurement; (c) soil moisture measurement; (d) groundwater level measurement; (e) snowpack measurement. Note that this answer will vary depending on the source of information and where the gauges are operational or not.

Q1.6   In-situ (ground-based) measurement of water cycle components is important for water management and has been the mainstay for much of the twentieth century and early twenty-first century. What do you think are the key challenges and limitations of the traditional way of measuring water today?

Q1.7   Name the key challenges to sharing or having access to water data for building robust water management solutions today that are global in scope. [Hints: hydropolitical; infrastructure; economic]

Q1.8   How do you think the vantage of space and using satellite remote sensing can help mitigate the current challenges to water data access for water management today?

Q1.9   For precipitation, search the internet and find out the first satellite mission that was used to estimate precipitation and at what scales (spatial and temporal). How many satellites are being used today concurrently to track precipitation globally and at what scales (spatially and temporally)?

Q1.10   Repeat question 1.9 for stream flow. Search the internet on the Surface Water and Ocean Topography (SWOT) mission. When did the mission start, and what type of data does it produce to facilitate water management?

Q.1.11 In 2022, the satellite mission called Landsat completed 50 years of monitoring the Earth's surface, after the first mission flew in 1972. Landsat data has been used to successfully track land cover change, deforestation, afforestation, wetlands, snow-covered areas, and surface water bodies (lakes and rivers). How do you think such data can help us manage water resources from space?

REFERENCES

Bose, I., F. Hossain, H. Eldardiry, et al. (2021). Integrating gravimetry data with thermal infra-red data from satellites to improve efficiency of operational irrigation advisory in South Asia. *Water Resources Research*, vol. 57, e2020WR028654. https://doi.org/10.1029/2020WR028654

Brockman, J. (2014). *What Should We Be Worried About? Real Scenarios That Keep Scientists Up at Night* (Edge Question Series). Harper Perennial.

Caldera, U. and C. Breyer (2019). Assessing the potential for renewable energy powered desalination for the global irrigation sector. *Science of the Total Environment*, vol. 694, 133598. https://doi.org/10.1016/j.scitotenv.2019.133598

Ceola, S., F. Laio, and A. Montanari (2015). Human-impacted waters: new perspectives from global high resolution monitoring. *Water Resources Research*, vol. 51, 7064–7079. https://doi.org/10.1002/2015WR017482

Durand, M., A. Barros, J. Dozier, et al. (2021). Achieving breakthroughs in global hydrologic science by unlocking the power of multi-sensor, multidisciplinary Earth observations. *AGU Advances*, vol. 2, e2021AV000455. https://doi.org/10.1029/2021AV000455

Giordano, M., A. Drieschova, J. A Duncan, et al. (2014). A review of the evolution and state of transboundary freshwater treaties. *International Environmental Agreements*, vol. 14, 245–264. https://doi.org/10.1007/s10784-013-9211-8

Grill, G., B. Lehner, M. Thieme, et al. (2019). Mapping the world's free-flowing rivers. *Nature*, vol. 569, 215–221. https://doi.org/10.1038/s41586-019-1111-9

Hossain, F., N. Biswas, M. Ashraf, and A. Z. Bhatti (2017). Growing more with less using cell phones and satellite data. *Eos*, vol. 98. https://doi.org/10.1029/2017EO075143

Hossain, F., S. Minocha, A. Alwash, and H. Eldardiry (2023). Restoring the Mesopotamian rivers for future generations: a practical approach. *Water Resources Research*, vol. 59, e2023WR034514.

Kondolf, G. M., Y. Gao, G. W. Annandale, et al. (2014). Sustainable sediment management in reservoirs and regulated rivers: experiences from five continents. *Earth's Future*, vol. 2, 256–280. https://doi.org/10.1002/2013EF000184.

Kummu, M., J. Guillaume, H. de Moel, et al. (2016). The world's road to water scarcity: shortage and stress in the 20th century and pathways towards sustainability. *Nature Scientific Reports*, vol. 6, 38495. https://doi.org/10.1038/srep38495

Lang, R. E., J. Lim, and K. A. Danielsen (2020). The origin, evolution, and application of the megapolitan area concept. *International Journal of Urban Sciences*, vol. 24, 1–12. https://doi.org/10.1080/12265934.2019.1696220

Liu, X., Y. Huang, X. Xu, et al. (2020). High-spatiotemporal-resolution mapping of global urban change from 1985 to 2015. *Nature Sustainability*, vol. 3, 564–570. https://doi.org/10.1038/s41893-020-0521-x

Minocha, S., F. Hossain, P. Das, et al. (2024). Reservoir Assessment Tool version 3.0: a scalable and user friendly software platform to mobilize the global water management community. *Geoscience Model Development*, vol. 17, https://gmd.copernicus.org/articles/17/3137/2024/gmd-17-3137-2024-assets.html

Orr, S., J. Pittock, A. Chapagain, and D. Dumaresq (2012). Dams on the Mekong River: lost fish protein and the implications for land and water resources. *Global Environmental Change*, vol. 22, 925–932. https://doi.org/10.1016/j.gloenvcha.2012.06.002

Richey, A. S., B. F. Thomas, M.-H. Lo, et al. (2015). Quantifying renewable groundwater stress with GRACE. *Water Resources Research*, vol. 51, 5217–5238. https://doi.org/10.1002/2015WR017349

Shiklomanov, A. I., R. B. Lammers, and C. J. Vörösmarty (2002). Widespread decline in hydrological monitoring threatens Pan-Arctic research. *Eos Transactions of the American Geophysical Union*, vol. 83, 13–17. https://doi.org/10.1029/2002EO000007

UN (2019). *World Urbanization Prospects: The 2018 Revision*. United Nations.

Union of Concerned Scientists (2021). UCS Satellite Database, www.ucsusa.org/resources/satellite-database (last accessed January 15, 2022).

Zarfl, C., A. E. Lumsdon, J. Berlekamp, et al. (2015). A global boom in hydropower dam construction. *Aquatic Sciences*, vol. 77, 161–170. https://doi.org/10.1007/s00027-014-0377-0

SUGGESTED READING

Douglas, E. A. and D. P. Lettenmaier (2003). Tracking fresh water from space. *Science*, vol. 301, 1491–1494. https://doi.org/10.1126/science.1089802

Thomas, B. F. and J. S. Famiglietti (2019). Identifying climate-induced groundwater depletion in GRACE observations. *Scientific Reports*, vol. 9, 4124. https://doi.org/10.1038/s41598-019-40155-y

2   An Introduction to Remote Sensing from Space

2.1   Chapter Overview

In the previous chapter, we introduced ourselves to the idea of satellite remote sensing for water management and why the technique is going to take on greater importance in years to come as challenges mount from climate change, competing needs, and lack of ground data. In this chapter, we will overview the basics of remote sensing, defining key concepts and terms. Using these concepts and terms, we will develop an understanding of the fundamental principle required for the success of remote sensing.

2.2   Introduction

To understand how satellite remote sensing works for water management, we must first understand how remote sensing from space works for estimating water cycle components. The first thing to recognize is that, unlike ground-based gauge measurements, remote sensing is an indirect method. Also, we are talking about fundamental differences between measurement techniques. Remote sensing often captures a variable or a phenomenon over an area or is spatially integrated rather than being point-based (refer to Chapter 1, section 1.5). For example, a traditional rain gauge directly measures the rainfall accumulated at the small location of the gauge footprint (probably over a 30-cm diameter) over a specified period of accumulation. Remote sensing will typically indicate rainfall that is spatially representative over a much larger area (spanning hundreds of meters to a few kilometers), and it can be instantaneous (based on when the satellite flew) or a temporal average. The same can be said of soil moisture and of river parameters such as width, slope, or even discharge. Whatever the differences, we should first build our understanding of satellite remote sensing with the humbling notion that it is indirect and therefore it can have both uncertainty (in terms of reproducibility) and error (in terms of comparison against a more trustworthy reference). Satellite remote sensing has endured despite these differences and limitations, and has continued to improve steadily over time, as we shall see in this book. What is even more exciting is that satellite remote sensing can often be a valuable qualitative indicator of increasing or decreasing trends in a water cycle component, without necessarily having to be quantitatively accurate. In many water management applications, just knowing whether streamflow or precipitation is on an increasing or decreasing trend over time at a region can be a valuable input for decision making.

2.3   What Is Remote Sensing?

A very common definition of remote sensing is as follows:

The process of seeking information using techniques that do not require actual contact with the object or area being observed (or measured).

So, basically, remote sensing is all about seeking information (sensing) on an object or phenomenon of interest without being in direct contact. This suggests an interesting question about Figure 2.1 – are the two people talking via remote sensing? We can argue both ways, but because the two ends are still physically connected with a wire, it is really not remote sensing in the strictest sense. So to further define the term, we must also define the essential requirement for remote sensing, which is energy. All remote sensing techniques work on the basis of energy. In this book and for water data, we will focus only on electromagnetic energy. However, readers should note that remote sensing can also be carried out using sound energy (sonar), as bats and submarines do to navigate.

Another way to look at remote sensing is through an everyday example that many of us have used during our childhood. Imagine there is a wall outside where you live, and you

Figure 2.1 Is this remote sensing? [Image by Sakib Chowdhury]

Figure 2.2 Using a ball to understand or to sense what a wall is made up of without getting close to or touching the wall. This action is active remote sensing.

have often wondered what the wall is made up of – steel, wood, or brick? You have not ventured close because you have heard rumors that the wall might be electrified. So, you devise the idea of throwing a ball at the wall (Figure 2.2). By observing how the ball bounces back and the sound it makes on contact with the wall (later, we will call this "target inter-action"), you can probably say with some confidence whether the wall consists of brick or flimsy wood. The very nature of throwing a ball at the target (the wall) and recording the interaction to seek information on the target is "remote sensing". In this case, this is active remote sensing (see later). A converse way of remote sensing, known as passive remote sens-ing, would have been to just observe closely how the wall looks naturally, compared with other types of walls.

2.4 Definition of Key Terms for Remote Sensing

To improve our understanding of remote sensing, it is important to formalize a few key terms that will be repeated throughout the book. The first three terms are "target", "foreground",

Figure 2.3 Target, foreground, and background, using the example of taking a picture in a studio. Here the background is the white wall behind the family being photographed. The foreground is the space between the family and the camera operator. The target is the family being photographed. [Image: Freepik.com. This figure has been designed using assets from Freepik.com]

and "background". The target is the object or the phenomenon we want to investigate via remote sensing. For any target, there is always a foreground (the medium in front of the target) and background (medium behind the target). Imagine you want to have your picture taken, standing in front of a whiteboard, by someone with a camera (Figure 2.3). You are the target, the space between you and the camera is the foreground. The background is the whiteboard.

It is important that readers try to think of every aspect of remote sensing broken down in terms of target (i.e. what it is that we are seeking information on), the foreground, and the background. However, not every type of remote sensing can always be broken down in that manner, as the reader will notice later in the book.

The other important terms to formalize our understanding are (Figure 2.4):

- Energy – the specific type of energy (electromagnetic) that is being used to seek information on the target
- Illumination – the phenomenon of energy beaming on the target
- Reflection/refraction
- Absorption/attenuation
- Diffraction/dispersion
- Scattering

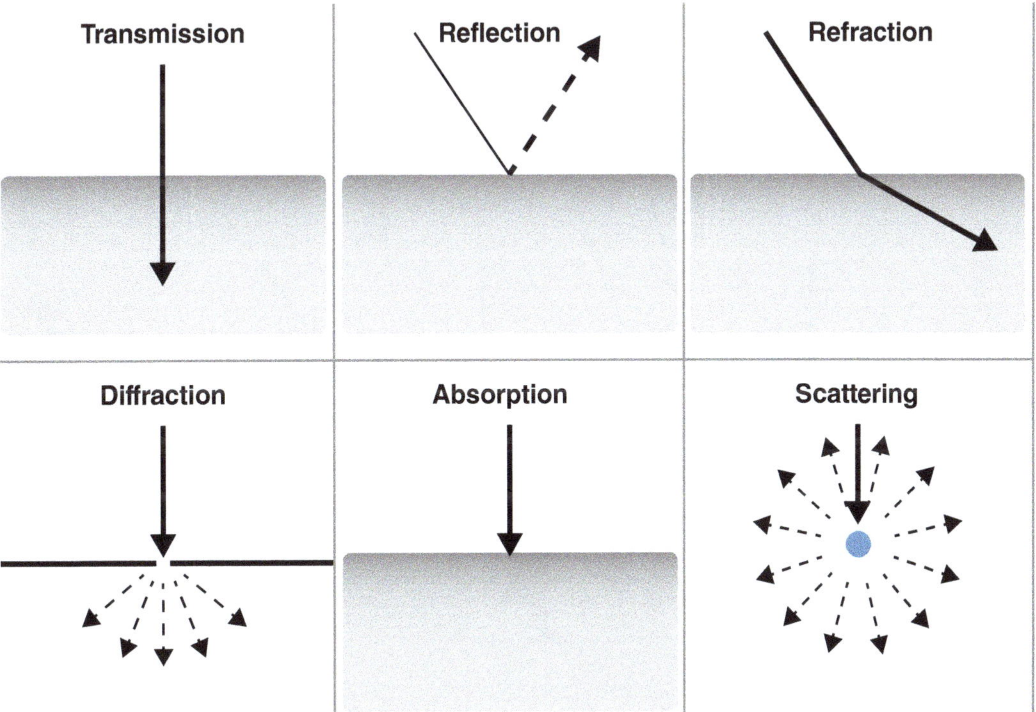

Figure 2.4 Electromagnetic energy can experience transmission, reflection, refraction, diffraction, absorption, and scattering (also known as dispersion/diffusion).

2.5   The Key Steps in Remote Sensing

There are a few key steps that happen during remote sensing from space. Let us look at Figure 2.5 and assume that the target (the object we want to seek information on from space) is at location C. It is daytime and the Sun is up. The satellite is at location D at a specified altitude from the ground surface. We may wish to take advantage of a specific type of the Sun's electromagnetic energy. In that case, the remote sensing would be called passive, as the energy source is not coming from the sensor or satellite carrying out the remote sensing. The energy from the Sun will interact with the target on the ground. This interaction may be reflection, refraction, dispersion, attenuation, or scattering (Figure 2.4). The satellite at D will record this interaction with the target, as some of the Sun's energy, after having interacted with the target, will reach the receiver of the satellite. The receiver's job is to record incoming energy from the target.

The nature of the interaction is critical here. We want an interaction that helps us to distinguish the target from the background and lets us quantify that behavior so that we can confidently attribute the energy interaction recorded by the satellite as being mostly due to the target. The satellite receiver will likely be receiving various kinds of energy, and it is important for it to know what energy was recorded uniquely due to the target interaction.

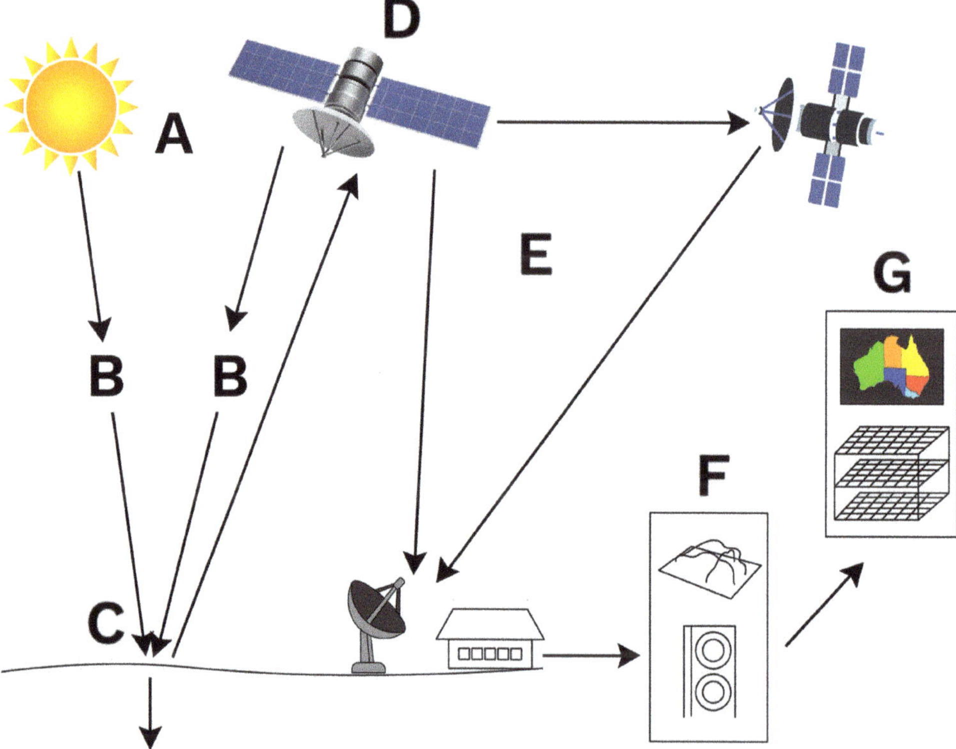

Figure 2.5 Key energy and process interactions during remote sensing from space for a target on the ground. [Image by Sakib Chowdhury]

Now the Sun's energy may not always be available: for example, it may be nighttime. Or, regardless of the Sun, the satellite may choose to send a more controlled form of energy to the target, repeat the process of target interaction, and record the interaction. This type of remote sensing is called active remote sensing, and although it is more expensive and energy-intensive, it affords more accurate and customizable remote sensing, owing to the ability to control the direction and nature of energy illuminating the target. We will talk more about this later to help the reader improve the fundamental concept behind this, using water as the target.

Note that the target interaction in Figure 2.5 can happen at C, but B, which is the propagation of the energy towards and away from the target, is also important to understand. If the energy from the Sun or satellite is severely weakened and almost dispersed by the time it reaches the target, the chances are that there will be very little target interaction left to characterize the target.

At D, the recording of the target interaction takes place. The rest of the remote sensing operation (E, F, and G) consists of quite standard logistical steps. Step E is for transmission of data to a ground station. This also includes reception and processing of the data. F is for interpretation and analysis of the satellite data to gather information on the target. Step G is application, once the data is ready, calibrated/verified for user consumption.

In all the above key steps, probably the most important step to improve our understanding of remote sensing vis à vis water management is "interaction with target". In future chapters, we will keep referring to this aspect to help us understand when water data from satellites can be skillfully inferred or not for the intended management application.

2.6 The Electromagnetic Spectrum

2.6.1 Electromagnetic (E-M) Waves

As mentioned in the introduction section of this chapter, remote sensing is based on energy. While various forms of energy can be used, for satellite remote sensing of water we will concern ourselves with only electromagnetic energy. It is therefore important to understand the electromagnetic energy spectrum.

In simple terms, electromagnetic energy can be defined as energy that travels at the speed of light (when in a vacuum). A specific energy type can be defined in terms of wavelength (lambda) and frequency, the product of which yields the speed of propagation. Such E-M waves are "transverse" in nature, which means that the wavelike motion of the energy is transverse to the direction of propagation. The best way to visualize a transverse wave is to throw a pebble into water and observe the ripples it creates – those are transverse waves. By contrast, sound waves are longitudinal, traveling through a process of rarefaction and expansion along the direction of travel. If you think of a French accordion being played while also being moved, the motion is analogous to what longitudinal waves would behave like.

E-M waves, as their name suggests, have an electric and magnetic component, at 90 degrees to each other (Figure 2.6; upper panel). The lower panel of Figure 2.6 shows the definition of wavelength that we use to define the type of E-M energy. The frequency is the number of waves (or cycles comprising crests and troughs) that the E-M energy is able to complete in a second.

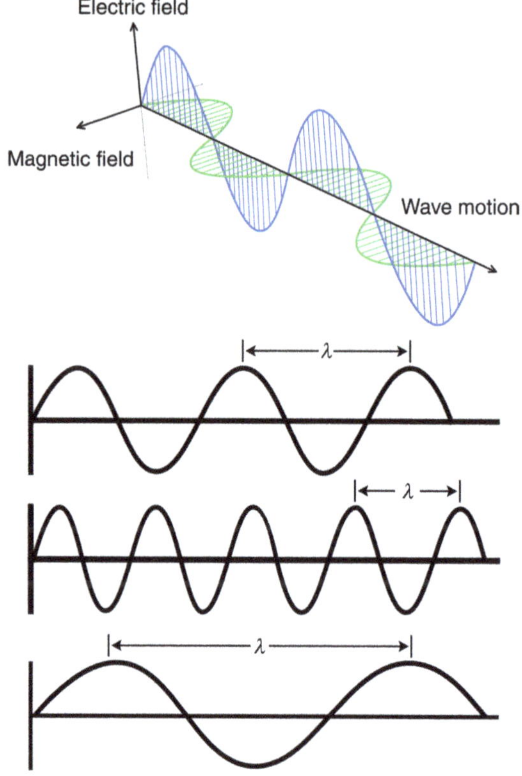

Figure 2.6 Example of an electromagnetic wave propagating in a transverse manner with varying wavelength lambda. [Image by Sakib Chowdhury]

Wavelength or frequency are very important parameters of an E-M wave if we are to understand target interaction and other issues such as spatial accuracy and sensitivity. We will talk more about this in Chapter 3.

In this book, wavelength rather than frequency will be emphasized to explain why certain E-M waves behave the way they do when interacting with certain targets on the ground. For now, readers can use a simple rule regarding E-M waves and keep that handy when trying to understand remote sensing of water:

1. A longer-wavelength E-M wave will have lower frequency because the product of the two must equal the speed of light. Such a wave can be considered as being less "excited" but "strong", with momentum to overcome obstacles in its course to reach a target or receiver. Think of a marathon runner.
2. Wavelength is generally correlated to the spatial resolution at which a target is remotely sensed. In other words, a shorter wavelength can reveal smaller spatial features of the target. The downside of this can be that a shorter wavelength may be easily weakened as it travels through a medium (which in our case is the atmosphere). There is also an altitude connection of the satellite here that we will cover later.

Now think of a marathon runner (long wavelength, lower excitation) versus a 100-meter sprinter (short wavelength, higher excitation). If you asked a marathon runner to run a 100-m

Figure 2.7 Using the example of rope exercise at the gym, we can understand longer versus shorter wavelength and comparative excitation energy as indicated by frequency. [Image: Freepik.com]

sprint or you asked a 100-m sprinter to complete a marathon, what is likely to happen along the way?

Before we close this discussion about wavelengths of E-M waves and how it affects the remote sensing of water, please think of the rope exercise shown in Figure 2.7. If the rope were very heavy, the waves would be longer but have much more force to overcome barriers on the other end. Vice versa, a light rope could be easier to operate for generation of smaller waves, needing less energy, but would likely dissipate quickly at the other end. Building such a conceptual understanding of what wavelength means relative to frequency can help us understand the skill that is associated with the remote sensing of a water-cycle variable for water management.

2.6.2 The E-M Spectrum

E-M waves come in all shapes and sizes, and together they form a spectrum. For remote sensing of water, we will concern ourselves with three parts of the spectrum (Figure 2.8): the visible (that humans can see with their eyes), infrared (near infrared, short infrared and long/thermal infrared), and microwave regions. The latter two regions cannot be seen by the human eye.

The visible part of the E-M spectrum (Figure 2.9) is really a very small part of the overall spectrum. But this part is important, as a lot of remote sensing of water and land cover can be carried out using these E-M waves during daytime when the Sun is out.

Figure 2.8 The E-M spectrum. [Image by Sakib Chowdhury]

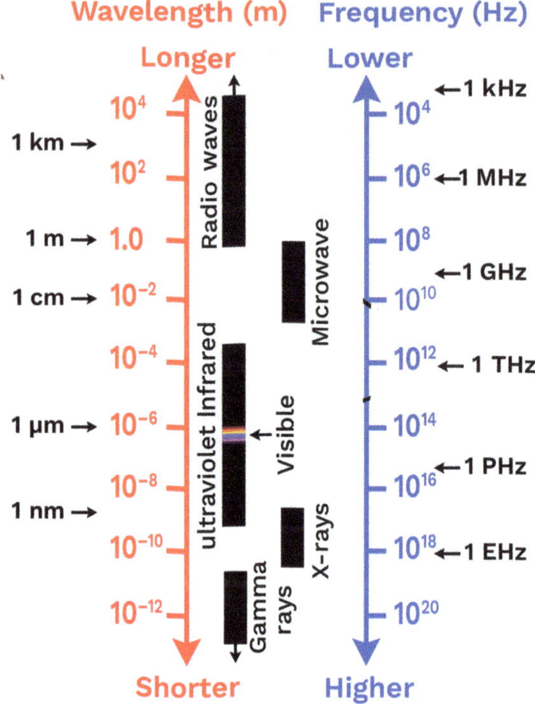

Figure 2.9 Close-up of E-M spectrum in the visible range. [Image by Sakib Chowdhury]

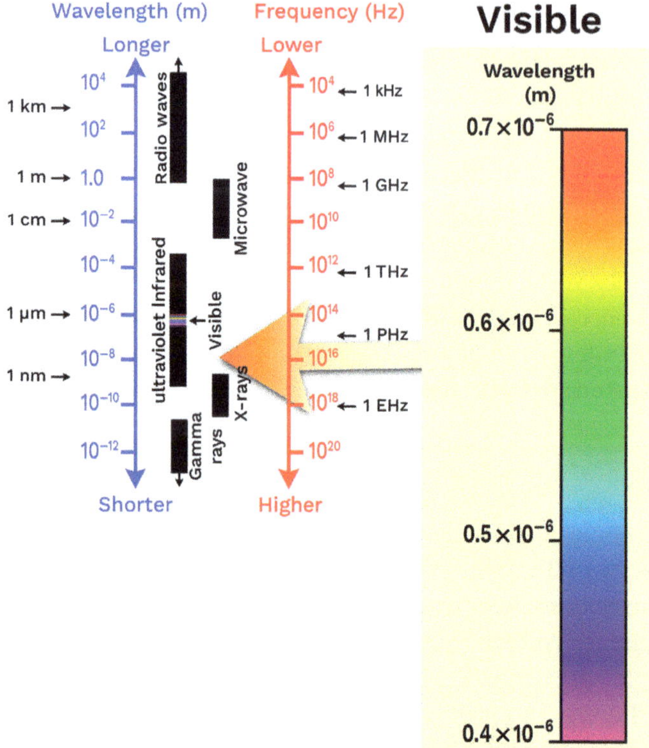

It is not important to memorize the specific wavelengths (as in how long each is) of each color in this visible spectrum. However, it is useful to remember how the colors are "stacked" from shorter (0.4 micrometer) violet color to longer (0.7 micrometer) red wavelengths (remember the mnemonic when you are not sure: VIBGYOR). One micrometer (μm) is one-millionth of a meter (10^{-6} m).

As the red E-M becomes a little longer than 0.7 micrometer, it starts to become less distinguishable to the human eye in the "infra" zone. In other words, infrared (IR for short) is this range of wavelengths longer than the red visible wavelength but typically shorter than 100 micrometers (Figure 2.10).

The IR range is broken down into three subgroups – near, mid, and far (Figure 2.10). The near IR (or NIR) is a wavelength for which many targets (such as vegetation) are reflective, and some are very absorptive (NIR absorbed completely), such as water. The Sun's E-M contains this NIR wavelength, so if we had a sensor that can detect NIR, we could clearly monitor how much of this NIR is being reflected and where. This NIR is also the wavelength used in many security cameras. Because it cannot be detected by the human eye, a security camera is easy to conceal from a robber at night while the camera secretly records the NIR reflection.

At the other end of the IR E-M spectrum, we have mid and far IR waves. These waves carry more thermal or heat energy than shorter wavelengths if the source producing the energy is

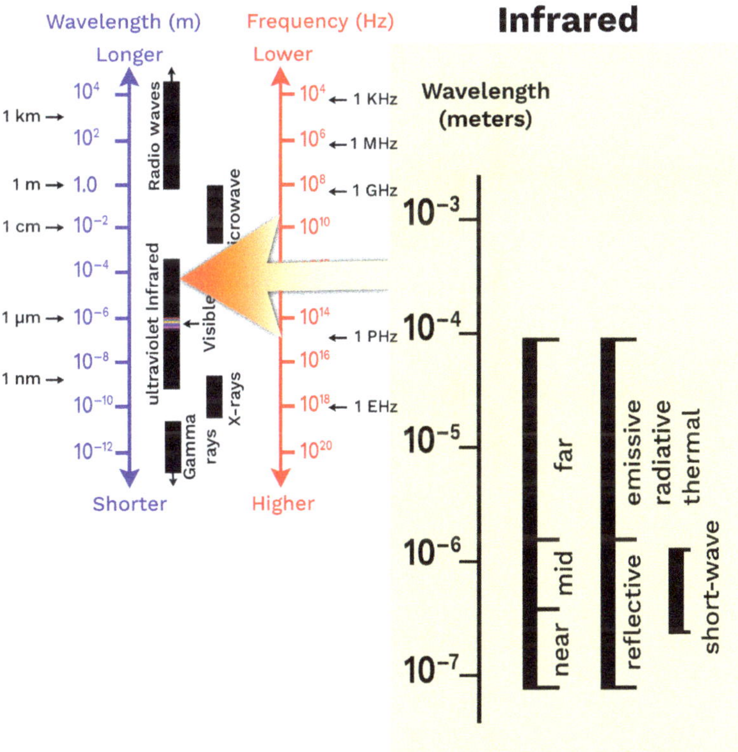

Figure 2.10 Close-up of the infrared region of the E-M spectrum. [Image by Sakib Chowdhury]

relatively lower temperature (than, say, a body at 1,000 kelvin, K). We will see later why that is the case, using basic E-M theory.

Any object that has a temperature higher than absolute zero (−273 degrees Celsius or 0 kelvin) is always radiating a combination of various E-M waves. As temperature increases, most of the energy radiated is in shorter wavelengths (also described by Planck's equation). For example, the surface of the Sun has a temperature of 6,000 K, and the wavelength that carries most of the energy is the blue wavelength. For Earth objects and targets, such as soil, trees, rocks, lakes, and cloud droplets, with much lower temperature in the range of 300–350 K (or 10–40 degrees Celsius), most of this energy is contained (and propagated) in the mid or far IR spectrum. This range is also known as the thermal IR range, as the energy is heat and can be felt.

The last and important range of the E-M spectrum used for water remote sensing is the microwave (MW) range of E-M waves, ranging from 1,000 micrometers (1 millimeter) to a million micrometers (or 1 meter) (Figure 2.11). This range is particularly suited for remote sensing of water in either active or passive form, owing to the unique relationship between MW energy and water. This relationship is due to the dipolar nature of water (H_2O), which makes water (as a target) respond differently to MW radiation compared with its background or foreground of non-water objects. The other unique property of MW is that cloud cover is

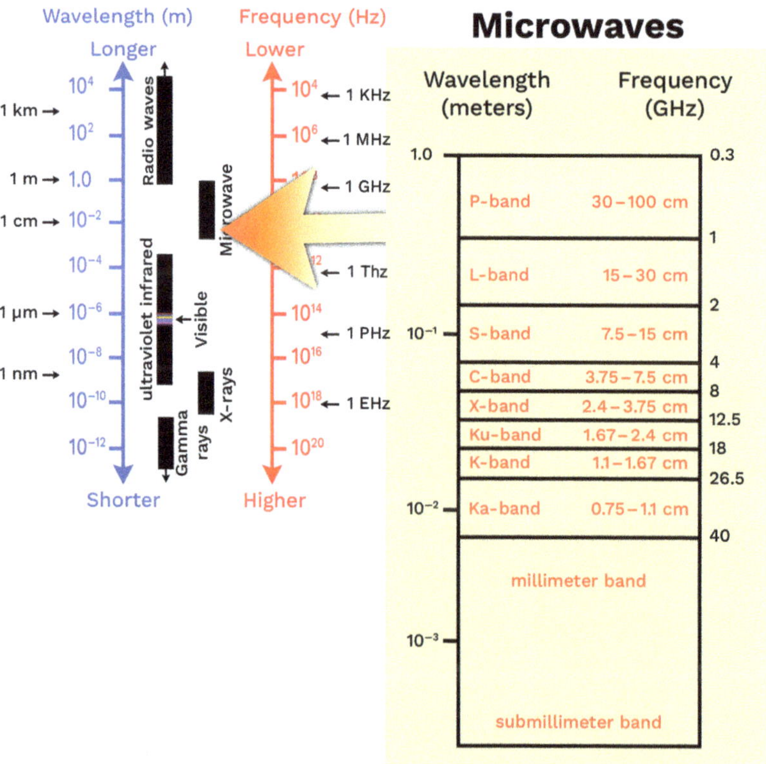

Figure 2.11 Close-up view of the microwave range of the E-M spectrum. [Image by Sakib Chowdhury]

mostly transparent to this wavelength range. In other words, the natural radiation in the MW range radiating from the surface (say, land, lake, or ocean surface) can propagate through the clouds and reach the satellite receiver. Vice versa, if MW rays are beamed from a satellite at high altitude towards a target on the ground, the presence of clouds will not matter much. The same cannot be said of IR waves, which are smaller in wavelength. If you are wondering why, then go back to the analogy of the marathon runner versus 100-m sprinter, or the heavy versus light rope (Figure 2.7). Both MW and mid/far or thermal IR E-M waves can also be used to estimate surface temperature of a target – a feature that we will utilize for tracking surface water quality of reservoirs, for plant evapotranspiration, and to explore management of irrigation water.

2.6.3 Target Interactions

As mentioned earlier, perhaps the most important step in remote sensing of water is to understand the interaction happening between the E-M wave and the target, which may be water or a water-relevant geophysical variable. First, let us imagine ourselves to be an E-M wave. We may be sent down to Earth from a satellite to "interact" (or "talk") to a target on the ground. Or we may have already been naturally interacting with the target and need to reach the satellite up above to record what that interaction is. Either way, we have to travel through the atmosphere. In the active mode, it is two-way with the return being "reflected" back to the satellite most of the time. In this propagation through the atmosphere, it is important to remember that the E-M wave will get "tired" because of attenuation, dispersion, scattering, absorption, or a combination of these phenomena (Figure 2.4). A lot of this has to do with the specific E-M wave and the specific atmospheric composition, as summarized in Figure 2.12.

When we want to carry out remote sensing, it is important to choose the E-M wavelength that can propagate through the atmosphere to reach the target, and/or propagate from the target to the satellite sensor. If we choose the wrong E-M wavelength, we will either not have any unique target interaction or be unable to record the target interaction by the satellite high above.

Figure 2.13 shows a typical target interaction with an E-M wave, where the target is a tree. Incident (I) is E-M wave/energy upon the target surface. Absorption (A), transmission (T), and reflection (R) will then follow in some combination. The total incident energy will interact with the surface in one or more of these three ways (A, T, and R). The proportions of each will depend on the wavelength of the E-M energy and the material and condition of the feature. For vegetation, NIR E-M waves are reflected much more off a tree or vegetation than MW E-M waves.

Now let's take this example a bit further to close the basic concept of remote sensing. Imagine our target is a wetland where there is surface water interspersed with vegetation. Our goal is to use remote sensing from satellite to detect the areas that are water and areas that are vegetation. For this exercise, we need to pick an E-M wavelength that will interact uniquely with each sub-target – vegetation and water. It turns out that the NIR E-M wave from the Sun is heavily absorbed by water, whereas it is reflected by vegetation. On the other hand, the blue

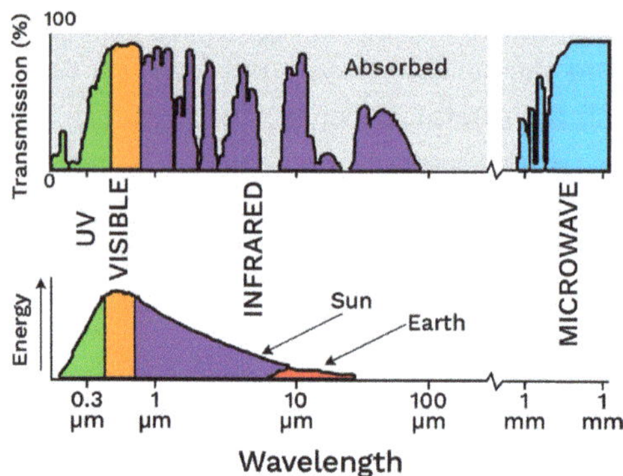

Figure 2.12 Upper panel shows the relative transmission or absorption of an E-M wavelength as it travels through the atmosphere. Notice how there are certain wavelength ranges where there is no transmission and other ranges with near-100% transmission. Lower panel shows the energy that is radiated from the Sun and Earth across the E-M spectrum according to Planck's equation (more about this in Section 3.2). [Image by Sakib Chowdhury]

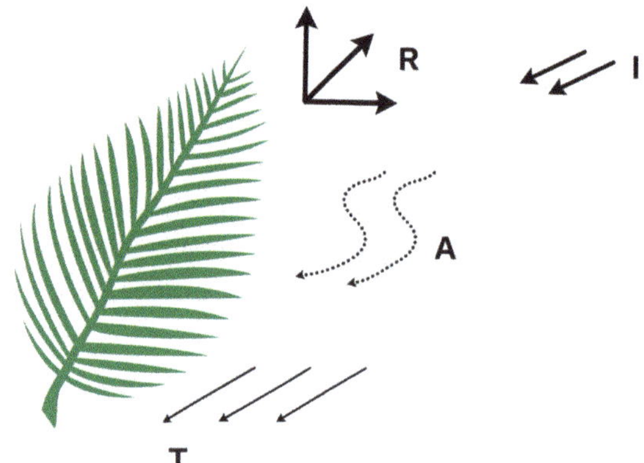

Figure 2.13 Typical interactions of an E-M wavelength with a leaf – reflection (R), transmission (T), absorption (A). I is the incident radiation.

wavelength of the visible E-M wave will be reflected quite extensively by a water surface but absorbed by vegetation (Figure 2.14). A green E-M wave from the Sun will get reflected off vegetation, but it may not be totally absorbed or reflected off water. Overall, it is the NIR wave that is a great discriminator of vegetation and water areas (Figure 2.14), and with the use of other visible wavelengths we can improve the confidence of areas detected as water or vegetated. Specifically, Figure 2.15 tells us that the NIR wavelength of about 0.75 micrometers will give us a map of target interaction where there is no reflection over waters and detectable reflection in vegetated areas.

So in essence, knowing where to "look" spectrally (i.e. on the E-M spectrum) and understanding the factors that influence the spectral response (i.e. target interaction) of the features of interest are critical to correctly seeking information on the target(s). These are best summarized as the *"Golden Rules of Remote Sensing"* (well, for most remote sensing applications). Figure 2.16 shows an example of this in the visible E-M wavelength.

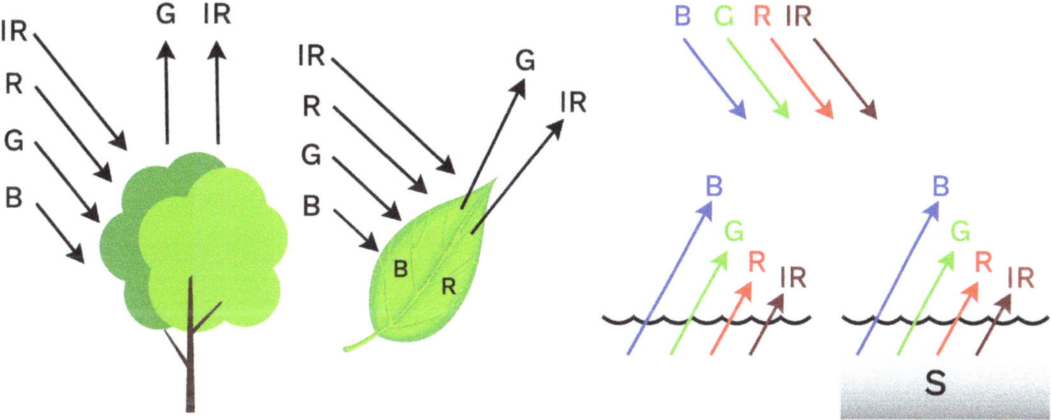

Figure 2.14 Differences in interaction of E-M waves of visible and IR wavelengths with vegetation and water. [Image by Sakib Chowdhury]

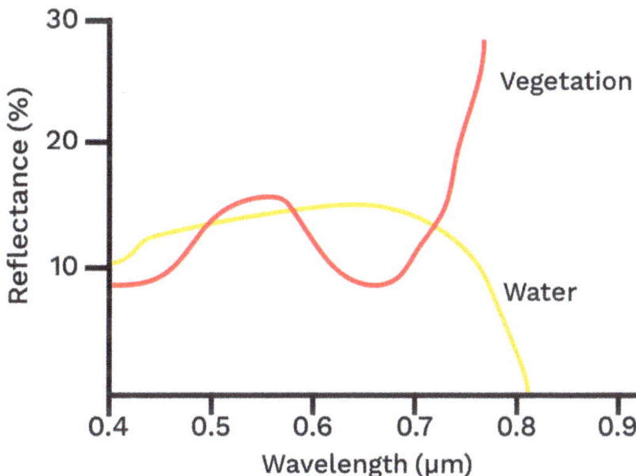

Figure 2.15 The reflectance of E-M wavelengths in the visible and NIR range for water and vegetation. [Image by Sakib Chowdhury]

Figure 2.16 The target must be distinguishable from the background in the particular E-M wavelength being used for remote sensing. [Image by Sakib Chowdhury]

2.7   Golden Rules of Remote Sensing

Rule 1: The target (object) must be distinct electromagnetically from the background.
Rule 2: The right form of E-M energy needs to be applied until the discrimination of the target from the background is maximum.

2.8   Types of Remote Sensing

It is worthwhile now to review the two major types of remote sensing, based on how the E-M energy is used (Figure 2.17).

Active remote sensing is where the satellite sensors direct E-M energy generated on board to "engage" the target and generate an interaction which is then recorded by the satellite's receiver.
Passive remote sensing is where the satellite does not direct any energy onto the target. In passive sensing, the satellite receiver records the natural (i.e. passive) radiation reaching the sensor at the top of the atmosphere.

These two types of remote sensing have their strengths and limitations. Passive remote sensing is obviously cheaper and easier to operate and maintain in space as it has a smaller energy consumption footprint. This is why most Earth-observing satellite missions in many fields have started out as passive missions. The downside of passive remote sensing is that the natural radiation propagating or reflected from the target can often be weak and hard to distinguish from the background radiation. This often requires the satellite to have a coarser spatial resolution to optimize the radiometric resolution (see later).

Figure 2.17 Active and passive remote sensing by satellites. [Image by Sakib Chowdhury]

Active remote sensing has a larger energy consumption footprint, and such satellites are not as widespread as passive satellite sensors. However, the method can yield more accurate information on the target than passive techniques, as there is greater control over the target interaction. Because of the energy requirement and the limited real estate in space, active satellites often have to compromise between lifespan (battery) and wavelength (longer wavelength would require larger antennae and more energy for generation).

2.9 Remote Sensing Platforms

Even though this book is about water remote sensing by satellites in space, it is helpful to review the various types of platforms that can be employed to carry out remote sensing. The word "platform" here refers the nature of "mounting" of sensors used by remote sensing instruments. The platform on which a particular sensor is mounted can determine several attributes of remote sensing such as altitude, swath, and revisit time (discussed later).

The three broad types of platforms are: ground-based; air-borne (or airplane); and satellite (or space) (Figure 2.18). A ground-based platform can be fixed or mounted on a mobile platform such as trucks. The distance between sensor and target is quite small, and such a platform affords highly accurate remote sensing. The limitation is that for global monitoring,

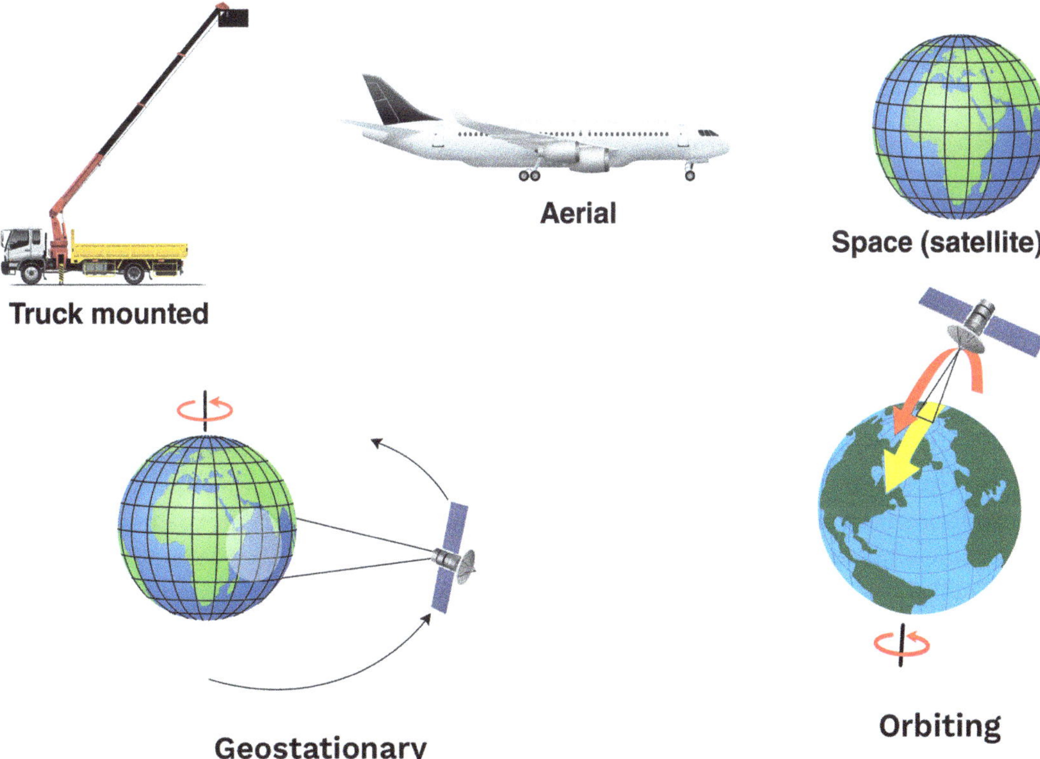

Figure 2.18 Various platforms for remote sensing. [Image by Sakib Chowdhury]

such a platform can only remote sense where there is land. Also, if the goal is to have highly frequent observations while covering a lot of ground globally, ground-based platforms would be severely limited.

In airborne remote sensing, sensors are mounted on an aircraft facing downwards or sideways to obtain images of the Earth's surface. Airborne sensing, like ground-based platforms, has the capability to offer higher spatial resolution images than satellites in space, owing to the lower altitude of sensing. The disadvantages are low coverage area and high cost per unit area of ground coverage. It is not cost-effective to map a large area frequently using an airborne remote sensing system. Airborne remote sensing missions are often carried out as one-time operations.

The most reliable and cost-effective platform for global and frequent monitoring of water is a satellite, which is spaceborne. Satellites can be classified by their orbital geometry and timing. Three common orbits commonly used are geostationary, Sun-synchronous (polar orbiting), and Sun-asynchronous. A geostationary satellite has a period of rotation equal to that of Earth (24 hours) so the satellite always stays over the same location on Earth. In Sun-synchronous orbit, the satellite orbits the polar region but crosses the equator typically at local noon. The solar panels of such satellites can be in the lit portion of Earth, thereby allowing power generation and longer lifespan. Also, the consistent timing of an overpass allows the study of climatology. In Sun-asynchronous orbit, the satellite crosses the equator at different times of the day. Such satellite platforms allow the study of diurnal variation of the target.

2.9.1 Altitude and Swath

As a satellite revolves around Earth, the sensor "sees" a certain portion of the Earth's surface. The area imaged on the surface is referred to as the swath. Imaging swaths for spaceborne sensors generally vary between tens and hundreds of kilometers in width. Look at the swath as the width of a paintbrush when painting a board. A wider brush would need fewer strokes to complete painting the board with the same color. Similarly, a wider-swath satellite can cover more ground in each revolution around Earth. Consequently, a wider swath can result in higher temporal frequency due to overlapping swaths (revisit time: time between consecutive satellite observations) for the same number of orbits in a day.

The altitude of the satellite is the vertical height from the ground to the position of the satellite in space (see Figure 2.19). Typically, Earth-observing satellites can have an altitude from a few hundred kilometers to more than 1,000 kilometers. In Chapter 3 we will see how the altitude affects the number of orbits possible in a day. As a rule of thumb, considering all things equal and maintaining the same wavelength, raising the altitude generally results in lowering the spatial resolution (increasing the footprint) of observation of the target. A way to mitigate this limitation would be to select smaller wavelength at higher altitudes, especially at geostationary orbits, which require more than 35,000 km of altitude (see Chapter 3 on why or how).

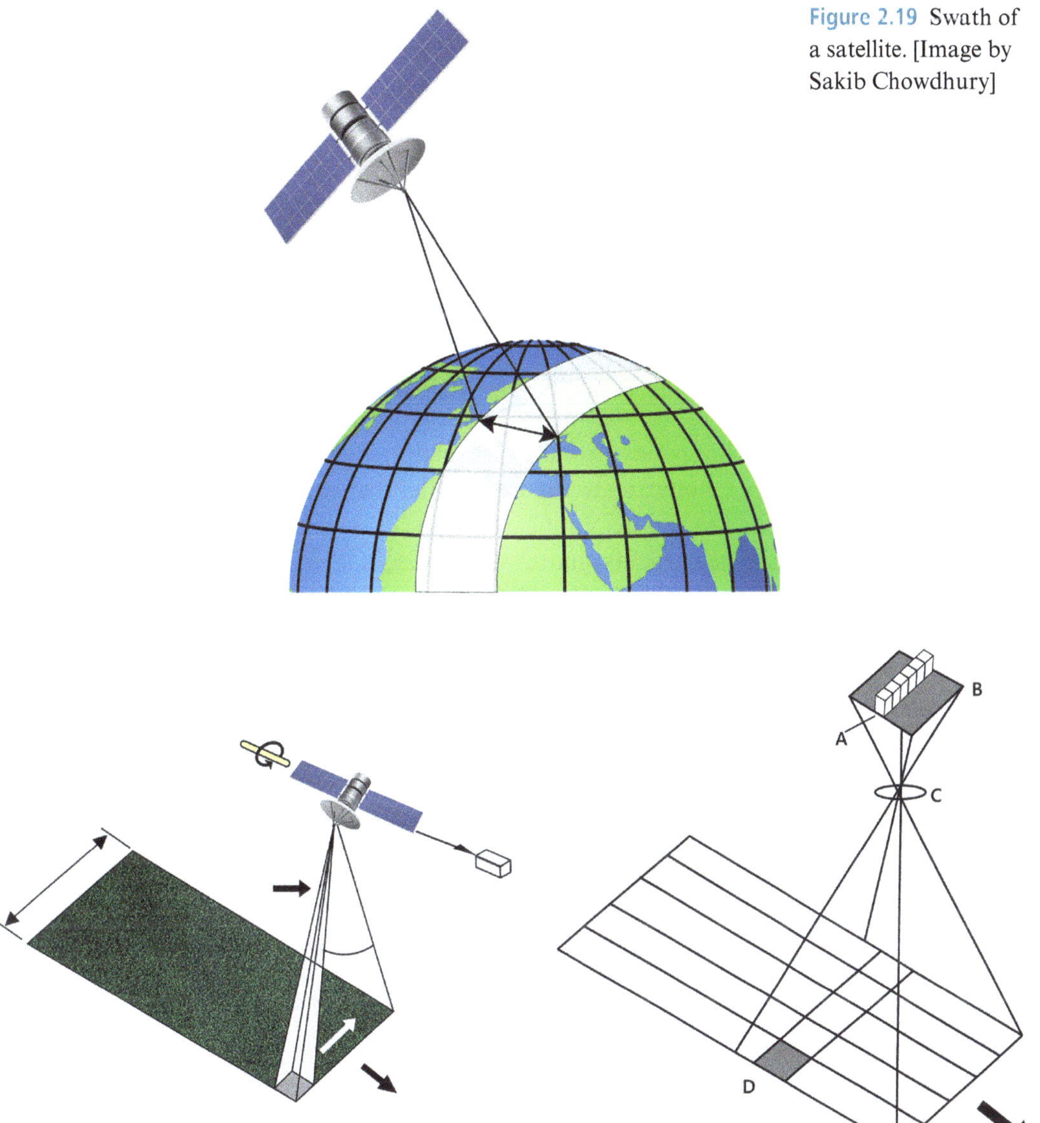

Figure 2.19 Swath of a satellite. [Image by Sakib Chowdhury]

Figure 2.20 Scanning types for remote sensing. Left: across-track scanning; right: along-track scanning. [Image by Sakib Chowdhury]

2.9.2 Remote Sensing Scanning Types

There are two main modes or methods of scanning used by satellites – across-track scanning, and along-track scanning (Figure 2.20). Across-track scanners scan Earth in a series of lines. The lines are usually perpendicular to the direction of motion of the sensor platform. As the platform moves forward over the Earth, successive scans build a spatial image along the orbit.

Along-track scanners also use the forward motion of the platform to record successive scan lines and build up a two-dimensional image, perpendicular to the direction of motion of satellite. However, scanning uses a linear array of detectors (A) located at the focal plane of the image (B) formed by lens systems (C), which are "pushed" along in the flight track direction (i.e. along track). These systems are also known as pushbroom scanners, as the motion of the detector array is analogous to the bristles of a broom being pushed along a floor (Figure 2.20).

2.9.3 Spatial Resolution and Pixel Size

The detail that is detectable in remotely sensed image is dependent on the spatial resolution of the sensor and refers to the size of the smallest possible feature that can be detected (Figure 2.21, right panel). The spatial resolution of satellites depends primarily on their instantaneous field of view (IFOV). Referring to Figure 2.21 (left panel), the IFOV is the angular cone of visibility of the sensor (A) and determines the area (B) on the Earth's surface that is "seen" from a given altitude at one particular moment in time. The size of the area viewed is determined by multiplying the IFOV by the distance from the ground to the sensor.

Most remote sensing images are composed of a matrix of pixels, which are the smallest units of an image. It is important to distinguish between pixel size and spatial resolution – they are not the same conceptually (Figure 2.21). If a sensor has a spatial resolution of 20 meters and an image from that sensor is displayed at full resolution, then each pixel should have an area of 20 m × 20 m on the ground. In this case the pixel size and resolution are the same. However, it is possible to aggregate the smallest features to a coarser resolution and display with larger-sized pixels, and, vice versa, disaggregate and display with smaller-sized pixels.

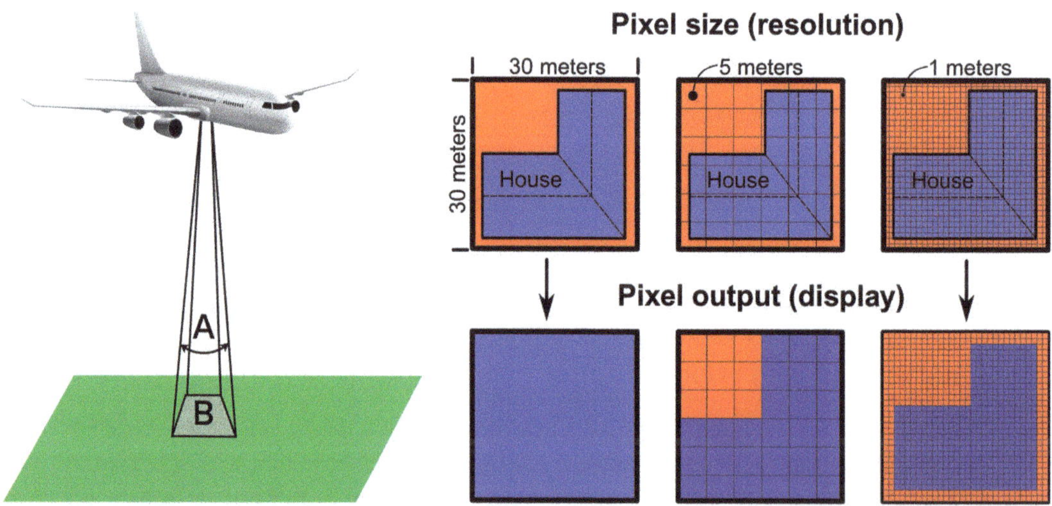

Figure 2.21 Pixel size and resolution. Left panel shows spatial resolution B with A as angle of view and C as altitude. Right panel shows an example of various levels of pixel aggregation using the example of a roof top. [Image by Sakib Chowdhury]

2.9.4 Other Remote Sensing Resolutions

Spectral resolution describes the ability of a sensor to define fine wavelength intervals. The finer the spectral resolution, the narrower the wavelength range for a particular channel or band. Black and white film records wavelengths extending over much or all of the visible portion of the electromagnetic spectrum. Its spectral resolution is therefore low, as the various wavelengths of the visible spectrum are not individually distinguished. Color film, on the other hand, is sensitive to the reflected energy over the visible portion of the spectrum and is able to record the various visible colors. Therefore a color film has higher spectral resolution than black and white film.

Radiometric resolution is defined as the sensitivity to the magnitude of the electromagnetic energy. The radiometric resolution of an imaging system describes its ability to discriminate very slight differences in E-M energy received at the sensor. The higher (or finer) the radiometric resolution of a sensor, the more sensitive it is to detecting small differences in reflected or emitted E-M energy.

2.9.5 Why We Cannot Maximize All Three Resolutions

We live in a world of tradeoffs when it comes to remote sensing. This means that we cannot maximize all three types of resolution (spatial, spectral, and radiometric) for remote sensing of a target. Something has to be sacrificed. For high spatial resolution, the sensor has to have a small IFOV, which reduces the amount of energy that can be detected, as the area of the ground resolution cell within the IFOV is smaller. This leads to reduced radiometric resolution, which is the ability to detect fine energy differences. Now, if we want to increase the amount of energy detected (and thus the radiometric resolution) without reducing spatial resolution, the wavelength range detected would have to be broadened for a particular channel or band. Doing this would, however, reduce the spectral resolution of the sensor. Conversely, coarser spatial resolution would allow improved radiometric or spectral resolution or both. Thus, the three types of resolution need to be balanced against the desired capabilities and objectives of the sensor and the goals of remote sensing. It is always a balancing act, as we shall see in later chapters.

Case Study 2.1: Earliest Form of Modern Remote Sensing

Aerial photography can be considered as the earliest form of modern-day remote sensing, which began with the invention of the camera. Soon after the development of photography, people became interested in taking photographs from the air. The earliest aerial photographs were taken from balloons.

In the 1850s, Gaspard-Félix Tournachon captured the first aerial photograph, using a hot air balloon (Figure 2.22). The picture taken remotely was of a French village in 1858. Unfortunately, none of these earliest aerial photographs exist today.

Case Study 2.1 (cont.)

Figure 2.22 Gaspard-Félix Tournachon, performing the earliest form of remote sensing from a hot air balloon. [Credit: Online Collection of Brooklyn Museum; photo: Brooklyn Museum, 2004]

Case Study 2.2: **Remote Sensing for Archaeology and Earth's History**

Using remote sensing, it is possible to identify loose soil that may have been prehistoric agricultural fields, or a site for burial. The Maya causeway was detected this way, using infrared radiation reflected at a different wavelength from surrounding vegetation. More advanced versions of such multispectral scanners (visible and IR) can detect irrigation ditches filled with sediment because they hold more moisture and thus have a temperature different from other soils. Also, the ground above a buried stone wall can be a touch hotter than the surrounding terrain because the stone absorbs more heat. These are all examples of the golden rule of remote sensing, to uncover human history long before modern-day remote sensing began as a subject.

Remote sensing has revealed ancient drainage patterns and rivers that once existed in the Sahara, when it was lush with verdant forests (Figure 2.23). Using radar images taken from a Japanese Earth observation satellite, researchers found ancient riverbeds running from the middle of the Sahara to the Mauritanian coast in West Africa, which appear to have originated in the Atlas Mountains to the north and the Hoggar Mountains to the east (Skonieczny et al., 2015).

Figure 2.23 Radar images reveal ancient rivers that once flowed through the Sahara desert, based on work reported by Skonieczny et al. (2015). [Photo credit: Philippe Paillou]

2.10 Conclusion

In this chapter, we have overviewed the basic concepts of remote sensing and the key principle based on which most remote sensing techniques work. This principle is that the "target" of interest has to appear distinct from the background. We learned that remote sensing requires energy, and in most cases, this energy is electromagnetic energy. Therefore, it is particularly important to understand the electromagnetic spectrum. In water management, the three regions of the spectrum that this book will cover are visible, infrared, and microwave. We also learned the meaning of terms such as background, foreground, transmission, scattering, attenuation, and reflection, among others. We learned about the three types of resolutions – spatial, spectral, and radiometric – and why it is necessary to achieve a balance among the three to get the best possible remote sensing of the target. In the next chapter, we will explore the physical theory behind remote sensing. The physical theory will help us understand the pros and cons, the advantages and limitations of using certain type of electromagnetic energy for the target of interest. Once we have built this basic understanding of the physical theory, we will be able to proceed to future chapters on remote sensing of specific targets, such as precipitation (Chapter 5), surface water (Chapter 6), and discharge (Chapter 7), and applications such as reservoir management (Chapter 8), irrigation management (Chapter 9), and water temperature management (Chapter 10).

EXERCISES: CHAPTER 2

Q2.1 Define "remote sensing" in two sentences max.

Q2.2 Itemize and then describe briefly the essential components and processes necessary for remote sensing (you should list at least three, and you should indicate how each one fits in the overall scheme through a drawing).

Q2.3 List a few everyday remote sensing techniques that you observe in life.

Q2.4 Qualitatively, show the components (microwave, infrared, visible, ultraviolet etc.) of the electromagnetic spectrum (i.e. show the relative position of each wave type as wavelength increases).

Q2.5 What are "reflected IR" and "thermal IR" radiation (in terms of their wavelength and behavior)? Explain briefly how each is beneficial for a given application of your choice.

Q2.6 What are the golden rules (fundamental principles) of remote sensing?

Q2.7a If you wanted to map the deciduous (e.g. maple, birch) and the coniferous (e.g. pine, fir, spruce) trees in a forest in summer using remote sensing data, what would be the best way to go about this and why? Use reflectance curves similar to Figure 2.15 illustrating the spectral response patterns of these two categories to help explain your answer.

Q2.7b What is the advantage of displaying various wavelength ranges, or channels, in combination as color images, as opposed to examining each of the images individually?

Q2.8 If the IFOV for all pixels of a scanner stays constant (which is often the case), then what will be the impact of resolution at nadir and off-nadir angles? At which look angle will the spatial resolution be coarse and why? Here the look angle is the angle made between the beam and the vertical line directly perpendicular to the ground (nadir).

Explain with a diagram and simple geometry.

Does this mean that spatial resolution will vary from the image center to the swath edge?

Q2.9 Explain why the following three resolution parameters – spatial, radiometric, and spectral – cannot be maximized together.

Q2.10 What advantages do sensors carried on board satellites have over those carried on aircraft? Are there any disadvantages that you can think of?

Q2.11 What are the various types of platforms for a remote sensing mission? Explain each one of them briefly.

Q2.12 What is the difference between Sun-synchronous and Sun-asynchronous orbiting satellites? Which one would you choose for performing climatologic studies and for studying the diurnal variation of a phenomenon?

Q2.13 What are "along-track" and "across-track" scanners? Explain the pros and cons of each scanning mechanism.

Q2.14 Between active and passive microwave remote sensing, which sensor would require a larger angle of view (or IFOV) and why?

REFERENCE

Skonieczny, C., P. Paillou, A. Bory, et al. (2015). African humid periods triggered the reactivation of a large river system in Western Sahara. *Nature Communications*, vol. 6, 8751. https://doi.org/10.1038/ncomms9751

SUGGESTED READING

Durand, M., A. Barros, J. Dozier, et al. (2021). Achieving breakthroughs in global hydrologic science by unlocking the power of multisensor, multidisciplinary Earth observations. *AGU Advances*, vol. 2, e2021AV000455. https://doi.org/10.1029/2021AV000455

Lettenmaier, D. P., D. Alsdorf, J. Dozier, et al. (2015). Inroads of remote sensing into hydrologic science during the WRR era. *Water Resources Research*, vol. 51, 7309–7342. https://doi.org/10.1002/2015WR017616

3   A Review of First Principles

3.1   Chapter Overview

In the previous chapter, we overviewed the basic concepts of remote sensing. In this chapter, we will build an understanding of the physical theory. A basic theoretical background will help us to understand the pros and cons, advantages and limitations of remote sensing of a specific target using a specific electromagnetic energy (or wavelength). This understanding is necessary for future chapters that are dedicated to specific targets such as precipitation, surface water, discharge, reservoirs, and crops.

3.2   Introduction

3.2.1   Black Body and Planck's Law or Planck's Equation

Earlier, in Chapter 2, we formalized our approach to remote sensing in the form of "target", background, and foreground. Now the first thing we need to focus on is the electromagnetic behavior of the target. This is best captured by the term "black body". From here on, we will try to think of the target in water management relative to a black body and understand how much it resembles a black body under certain circumstances.

So, what is a black body?

A black body is an object that absorbs all incident radiation on it. In other words, if E-M energy were directed at a black body, it would absorb all. The name "black body" is given because it absorbs all colors of light. For thermodynamic equilibrium, an object cannot go on absorbing without causing thermal instability, so a black body also emits a maximum amount of radiative energy for maintaining equilibrium, i.e. more than any other type of body at the same temperature. Thus, a black body also emits black-body radiation (see Figure 3.1).

There are nuances to this definition of a black body. An object or target can be a black body at a specific wavelength of E-M energy, or across a range of wavelengths, or at all E-M wavelengths. The goal with most targets in remote sensing of a water target (or variable) is to use the type of E-M energy that ensures near black-body behavior compared with its surroundings. If we know that a target behaves as a perfect black body at a given wavelength, or if we happen to know quantitatively the closeness to a black body, then we can use well defined mathematical relationships to estimate its temperature or estimate energy radiating from it at a given wavelength. This mathematical relationship is called Planck's law or Planck's equation, shown below (Equation 3.1).

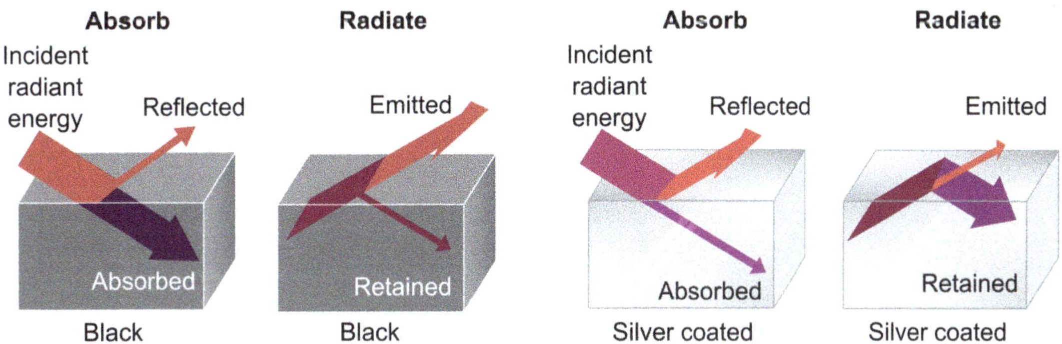

Figure 3.1 Demonstration of the concept of a black body. A black object is a good absorber and a good radiator, while a white (or silver) object is a poor absorber and a poor radiator. [Source: OpenStax College. Located at: http://cnx.org/contents/031da8d3-b525-429c-80cf-6c8ed997733a/College_Physics.]

$$i_\lambda\left(T\right) = \frac{2hc^2}{\lambda^5\left(e^x - 1\right)} \quad \text{where} \quad x = \frac{hc}{k\lambda T} \tag{3.1}$$

c, speed of light, 3.00×10^8 m s^{-1}

h, Planck's constant, 6.63×10^{-34} J s

k, Boltzmann's constant, 1.38×10^{-23} J K^{-1}

The i on the left hand side of Equation 3.1 indicates the E-M energy as flux (intensity over an area the energy passes through), as a function of the temperature and for a given wavelength. Of course, this equation applies to an object that is a perfect black body (absorbs all and radiates all at that wavelength).

If we plotted the equation for various temperatures to see how i behaved as E-M waves became longer, then our plot would look like that in Figure 3.2.

The topmost line, at 5,000 K, can be regarded as one from the surface of the Sun, which has such a high surface temperature. An object that hot would be radiating most of its E-M energy at shorter wavelengths (peaking in the visible range) and much less at longer wavelengths, if it was a black body. On the other hand, an object at a temperature similar to the environmental range we live in on planet Earth (say 300 K) would have radiate most of its E-M energy at very long wavelengths (the 300 K line is not shown here, but you can do the math by using Planck's equation).

In short, what the Planck's equation tells us is that the hotter the black body, the shorter the wavelength carrying most of the radiative energy. Recall the earlier discussion in Chapter 2 about wavelengths and the exercise rope (Figure 2.7 in Chapter 2). Short-wavelength E-M energy will have high frequency (or excitation) but also very high energy (just see how high the peak is at 5,000 K). This allows the Sun's energy, mostly short wavelength, to penetrate Earth's atmosphere easily and reach the ground to illuminate the planetary boundary layer we live within. On the other hand, the Earth is much cooler (300 K), and the wavelengths radiating are much longer (infrared or longer) but at low energy (just see how the peak decreases as the wavelength increases on the horizontal axis in Figure 3.2). Because of its low energy, such radiation generated by Earth through warming up by the Sun does not pass through the atmosphere

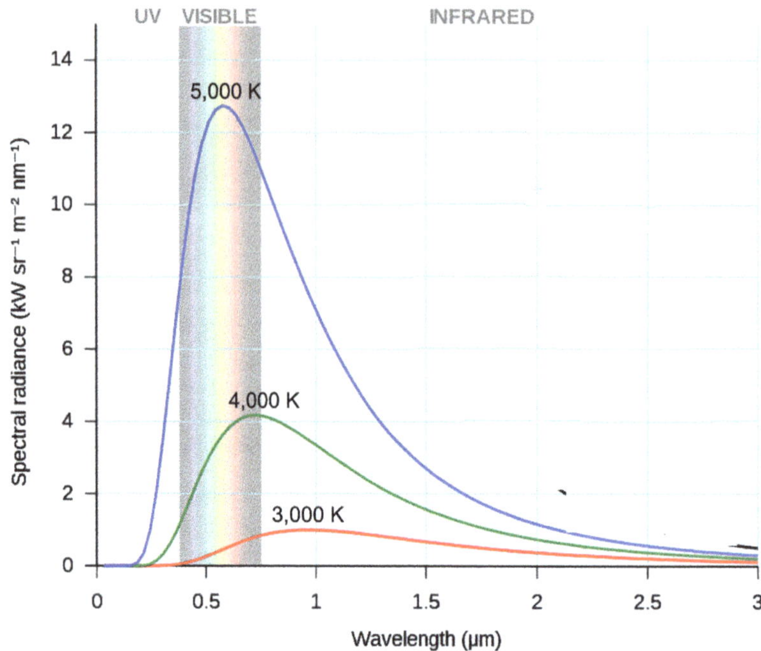

Figure 3.2 Planck's law or equation in graphical form, showing how the intensity of electromagnetic radiation from a black body varies as a function of wavelength of the radiation and temperature of the black body. [Source: Darth Kule, public domain, via Wikimedia Commons]

as easily as the shorter-wavelength, high-energy Sun's rays. Thus, the low-energy radiation from the Earth remains trapped – creating a more livable environment for us with thermal equilibrium (Figure 3.3). The trapping of the energy is due to an effect called the greenhouse phenomenon – and it is exactly as the name suggests, if you have seen a greenhouse that keeps the inside warmer than the outside during winter.

There is a downside to too much trapping of the Earth's longer-wave radiation, and this is now happening because of excessive greenhouse gas emissions from anthropogenic activities. Too much trapping of the Earth's heat means a rise in the Earth's planetary temperature (and it has been rising since the later nineteenth century). It is important to stress, however, that the greenhouse effect is what makes our Earth habitable for humans and many other lifeforms. Without such an effect, the Earth's planetary temperature would theoretically be around 255 K (or about −18 degrees Celsius).

3.2.2 Stefan–Boltzmann Law

If we mathematically integrate Planck's law across all possible wavelengths (from zero to infinity) to get the total E-M energy summed over all wavelengths rather than the flux for a given wavelength, we get an elegant equation known as Stefan–Boltzmann's equation (Equation 3.2).

The greenhouse effect

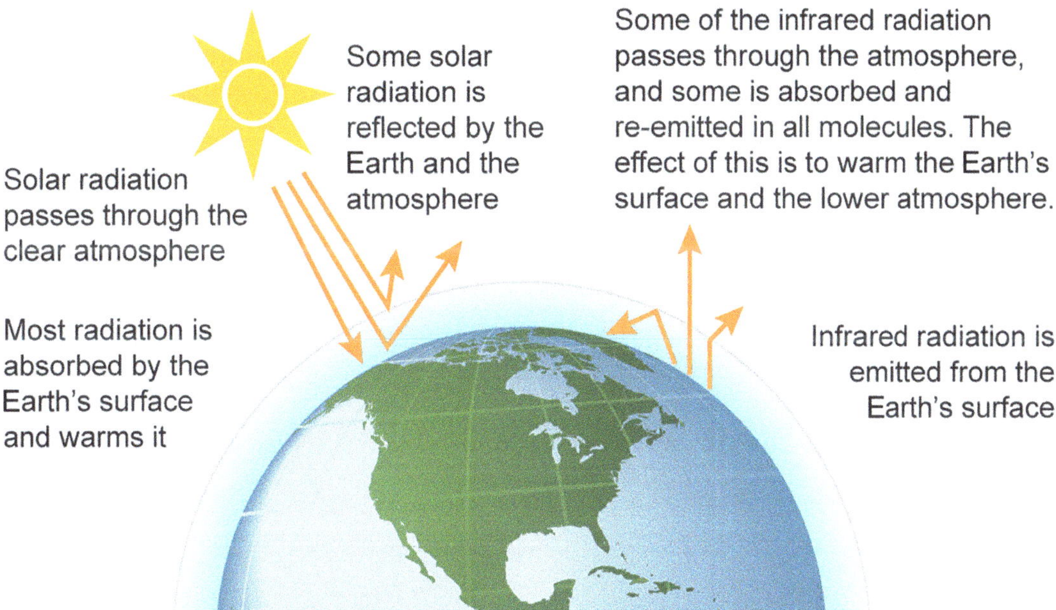

Figure 3.3 Greenhouse effect, and how the planet Earth is kept warmer. [Source: US Energy Information Administration. Image adapted from US Environmental Protection Agency]

$$E = \sigma T^4 \qquad (3.2)$$

E, power per unit area, $\mathrm{W\,m^{-2}}$

T, temperature, K

σ, Stefan–Boltzmann constant, $5.67 \times 10^{-8}\,\mathrm{W\,m^{-2}\,K^{-4}}$

In addition to tracking the total E-M energy from a blackbody target as a function of temperature, the Stefan–Boltzmann equation (Equation 3.2) helps us carry out some interesting academic exercises on how planetary temperature is maintained by the Sun's energy. The Sun's energy is incident on a circular area with the radius of the planet. However, the planet absorbs all that and becomes warmer in all directions (assuming thermodynamic equilibrium) over a spherical surface area. For a given albedo (which is the fraction of total incident solar radiation that is reflected back to space), we can actually calculate the planetary temperature if none of that longer-wavelength radiated E-M energy was trapped via greenhouse gases (Figure 3.3). For Earth, with an albedo of 0.3, the planetary temperature with no long-wave radiation trapping by greenhouse gases would be 255 K (−18 °C). For Mercury, with an albedo of 0.06, it is 442 K; and for Venus, it is 227 K. Venus has a runaway greenhouse gas problem due to its high concentration of methane.

Example (adapted from *Atmospheric Science: An Introductory Survey* by Wallace and Hobbs, 2006): Calculate the equivalent blackbody temperature of Earth, assuming a planetary albedo of 0.32. The planetary albedo is the fraction of the total incident solar radiation that is reflected back into space without absorption. Assume that Earth is in radiative equilibrium, so that there is no net energy gain or loss due to radiation.

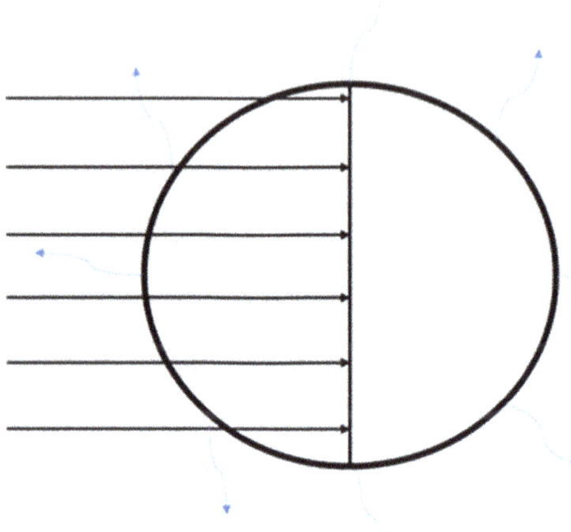

Figure 3.4 Incoming arrows – solar radiation; outgoing arrows – planetary radiation.

Let S be the irradiance of solar radiation incident upon Earth ($1{,}380 \ \text{W m}^{-2}$); E the irradiance of planetary radiation emitted to space; R_E the radius of Earth; and A the planetary albedo. For radiative equilibrium, incoming flux = outgoing flux (see Figure 3.4):

$$(1-A)S\pi R_E^2 = E 4\pi R_E^2$$

Therefore,

$$E = (1-A)S/4 = 234.6 \ \text{W m}^{-2}$$

Now applying the Stefan–Boltzmann equation, we can find the blackbody temperature of Earth for the given radiation:

$$T_E = 253.3 \ \text{K} \ \left(\text{or } 19.3 \text{ degrees Celsius below freezing}\right)$$

3.2.3   Wien's Displacement Law

Wien's displacement law (Equation 3.3) is derived from Planck's equation to find the wavelength λ at which the maximum intensity of E-M waves is radiated for a black body at a given temperature T:

$$\lambda_{\text{max}} \left[\mu\text{m}\right] = \frac{2{,}897}{T} \tag{3.3}$$

T, blackbody temperature (K).

Example: Calculate the power radiated by a 10-cm-diameter sphere of aluminum at room temperature (20 °C). Assume it is a perfect radiator (black body).

Power radiated per unit area (from the Stefan–Boltzmann equation), J

$$J = \left(5.67 \times 10^{-8}\, \text{W m}^{-2}\, \text{K}^{-4}\right) \left(293\ \text{K}\right)^4$$

$$= 418\ \text{W m}^{-2}$$

$$\text{Area} = 4\pi r^2 = 4\pi \left(5 \times 10^{-2}\, \text{m}\right)^2 = 3.14 \times 10^{-2}\, \text{m}^2$$

$$\text{Power radiated} = J \times \text{area} = 13\ \text{watts}\ \left(\text{Answer}\right)$$

What is the wavelength of peak emission for the surface of the Sun at 6,000 K? Assume the Sun is a perfect black body.

Apply Wien's displacement law here.

$$\lambda_{\max} = 2{,}897 / 6{,}000\ \left(\text{in micrometers, } \mu\text{m}\right)$$

$$= 0.4828\ \mu\text{m or } 483\ \text{nm (which is the wavelength for blue light)}$$

3.3   Properties of Atmosphere and Surface

Let us talk now about some metrics that quantify certain properties of the atmosphere when an E-M wave is traveling through it. We can look at these as "travel" metrics. Remember, the atmosphere is the medium of propagation here when we are remotely sensing water from space. As an E-M wave travels, whether it is from the target or towards the target, it experiences transformation that is important for us to understand. When the E-M radiation strikes a surface, part of it is absorbed, part of it is reflected, and the remaining part, if any, is transmitted, as illustrated in Figure 3.5. The fraction of irradiation absorbed by the surface is called the absorptivity, α, the fraction reflected by the surface is called the reflectivity, ρ, and the fraction transmitted is called the transmissivity, τ. That is,

$$\text{Absorptivity}: \quad \alpha = \frac{\text{Absorbed radiation}}{\text{Incident radiation}} \quad 0 \leq \alpha \leq 1 \tag{3.4}$$

$$\text{Reflectivity}: \quad \rho = \frac{\text{Reflected radiation}}{\text{Incident radiation}} \quad 0 \leq \rho \leq 1 \tag{3.5}$$

$$\text{Transmissivity}: \quad \tau = \frac{\text{Transmitted radiation}}{\text{Incident radiation}} \quad 0 \leq \tau \leq 1 \tag{3.6}$$

In general, these travel metrics are dependent on the wavelength and composition of the atmosphere. As they are basically fractions of the total incoming radiation, their sum is 1.

$$\rho(\lambda) + \alpha(\lambda) + \tau(\lambda) = 1 \tag{3.7}$$

Note also that reflectance over the entire wavelength range is called *albedo*.

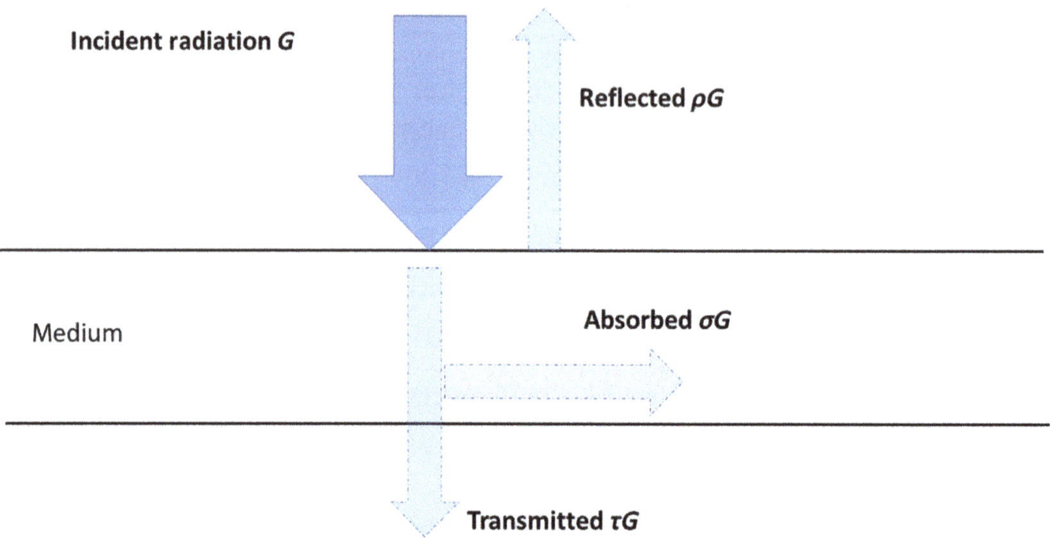

Figure 3.5 E-M radiation as it propagates through a material or medium. Here ρ, σ, and τ are fractions that add up to 1.

3.4   Emissivity

As mentioned earlier, all objects at temperatures above absolute zero (>0 K) emit E-M radiation. However, for any particular wavelength and temperature, the amount of thermal radiation emitted depends on the emissivity of the object's surface. This is where the metric emissivity comes in, which is perhaps the most important property of a target and the background for understanding how effective remote sensing of water will be based on the golden rules (see Chapter 2).

Emissivity is defined as the ratio of the energy radiated from a material's surface to that radiated from a perfect emitter, known as a black body, at the same temperature and wavelength and under the same viewing conditions. It is a dimensionless number between 0 (for a perfect reflector) and 1 (for a perfect emitter).

$$\text{Emissivity, } \varepsilon = \frac{\text{Energy emitted from a surface}}{\text{Energy emitted by a black body at same temperature}} \tag{3.8}$$

The emissivity of a surface, whether it is the target or background, depends not only on the material but also on the nature of the surface. For example, a clean and polished metal surface will have a low emissivity, whereas a roughened and oxidized metal surface will have a high emissivity. The emissivity also depends on the temperature of the surface as well as wavelength and angle of the incident E-M radiation.

3.4.1   Emissivity of Targets on Land and Oceans

For remote sensing of water from space, our key target will often be water (in its various forms – cloud droplets, snow, liquid water, frozen ice). And this target may be over land or

Table 3.1 Emissivity of various objects in the thermal infrared wavelength range. Notice how water acts as a near-black body

Material	Emissivity
Polished silver	0.02
Polished copper	0.03
Polished gold	0.03
Aluminum foil	0.07
Wood	0.85
Asphalt pavement	0.90
White paint	0.90
Vegetation	0.94
White paper	0.94
Water	0.95
Black paint	0.98

Collected from various sources on the internet such as www.thermoworks.com/emissivity-table/ and www.transmetra.ch

over ocean as the background. So, it is important for us to be aware of the emissivity of the water or water-sensitive targets in various forms, as well as emissivity of the background. Table 3.1 shows emissivity values of some standard materials in thermal infrared wavelength range.

A few other factors that can alter the emissivity of a target are as follows:

Moisture content – the more moisture an object contains, the greater is its ability to absorb energy and become a good emitter. Wet soil particles have a high emissivity similar to water.

Compaction – the degree of soil compaction can affect emissivity.

Field-of-view – the emissivity of a single leaf measured with a very high-resolution thermal radiometer will be different from an entire tree crown viewed using a radiometer with coarse spatial resolution.

Wavelength – the emissivity of an object is generally considered to be wavelength dependent.

Viewing angle of the satellite sensor.

A few rules of thumb that we can follow to help us out here are as follows:

• In the thermal IR (4 to 100 micrometers), nearly all natural surfaces are efficient emitters with emissivity >0.8 (unless they are polished with a very smooth surface).
• In the shortwave region (0.1–4 micrometers), emissivity is negligibly small.
• In the microwave region (0.1–100 centimeters), emissivity depends on the type and state of the surface.
• Emissivities of some water-relevant surfaces in the IR region (10–12 micrometers) are shown in Table 3.2.
• Figure 3.6 shows how emissivity varies in the microwave wavelength ranges for various water surfaces.

Table 3.2 Emissivity for various water related variables in the thermal IR wavelength

Surface	Emissivity
Water	0.993–0.998
Ice	0.98
Green grass	0.975–0.986
Sand	0.949–0.962
Snow	0.969–0.997

Collected from various sources on the internet, such as www.thermoworks.com/emissivity-table/ and www.transmetra.ch

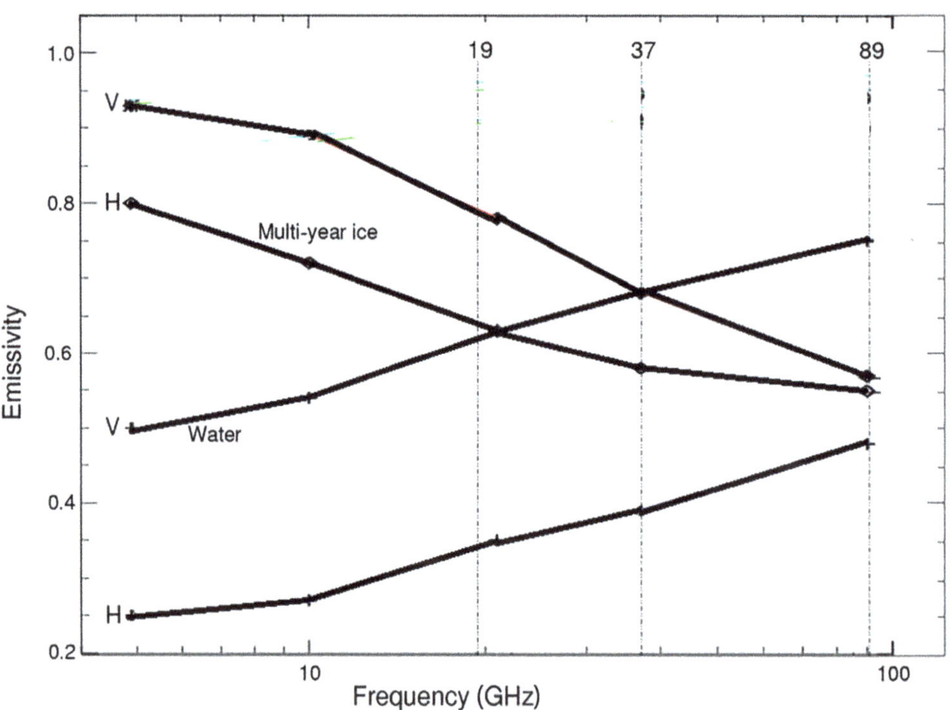

Figure 3.6 Variation of microwave emissivity for various surfaces as a function of frequency. Here H is for horizontal emissivity and V for vertical emissivity. [Adapted from Spreen et al., 2008]

3.5   Kirchhoff's Law

Kirchhoff's law was postulated by the German physicist Gustav Robert Kirchhoff, and states that the emissivity and the absorptivity of a surface at a given temperature and wavelength are equal. In other words,

For an arbitrary body emitting and absorbing thermal radiation in thermodynamic equilibrium, the emissivity is equal to the absorptivity.

$$\text{emissivity } \varepsilon = \text{absorptivity } \alpha \qquad (3.9)$$

One way to look at this is to remember the following rule of thumb:

(1) Good absorbers are good emitters but poor reflectors.
(2) Poor absorbers are poor emitters but good reflectors.

The law is based on the Second Law of Thermodynamics and the common sense of energy balance. If a target is not reflecting the E-M energy incident on it and if it is also not dispersing, then it must be absorbing; whatever it is not able to absorb should be getting reflected or dispersed. Also, a target cannot go on absorbing indefinitely and not respond in kind, as it has to maintain equilibrium. To maintain this thermodynamic equilibrium, it must also emit what it absorbs.

$$\text{For a } \textbf{black body}: \qquad \varepsilon(\lambda) = \alpha(\lambda) = 1 \qquad (3.10)$$

$$\text{For a } \textbf{gray body}: \qquad \varepsilon(\lambda) = \varepsilon = \alpha(\lambda) = \alpha < 1 \qquad (3.11)$$

$$\text{For a } \textbf{non-black body}: \quad \varepsilon(\lambda) = \alpha(\lambda) < 1 \qquad (3.12)$$

If reflectivity increases, then emissivity must decrease. If emissivity increases, then reflectivity must decrease.

For example, water absorbs almost all incident energy and reflects very little at many wavelengths. Blue is an exception where reflectivity is quite high for pure water. Therefore, water is a very good emitter and has a high emissivity close to 1, especially at thermal infrared and microwave wavelengths. Conversely, a sheet metal roof reflects most of the incident energy and absorbs very little, yielding an emissivity much less than 1. Therefore, metal objects such as cars, aircraft, and metal roofs almost always look very cold (dark) on thermal infrared imagery – they are poor emitters.

3.6 Brightness Temperature

Satellite remote sensing sensors record E-M energy that is either reflected from the target or radiated/emitted by the target. They do not record or measure directly the specific water-related geophysical variable we are interested in for water management, such as precipitation rate, discharge, or soil moisture. Assuming the target is a perfect black body, this E-M energy can be redefined in terms of brightness temperature, which makes for a more convenient way of understanding remote sensing of the target vis à vis its background.

Brightness temperature (T_b) is defined as the temperature of a black body that emits the same intensity as measured for that object. Brightness temperature is found by inverting the Planck function (equation). For a black body, brightness temperature = kinetic temperature. Here, the kinetic temperature is basically the temperature we can feel or measure with a thermometer.

$$T_b{}^4 = \varepsilon T^4 \qquad (3.13)$$

3.7   Atmospheric Windows and Absorption Band

Let us now talk about atmospheric windows. A window is an opening that allows us to see the outside world. A window also allows the Sun's rays to enter a room to illuminate the inside. In a similar way, an atmospheric window is a range of wavelengths of the E-M spectrum that can propagate through the Earth's atmosphere and reach its surface. If the target is on the ground and the satellite is up in space, we need to use wavelengths that are atmospheric windows. The optical, infrared, and radio windows comprise the three main atmospheric windows (Figure 3.7a). We can see from Figure 3.7a that there are certain ranges of wavelengths for which transmission is zero or negligible. On the vertical axis, 100% opacity means no transmission. These ranges are often called absorption bands and should be considered as a closed window. However, such absorption bands can have value in inferring the make up or composition

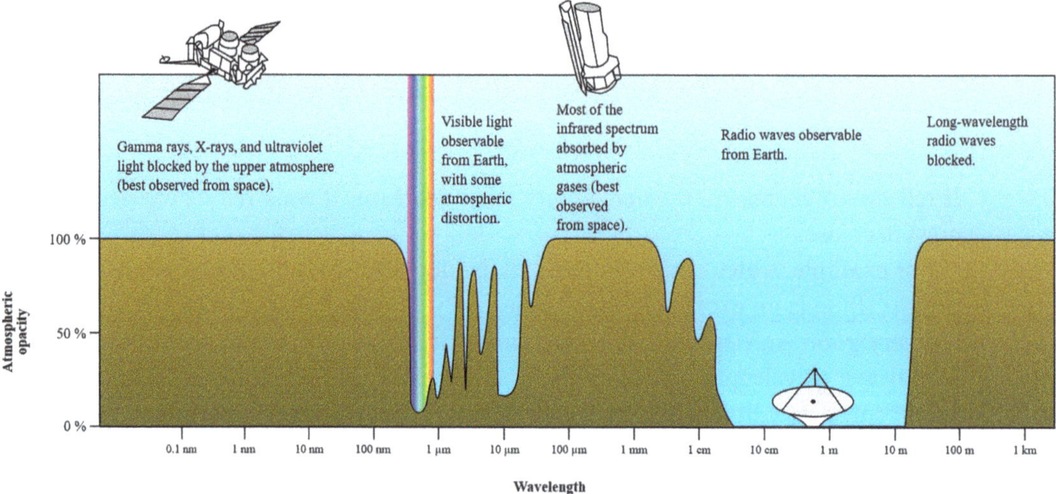

Figure 3.7a A diagram of the E-M spectrum with the Earth's atmospheric opacity and the types of sensors used to image parts of the spectrum. [Source: NASA]

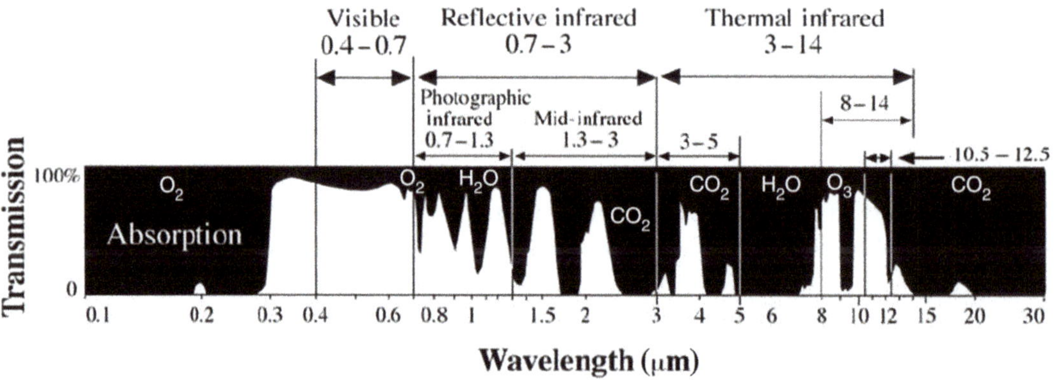

Figure 3.7b Atmospheric windows in the E-M spectrum, showing the role that gases play in absorption.

of the atmospheric column, as the absorption at specific wavelengths is due to specific gases (Figure 3.7b). Although not the subject of this book, such bands or "closed windows" are used widely in the remote sensing of atmospheric chemistry, tracking methane and other gas compositions.

3.8 Thermal Infrared Imaging

Thermal infrared (TIR) energy is emitted from all objects that have a temperature greater than absolute zero. However, we should keep in mind that according to Planck's equation, when the temperature is very high (say 6,000 K, which is roughly the temperature of the Sun's surface), most of the E-M energy generated is in the shorter nonthermal IR range, and the thermal IR energy is almost negligible.

Humans and most animals can detect thermal energy only through the sense of touch. Our eyes are not sensitive to the reflective infrared (0.7–3 μm) or thermal infrared energy (3–14 μm). Some animals, such as snakes, rattlesnakes in particular, appear to have an innate ability to "see" the heat signature of a target in the TIR range. Engineers have developed detectors that are sensitive to thermal infrared radiation. These thermal infrared sensors allow humans to sense a previously invisible world as they monitor the thermal characteristics of the landscape (Figure 3.8).

3.9 The Retrieval Problem in Satellite Remote Sensing

As mentioned earlier, remote sensing of a geophysical variable related to water is always an indirect method. We are not measuring the water variable directly in the way we would measure precipitation using a bucket to collect rain and find out the accumulation over a period of

Figure 3.8 Thermal infrared imagery of a landscape as sensed by the GOES (Geostationary Orbiting Environmental Satellite) TIR sensor. [Source: NOAA]

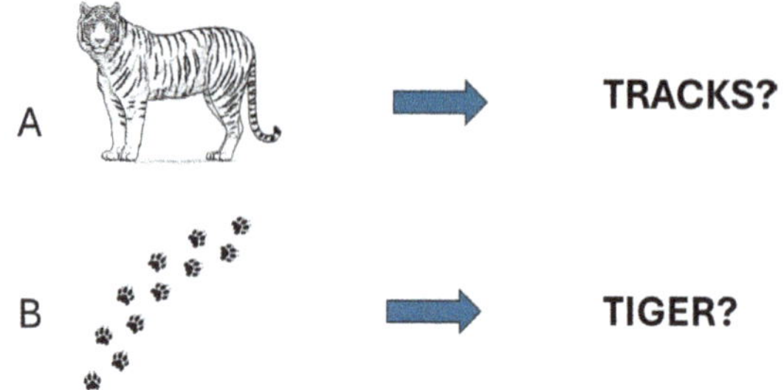

Figure 3.9 The retrieval problem in remote sensing. (A) represents the forward problem by directly measuring the variable of interest; (B) represents the inverse problem using indirect measurements made by the satellite. [Modified from Bohren and Huffman, 1983]

time. Rather, in remote sensing, we are tracking or measuring the E-M energy either radiated, reflected, dispersed, or absorbed by the target representing the geophysical variable. Often, this E-M energy is defined in the form of brightness temperature or reflected energy. This indirect approach poses to us a retrieval problem, which means that we have to reconstruct from the measurements the unknown quantity that we are interested in. This reconstruction is called "retrieval" or "inversion" (Figure 3.9).

Mathematically, we can present the retrieval problem as inverting the forward model (see below). Here, the forward model presents a relationship between satellite remote sensing data and the variable (target) of interest via a function F. Y is what the satellite measures, but x is really what we are after. For example, Y could be brightness temperature at a certain wavelength for a cloud top; x may be the rainfall rate expected from that cloud.

$$\text{DATA} = \text{FORWARD MODEL (STATE)} + \text{NOISE}$$

$$Y = F(x) + n \tag{3.14}$$

where Y is data (radiance/reflectivity)

F is forward model – for example a radiative transfer model or instrument response

x is state vector variables: physical/chemical properties

n is noise: instrument noise

To get to x, we have to invert the forward model as follows:

$$X \pm \sigma_x = F^{-1}(Y, n) \tag{3.15}$$

Using E-M theory, it is sometimes possible to set up a physical and mathematically robust equation between x and Y, often known as a radiative transfer model, with function F.

The challenge is that often there may not be a well defined mathematical and physically unique radiative transfer relationship between x and Y in Equation 3.15. Thus, we sometimes have to solve this problem by collecting sufficient information on Y and corresponding x and creating either look-up tables (like nomographs) or statistical regression equations. Other methods to solve the inverse model are by machine learning or optimal estimation theory. Owing to a lack of a physically tight and unique relationship between the E-M measurement and our water variable of interest, we often run into remote sensing problems becoming ill-posed. In other words, multiple solutions may produce similar outcomes, or the solution may be very sensitive to slight quantitative changes in the satellite remote sensing observations.

Despite these challenges, satellite remote sensing has evolved over the past many decades to yield fairly reliable and often skillful and actionable information for water management by using a teamwork of satellite sensors, wavelengths, and methods, as we shall see in later chapters. Such multi-sensor and multi-algorithm methods have now helped unleash the full potential of satellite data for water resources management. The good news is that the quality of remote sensing data for water management keeps getting better each year.

3.10 Orbital Mechanics

Earth-observing satellites orbit around Earth at a given altitude or try to maintain the same rotational speed as the Earth if they are in geostationary orbit. In this section, we will explore how altitude and rotational speed play a role in the design or choice of a satellite platform or orbit.

First, let us review the basic laws of physics as Isaac Newton first formulated a few hundreds of years ago.

Laws of Motion

1. Every body continues in its state of rest or of uniform motion in a straight line unless it is compelled to change that state by a force applied upon it.
2. The rate of change of momentum is proportional to the applied force and is in the same direction as that force. Momentum = mass × velocity, so law (2) becomes

$$F = \mathrm{d}(mv)/\mathrm{d}t = m\mathrm{d}v/\mathrm{d}t = ma \tag{3.16}$$

3. For every action, there is an equal and opposite reaction.

Law of Gravitation

The force of attraction between any two particles is

- proportional to their masses m_1 and m_2
- inversely proportional to the square of the distance between them

$$F = Gm_1m_2/r^2 \tag{3.17}$$

$$G = \text{gravitational constant} = 6.673 \times 10^{-11}\,\mathrm{N\ m^2\ kg^{-2}}$$

These laws explain how a satellite stays in orbit.

Law (1): A satellite would go off in a straight line if no external force (such as gravity) were applied to it.

Law (2): An attractive force makes the satellite deviate from a straight line and orbit Earth.

Law of Gravitation:

This attractive force is the gravitational force between Earth and the satellite. Gravity provides the inward pull that keeps the satellite in orbit.

Assuming a circular orbit, the gravitational force must equal the centripetal force.

$$mv^2 / r = Gmm_E / r^2 \qquad (3.18)$$

where

v = tangential velocity
r = orbit radius = RE + h
RE = radius of Earth
h = altitude of orbit = height above Earth's surface
m = mass of satellite
m_E = mass of Earth

$$v = \sqrt{\frac{Gm_E}{r}} \qquad (3.19)$$

so v depends only on the altitude of the orbit (not on the satellite's mass).
The period of the satellite's orbit is

$$T = 2\pi r / v = 2\pi r \sqrt{r / Gm_E} \qquad (3.20)$$

For example, if a satellite orbits Earth at 600 km altitude, then,

$$r = 600 + RE = 600 + 6{,}378 \text{ (Earth's equatorial radius)} = 6{,}978 \text{ km}$$

Using Equation 3.19, the velocity of the satellite is $7{,}558 \text{ m s}^{-1} = 7.6 \text{ km s}^{-1}$
From Equation 3.20, the time to complete one revolution around Earth is

$$T = 96.7 \text{ min}$$

Case Study 3.1: From the American Museum of Natural History...

As humans, our eyes cannot process "infrared" wavelengths as we can only see the colors of the rainbow. But some animals are able to detect thermal infrared waves which radiate from warm objects.

That includes the venomous snakes that are commonly known as pit vipers. Such snakes have a pair of heat-sensing organs located in "pits" between their eyes and nostrils. Scientists have found that there exists a connection between these sensing organs and the brain via an optic nerve. Two of these front-facing organs help such snakes "triangulate" in total darkness the direction and distance of the source of heat, which is often a warm-blooded prey radiating thermal infrared heat (Figure 3.10).

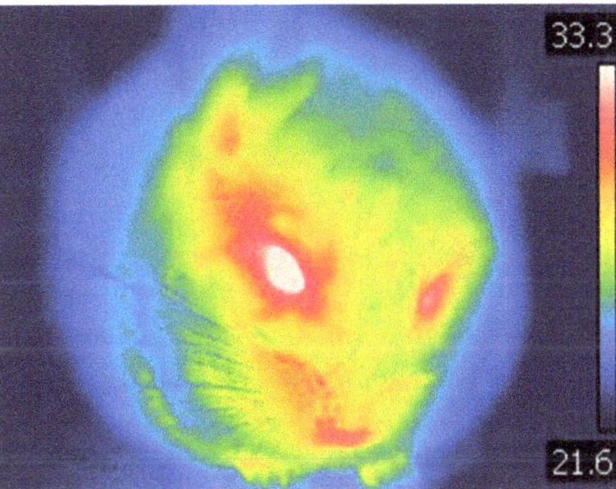

Figure 3.10 How the image of a meal in the form of a chinchilla would look to a rattlesnake moving in the dark. In the thermal IR imagery, red regions indicate higher temperature.

3.11 Conclusion

In this chapter, we have now overviewed the basic physical theory of remote sensing. We have learned what some of the key equations mean, such as Planck's equation, Stefan–Boltzmann's equation, and Wien's displacement theory. We have defined what a black body is, as well as other key terms such as gray body (Equation 3.11), emissivity, atmospheric window, and brightness temperature. This chapter also overviewed the basic theory of motion to understand the orbital mechanics of satellites in space. Before we go on to apply this physical understanding for water-relevant targets, we will take a pause in the next chapter and introduce ourselves to the concept of cloud computing. This is important because all future chapters on remote sensing of specific targets such as precipitation, surface water, and discharge, and applications such as reservoir and irrigation management, will include tutorials and exercises built around cloud computing. You can skip the next chapter if your goal is to develop only a theoretical and conceptual understanding of satellite remote sensing for water management. However, to build hands-on data literacy in using currently available satellite data, the next chapter is important. The specific cloud-computing platform we will introduce our readers to is Google Earth Engine, which is publicly available (at the time of publishing this textbook). We recognize that readers may prefer an alternative "back up" to Google Earth Engine, so we also provide the traditional way of using codes/scripts and data locally.

EXERCISES: CHAPTER 3

Q3.1　What is the "atmospheric window" (for absorption and transmission) and how is it leveraged for remote sensing features of Earth's surface? Give a specific example.

Q3.2   If you are performing active remote sensing of Earth's ground features, would you design your channels in the absorption band or atmospheric window ranges? Why?

Q3.3   What are (a) a perfect black body, (b) a gray body, and (c) a non-black body? Explain this in terms of emissivity.

Q3.4   Environmental satellites (or the remote sensing technique) observe radiative or electromagnetic properties of the target as opposed to the actual phenomenon (e.g. precipitation, soil moisture) that is estimated. Describe in basic steps (i.e. just the overall mathematical formulation) how the physical phenomenon is "retrieved".

Q3.5   If you knew the emissivity, how would you calculate the surface (kinetic) temperature of a surface using a radiometer that measured the brightness temperature at long (microwave) wavelengths?

Q3.6   You are given three equations: Planck's equation, Wien's displacement law, and the Stefan–Boltzmann law. (a) Which one describes the radiative flux (power per area per wavelength) as a function of surface temperature? (b) Which one describes the wavelength at which the radiative flux is maximum for a given surface temperature? Explain how that is derived from (a).

Q3.7   Calculate the time taken for a satellite to revolve around the Earth once from an altitude of 600 km from the ground surface.

Q3.8   If a satellite was orbiting at 250 km from the ground, how many orbits would it be able to complete in a day?

Q3.9   What is the difference between brightness temperature and kinetic temperature?

Q3.10   For a black body with a surface temperature of 35,000 kelvin, which wavelength transmits the maximum amount of E-M energy? What is the total energy emitted by such a black body?

Q3.11   If Earth were flat with a perfectly circular shape (not a sphere), what would be the equivalent blackbody temperature of Earth? Assume planetary albedo of 0.3 and Earth as a black body.

REFERENCES

Bohren, C. F. and D. R. Huffman (1983). *Absorption and Scattering of Light by Small Particles*. Wiley.

Spreen, G., L. Kaleschke, and G. Heygster (2008). Sea ice remote sensing using AMSR-E 89-GHz channels. *Journal of Geophysical Research*, vol. 113, C02S03. https://doi.org/10.1029/2005JC003384

Wallace, J. and P. Hobbs (2006). *Atmospheric Science: An Introductory Survey*. Academic.

SUGGESTED READING

Rees, W. G. *Physical Principles of Remote Sensing*. Cambridge University Press.

4   Introduction to Cloud Computing[1]

4.1   Chapter Overview

In the previous chapters, we built the basic foundation of satellite remote sensing. In this chapter, we will explore cloud computing, a relatively recent innovation in information technology that has dramatically improved data accessibility and the practicality of applying large satellite remote sensing datasets for water management. Future chapters on specific targets and water management themes will have hands-on examples and assignments based on actual satellite data. Most of these chapters will assume prior knowledge of cloud computing for understanding and completing assignments. Since cloud computing is gradually proliferating in all walks of water management practice, the aim of this chapter is to introduce readers to cloud-computing concepts and specific tools currently available for dealing with the very large satellite datasets on water.

4.2   What Is Cloud Computing?

Cloud computing is defined as the delivery of different services through the internet. These resources include tools and applications such as data storage, servers, databases, networking, and software. In simple terms, it can be considered as seeking "services" or outputs by fetching them from the "cloud" where all the analysis and service rendering are carried out, rather than doing everything locally in-house at the user end. More analogously, cloud computing is like seeking services that are located in a distributed manner elsewhere and all that a user needs is "connectivity" to the "grid." For example, if we need electricity to power our homes and offices, we usually do not generate the electricity locally, but plug into the wall's electricity outlet, which is connected to the electricity grid (or electricity cloud) where the energy is delivered to us from the power station. The same could be said for drinking water if you are living in a city or town with a distributed system for pressurized clean water. These days, we do not maintain a groundwater well in our backyard, or fetch water manually from the nearest river or lake ourselves, if we are living in a functional city. Rather, our drinking water is provided by the city's utility service or water district, and we need local plumbing to connect our end to the city's water distribution network (i.e. the "cloud").

When we think about it, the notion of seeking services that are prepared elsewhere and then delivered to the demand point has been something humankind figured out a long time ago, ever since farming began. For example, instead of each individual trying to grow all types of food

---

[1] Contributed by Dr. Matthew Bonnema, NASA Jet Propulsion Laboratory, and Pritam Das, University of Washington.

locally, we often pay to collect food grown more centrally and elsewhere. With this in mind, let us look at cloud computing a little more closely to understand its value for water management in the context of satellite remote sensing.

4.2.1 Why Use Cloud Computing for Satellite Data?

Chances are that we have all been exposed to or affected by cloud computing if we have had a touch-screen phone or used smartphone "apps". These apps send a request to the "cloud" where the massive datasets are kept (for example, checking today's weather and forecast for the week) and can be analyzed. Based on the specific request, analysis is carried out, and the information based on user-specified location is delivered. The beauty of this cloud-computing approach is that connectivity is all that we need; we do not need to store the very massive datasets ourselves. This particularly applies to satellite remote sensing datasets. Recall from Chapter 1 that in 2014, the total amount of remote sensing data on water NASA was generating was in the tens of terabytes (TB) a day. This has exponentially increased each year and took a dramatic upward trajectory with new satellite missions relevant for water management. In Figure 4.1, the raw data generated by five of the major Earth observing satellites for just one region (Australia) can be seen to have increased exponentially, while the total number of satellites launched each year continues to rise (Figure 4.1, left panel).

The broader community has come to the realization that it would be inordinately expensive and also impractical if we always had to download all the raw data over the internet and store it before analyzing it to generate the much smaller-sized outputs, in kilobytes or megabytes, that are relevant for water management decision making. For example, by processing 40 years of data in the visible wavelength from the Landsat satellite mission over a region, we can generate a time series of surface area for a selected reservoir. This time series will be a few orders of magnitude smaller in data size than the main input data. So why not generate the time series in the cloud to spare ourselves from having to download the massive amounts of datasets over the internet (which would be costly and also time-consuming) and just download the time series?

Today, there are satellites that can generate water or land-cover maps, both historical and current, based on massive observation datasets. The user-ready product or service that the water manager needs is an image, probably geolocated (think of a GeoTIFF) and would probably have a size of a few megabytes (MBs). The processing of the raw satellite data would likely involve a few orders more data. When time is of the essence or when simplicity and clarity are more impactful for water management coordination, it makes little sense to have to download the terabyte- or petabyte-scale data routinely, store it, and analyze it locally, assuming all the required CPU facilities existed. A far more powerful approach would be to do all that in the cloud where the satellite data is kept and rapidly extract the inundation maps (just like a mobile phone app would do to fetch such data).

We can come up with many other examples when it comes to water management, such as irrigation advice for farmers (see Case Study 1.3 in Chapter 1), flood forecasting, estimating reservoir storage change, and tracking reservoir operations.

Cloud storage has become increasingly popular and necessary among satellite data agencies, such as NASA and their distributed array of archives center (called DAAC – Distributed Active Archive Center), that need increasingly larger storage space. As mentioned earlier, such cloud-based storage makes it possible to save files to a remote database and retrieve them on demand.

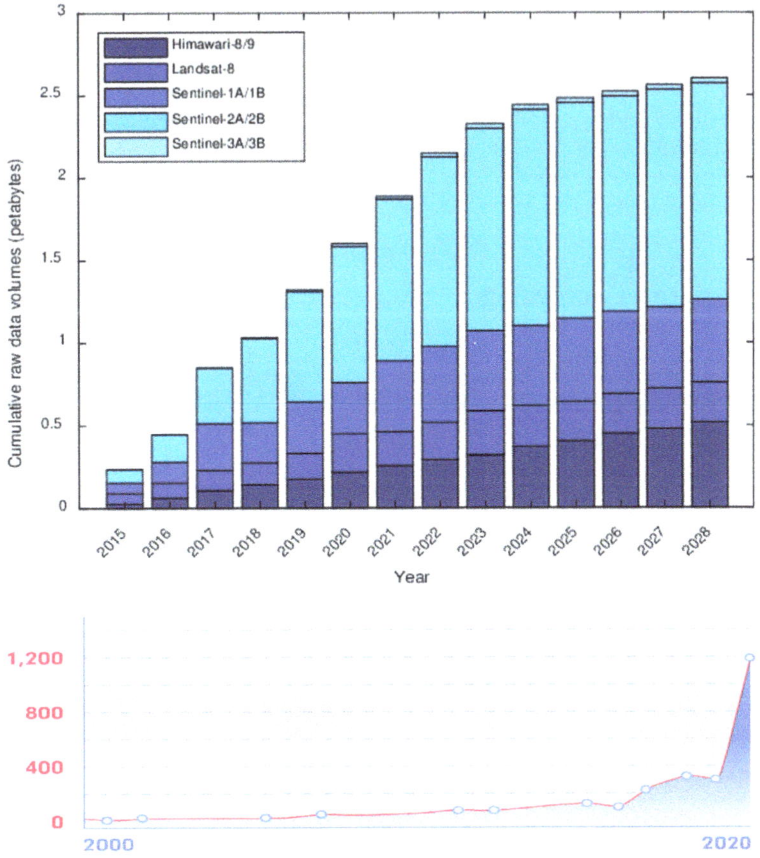

Figure 4.1 Left panel: estimated volumes of Earth observation data produced by the Landsat 8, Sentinel-1, -2, and-3, and Himawari-8/9 missions from 2014 to 2029, for Australia alone. [After Lewis et al., 2017.] Right panel: total number of satellites launched each year since 2000. [Source: Union of Concerned Scientists]

4.2.2   Why Use Cloud Computing for Water Management?

Recall from Chapter 1 that the landscape of water management is changing. It has been changing for the past 20 years. Like other fields, water management today needs to adopt the latest technology to solve emerging challenges, or current problems whose solution is made challenging using traditional water management based on in-situ data only. So we now see water management adopting environmental sensing, Internet of Things, remote sensing, "Big Data", artificial intelligence, and crowd-sourced data, all out of necessity. If you remember from Chapter 1, there is the hydropolitics of water sharing, or not having enough water to share, or ground-based water data being treated as an issue of national security. It is therefore becoming fundamentally impossible to track and manage our water where we live, and in places we need to apply it, using traditional methods (a gauge or sensor in the ground), owing to costs, high need for manpower, hydropolitics, or lack of ground infrastructure.

As innovations continue to explode in water management due to advances made in information technology (see two examples in Chapter 1: Case Study 1.1 and 1.2), today's and tomorrow's water managers need to be exposed to how the field is changing in terms of information

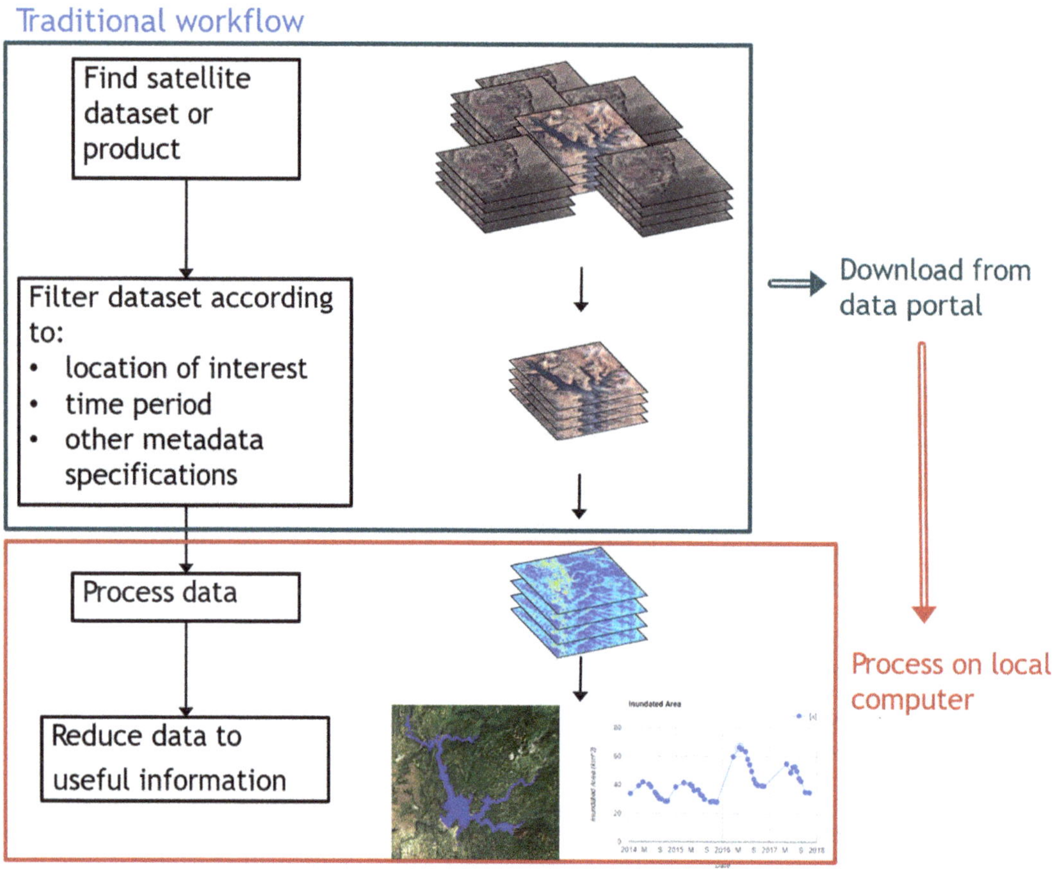

Figure 4.2a Traditional workflow process for satellite data in the pre-cloud-computing environment. [Source: Dr. Matthew Bonnema, NASA Jet Propulsion Laboratory]

technology and the easier access to large datasets on satellite remote sensing of water. The goals of this chapter will therefore be the following:

1. To learn the basic concepts of cloud computing for satellite data geospatial analysis;
2. To learn how to navigate satellite water data in the Google Earth Engine cloud-computing environment.

4.2.3 Visualizing Cloud Computing for Satellite Data

The traditional (non-cloud-computing) process of handling satellite data for water management can be summarized in Figures 4.2a–c.

4.2.4 Why Use Cloud Computing for New Satellite Water Missions such as SWOT and Other Water Data?

As seen from Figure 4.1, newer Earth-observing satellites, measuring water or other variables important for water management, will create terabytes of data every day and therefore the

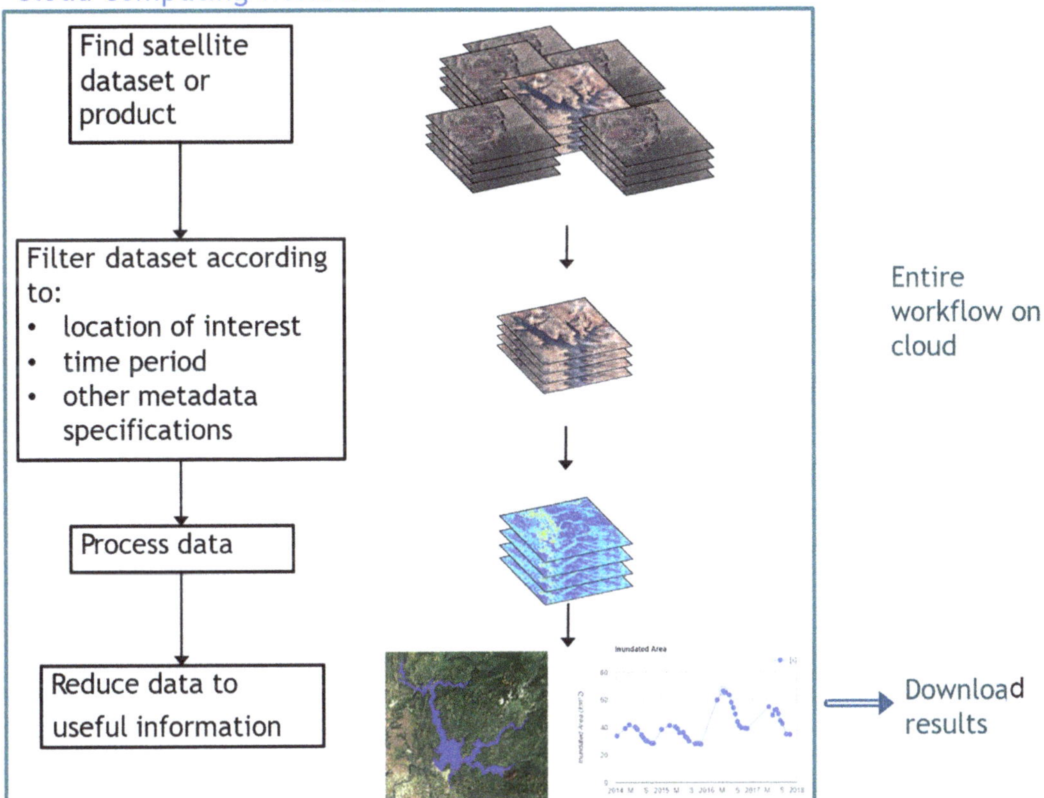

Figure 4.2b The cloud-computing workflow, showing where the key differences lie with traditional at-user computing. [Source: Dr. Matthew Bonnema, NASA Jet Propulsion Laboratory]

download sizes could be prohibitive for use. Also, working with Big Data on local hardware can be challenging and costly owing to the high internet bandwidth requirement. Specifically for satellite missions tailored for surface water, such as the Surface Water and Ocean Topography (SWOT) satellite mission, the data size will experience a dramatic increase (Figure 4.3). SWOT will be covered in a little more detail in Chapters 6 and 7. Major space agencies and satellite data centers, and open-access private companies (such as Google Earth Engine and Microsoft Planetary Computer), are therefore investing in cloud-computing infrastructure particularly for satellite water data accessibility.

4.2.5 How Will Cloud Computing Work?

In order to understand how to make best use of cloud computing for processing and analyzing satellite data, we first need to identify how satellite data is handled in general. The traditional workflow for using satellite data, as seen in Figure 4.2a, to inform water management follows the same basic steps:

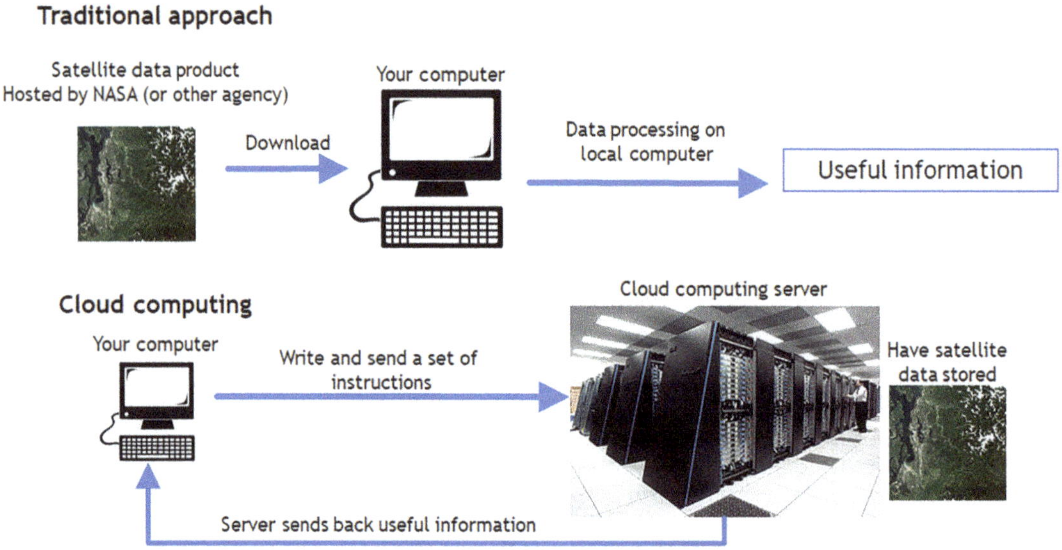

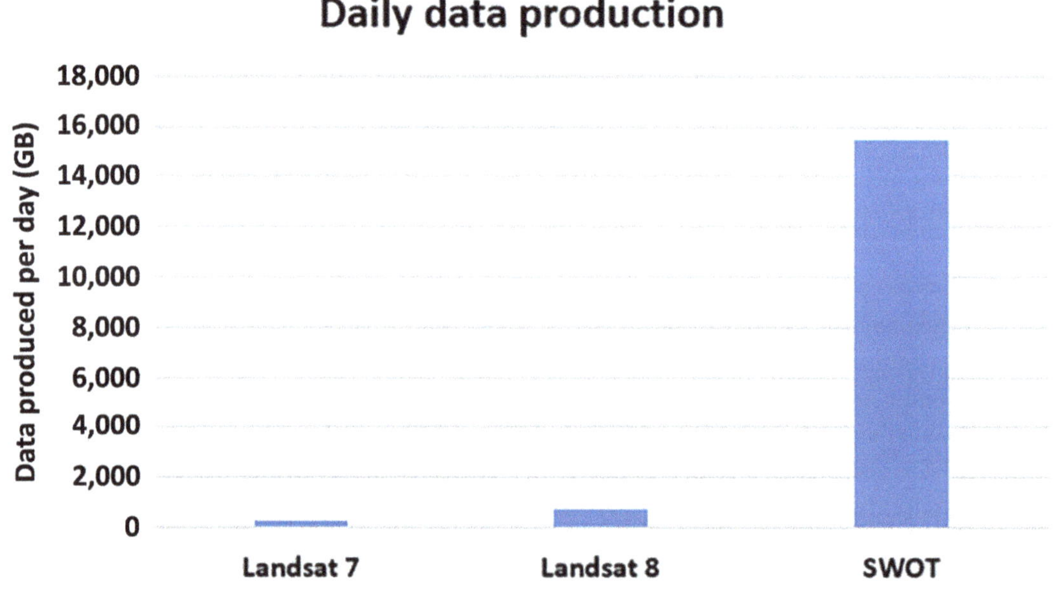

Figure 4.2c Overall summary showcasing cloud-computing versus traditional non-cloud-computing approach. [Source: Dr. Matthew Bonnema, NASA Jet Propulsion Laboratory]

Figure 4.3 Comparison of data production rate for SWOT and Landsat missions.

1. Search for a satellite data product collection (for example at Earth Data Search: https://search.earthdata.nasa.gov/search) and subset the collection to the specific product files located in the space and time of interest.
2. Download the identified data files to a local computer.
3. Process the data using code scripts or tools on the local computer.

We already know from Chapter 3 that remote sensing data rarely provides the exact geophysical quantity needed for a given use and instead must undergo some kind of mathematical translation. In some sense, step 1 has always occurred in the cloud, where the computations needed to find and filter satellite data occur not on a local computer, but rather on a server provided by the distributor of the satellite data. Steps 2 and 3, however, are where significant bottlenecks can occur in this traditional model: not enough computation resources to perform the processing, and/or not enough internet bandwidth and local storage to effectively download and store satellite data. When using cloud computing to handle remote sensing data, it can be useful to identify where such bottlenecks occur in traditional local-computer workflows, and this can guide how you implement your cloud-computing solution.

To illustrate this point, let us take the example of flood management. Processing satellite imagery and using it to identify flooded regions is a relatively computationally light task. It is not a task that you would do on your phone, but even a modest laptop can handle simple water classification algorithms. The real bottleneck is instead in the downloading of the raw satellite data. Satellite imagery consists of many bands, so every image downloaded can be a composite of over 10 individual rasters (Figure 4.4). Effective flood management tends to require higher spatial resolution, and datasets with higher spatial resolution have larger file sizes. Rapidly responding to floods necessitates rapid access and analysis of satellite imagery. Thus, the objective of our cloud workflow is to reduce the amount of data we need to download at the end. This does not necessarily mean completing the entire end-to-end workflow in the cloud, as there are limitations to using the cloud (cost, interoperability with existing local tools or systems). Instead, the solution might be to access and classify the relevant satellite imagery, merge individual water maps across some spatial or temporal domain, download the map, and then feed

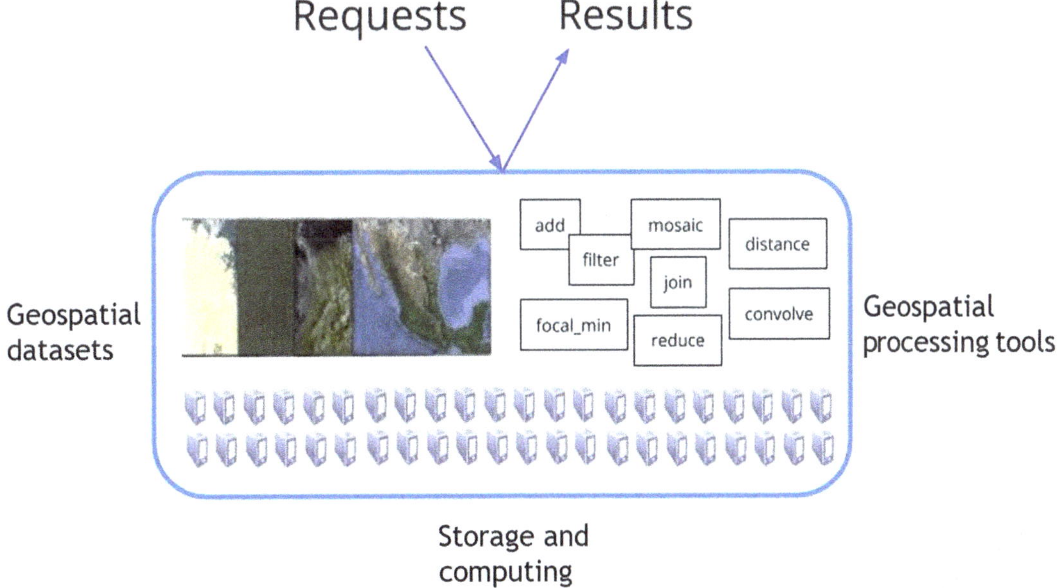

Figure 4.4 Schematic of cloud computing. Users send requests to a cloud-computing platform, and results are sent back to users. [Source: Dr. Matthew Bonnema, JPL, NASA]

the map into a decision support tool or further analyze the map on local hardware to identify affected regions. This approach turns potentially hundreds of gigabytes into a single map of manageable file size that can be downloaded near-instantly. At the point in the workflow when the map becomes a manageable size for rapid download, there is much less benefit to keeping the workflow in the cloud because the major local-computational bottleneck has already been solved. At this point, it is up to the user to weigh the costs and benefits of performing the rest of the process in the cloud, and the answer can sometimes be very specific to use case.

4.3   Providers of Cloud Computing for Satellite Water Data

When implementing cloud computing to solve a water management problem, it is important to be aware of the different cloud-computing platforms available. Before we get into the major players in the cloud field, we first need to acknowledge that the landscape of cloud computing, like all other technology sectors, evolves over time, with new companies trying new approaches, and old companies adding services or changing policies and pricing. The overview provided here cannot account for such changes in the future but will try to cover only the aspects of cloud-computing providers that appear relatively stable for the near future. That said, there are two major players in cloud computing that must be discussed when outlining the options for water management, Amazon and Google.

Amazon Web Services (AWS) is the largest cloud platform in the world (at the time of writing) and has become an essential service for a wide variety of internet applications, corporations, and government agencies. As the name implies, AWS provides a number of cloud services (e.g. tools for making use of many aspects of the cloud). The sheer number and complexity of these services can be overwhelming, so let us start with two of the simplest and most general services, storage and computing. The AWS storage service (called "S3") is similar in concept to any online data storage service (Dropbox, Google Drive, Microsoft One Drive, etc.), in that it is a place to store data on the internet. Users can create storage "buckets" and upload and download files to and from those buckets. This may sound mundane, but the real power of S3 buckets (or any cloud storage service) is how they interact with other cloud services, such as computing services. AWS's basic computing service (called "EC2") is similar in concept to a computer sitting in an office somewhere.

When you want to use it, but don't want to go to the office, you connect to it from your computer at home (or wherever you are with an internet connection) and send instructions for what you want the computer to do, often in the form of command line commands or scripts. The server performs those instructions and sends back whatever you asked for. EC2 computing instances are like this, except instead of a computer, you are given a certain portion of a massive server at an AWS computing center. This EC2 instance can be configured with a variety of hardware options that can be customized specifically for a given task. A key difference between a remote computer and an EC2 instance is that EC2 instances can access data stored in an S3 bucket much faster than local computers can. This means that using an EC2 instance to read and process big satellite data can be dramatically faster doing the same workflow locally, even if the EC2 instance has the same processing power as your local computer. After processing is complete, results can be saved to an S3 bucket, where they can then be accessed by local machines for offline analysis or distributed amongst other users. AWS offers many other cloud services beyond S3 and EC2, but leveraging these services to their fullest potential requires a deeper dive into their intricacies and caveats than can be provided here. Our recommendation for implementing a workflow in the cloud is to first use these two relatively simple and general-use services, and only explore more complex and specific services when use-case-specific issues or deficiencies arise.

A major benefit of using AWS for water management is that all Earth observations collected and distributed by NASA are now hosted in AWS S3 buckets. This means that there is a deep catalog of remote sensing data readily available to access in the cloud, including hydrology data. PO.DAAC (Physical Oceanography Distributed Active Archive Center) is the NASA DAAC that hosts a majority of NASA's hydrologic remote sensing data, such as the SWOT satellite mission, Dynamic Surface Water Extent (DSWx), and Soil Moisture Active and Passive (SMAP) satellite mission (see Table 1.1 for details). The scientists and engineers who work for PO.DAAC acknowledge the difficulties in learning how to best make use of the cloud to access their datasets and have created an extensive series of tutorials, wrapped up in one location titled the "PO.DAAC Cookbook" (https://podaac.github.io/tutorials/). Most of these tutorials are written in Python, but they also include examples of GIS (Geographical Information System) workflows. The entire cookbook has useful tutorials, tailored for different types of remote sensing data, but we will recommend several tutorials below that are especially relevant for water resources management. Before we do, though, it must be acknowledged that while access to NASA data in the cloud is freely available to everyone, performing workflows in EC2 and storing data in your own S3 bucket requires an AWS account and is subject to expenses. There are free-tier options for AWS beginners, but these may not be available to everyone, so it may not be possible to complete the recommended tutorials (if you are looking for a free option without much setup, keep reading). Still, many of these tutorials provide both in-cloud and local versions so that users can see the adaptations needed to migrate to the cloud.

PO.DAAC AWS Cloud Tutorials (note: many of these links may evolve or the contents be relocated over time, so it is important to search based on the title of the page):

- Tutorial 0: Cloud cheat sheet: https://nasa-openscapes.github.io/earthdata-cloud-cookbook/glossary.html#cheatsheets-guides

   While this isn't a tutorial, it's a good primer on NASA EarthData in the cloud and is a good reference for cloud-computing terminology.
- Tutorial 1: Access data in the cloud: https://podaac.github.io/tutorials/external/access-cloud-python.html

   This simple example shows how to read NASA Earth data into an EC2 instance from S3.
- Tutorial 2: Access and visualize SWOT hydrology datasets: https://podaac.github.io/2024-SWOT-Hydro-Workshop/Tutorials/SWOTHR_s3Access_real_data_v11.html

   This tutorial takes you on a visual tour of all the different datasets produced by SWOT for monitoring surface water in the cloud.
- Tutorial 3: https://podaac.github.io/tutorials/quarto_text/CloudvsLocalWorkflows.html

   This tutorial is actually a suite of tutorials that compare and contrast cloud and local versions of simple hydrology workflows using a variety of hydrologic datasets: GRACE, MODIS, DSWx, and Harmonized Landsat-Sentinel Imagery.

Another cloud computing platform is by Google and is called **Google Cloud Services** (GCS). On the surface, GCS functions similarly to AWS, offering similar computational and storage services. However, one service provided by GCS sets it apart and is important to discuss in the context of water management: Google Earth Engine (GEE). In this chapter, we will provide an example tutorial on GEE for first-time users. The tutorial is on a relatively simple application for viewing satellite imagery in an interactive way. GEE takes this idea and adds access to a deep catalog of geospatial data, a powerful suite of analysis tools, and cloud computation resources to read the data and execute the computations. According to Google:

Google Earth Engine combines a multi-petabyte catalog of satellite imagery and geospatial datasets with planetary-scale analysis capabilities. Scientists, researchers, and developers use Earth Engine to detect changes, map trends, and quantify differences on the Earth's surface. Earth Engine is now available for commercial use, and remains free for academic and research use.

That last statement, "free for academic and research use", is what makes GEE an excellent starting point for learning how to use the cloud for water management. Beyond that, the ability to access multiple (sometimes decades-long) datasets rapidly and in a single workflow can be a powerful tool for solving Big Data problems that would be otherwise impossible.

GEE can be thought of as one part cloud-computing platform, one part coding environment, and one part GIS application. In general, using GEE involves writing scripts in a web browser application that search, access, and process one or more geospatial datasets. When these scripts are run, the web application sends a request to GCS and waits for the results to come back. These results can then be visualized on an interactive map or exported to a user's Google Drive. This platform model is useful for a wide variety of tasks, from rapidly visualizing a small dataset to learn how it works, to massive Big Data workloads (e.g. classifying ~40 years of satellite imagery). Another useful feature of GEE is the capability to share analysis and results easily and rapidly, which is important for water management as coordination is frequently required among multiple agencies, various types of datasets and stakeholder needs to find solutions. Users can even make customized graphical user interfaces (GUI) for exploring datasets or results, allowing people who are not familiar with GEE to explore the output of a given workflow. This utility is very powerful for many water management problems, as it can be empowering for stakeholders to have the ability to analyze big datasets, even if they do not have cloud-computing skillsets.

For first-time users, the following link on how to get started in GEE (prepared by Dr. Matthew Bonnema) https://mbonnema.github.io/GoogleEarthEngine/ is an excellent starting point. Building on this resource, readers can later use the GEE developer guide at https://developers.google.com/earth-engine/. When coming from traditional GIS workflows, GEE can seem like magic. However, it is important to understand its limitations. GEE is designed to process gridded raster datasets (e.g. satellite imagery, precipitation, soil moisture, etc.) and is extremely efficient at handling such datasets. Many remote sensing datasets, however, come in different formats. SWOT is a prime example of this, where many of the products it produces are shapefiles. While GEE can load and work with this type of data, it is not designed to perform large-scale processing of such data. For working with SWOT data and other non-raster type data, GEE may not be the appropriate platform. Another key limitation of GEE is its "walled garden" approach to data processing. Users can only use functions included in the GEE software library. This library is fully featured and has many tools that, working together, can accomplish many geospatial tasks. However, if a given workflow requires something outside of what these libraries can do, then implementing it in GEE will be difficult. Lastly, while the GEE data catalog is deep, it does not include every public remote sensing dataset. Users can bring their own datasets into GEE and doing this on small scales is relatively simple, but loading a large dataset from outside GEE will require some development time to solve.

4.4   Conclusions

To summarize, there are two key cloud platforms with relevance to water management, AWS and GEE. AWS offers access to the entire Earth data catalog hosted by NASA and offers flexibility to handle any dataset of any type. However, computations and storage in AWS are subject to fees, and setting up and managing your own cloud services in AWS can be challenging. GEE, on the other hand, offers a narrower data catalog and set of data types that can be processed, but is free for educational and research purposes, and eliminates the need for users to worry about management of cloud services. Ultimately, which cloud platform is right, or whether the cloud is even appropriate, is dictated by the needs of any given water management problem.

Cloud computing is a powerful tool for accessing and processing remote sensing data for water management and can help overcome many challenges with traditional local computing. Big datasets that were previously unusable, owing to their large file size taking too long to download and/or process, are now accessible by a wider range of end users. Smaller datasets can now be accessed and processed even faster, lowering latency caused by local workflows and enabling users to more quickly get their hands on vital information. With these improvements come other barriers to entry. Working in the cloud requires new skills and work to adapt local workflows to be compatible with cloud computing.

This chapter has introduced readers to the idea of cloud computing. As satellite data on water becomes more and more abundant from a growing number of satellite missions, and as information technology continues to advance, analyzing all that data in the cloud rather than locally is a more cost-effective way to manage our water. We believe that future water management strategies will need to make cloud computing an integral part of improving decisions on how to manage our water. With that in mind, we have developed a tutorial and a hands-on exercise for readers to begin building the necessary cloud-computing data skills. In the next chapters covering various topics on water, more detailed and water-management-specific tutorials will be provided to help readers build on this basic cloud-computing and data handling skill. These tutorials will showcase how to leverage cloud computing and also local computing to harness large satellite remote sensing datasets for water management.

A TUTORIAL ON CLOUD COMPUTING USING GOOGLE EARTH ENGINE[2]

In this tutorial, we will be looking at the use of satellite data in investigating the effect of natural disasters such as floods, which is very important in water management. The goal of this tutorial is to develop a basic understanding of how cloud computing works in Google Earth

---

[2] Contributed by Sarath Suresh, University of Washington

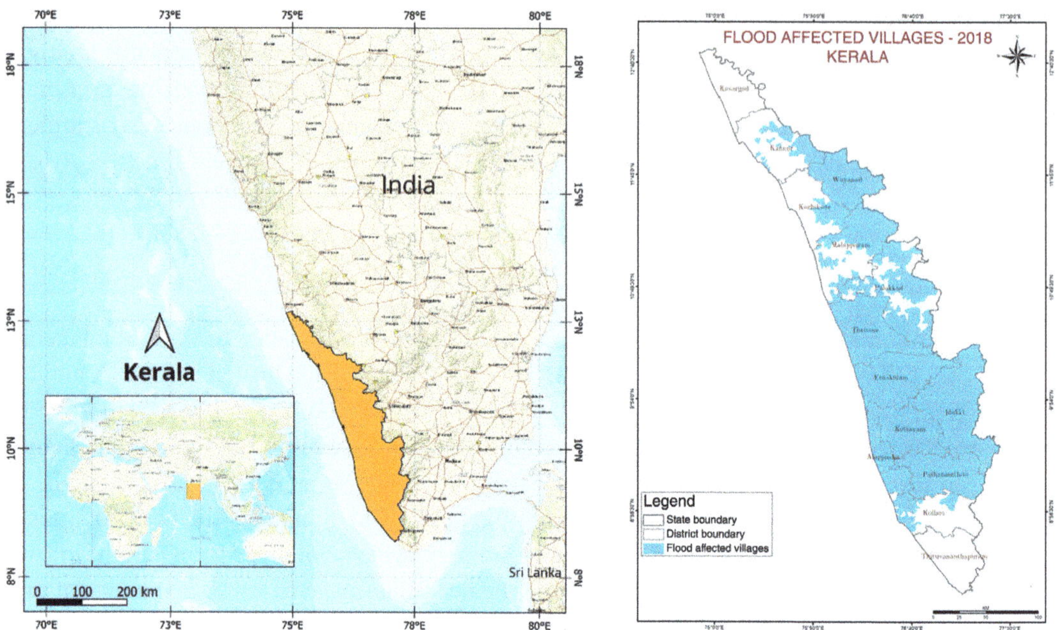

Figure 4.5 Kerala: location of Kerala in India and world map (left panel), and flood map during 2018 August floods in Kerala (right panel). [Flood map prepared from information reported by Kerala State Disaster Management Authority]

Engine rather than how investigation of flood inundation is actually carried out in detail. In future chapters (Chapter 6), we will cover the methodology for inundation and surface water detection in much greater detail.

In the month of August 2018, the Indian state of Kerala experienced one of the worst cases of flooding in history. Unexpectedly high precipitation triggered by a propagating low-pressure system from the Bay of Bengal forced 35 dams to simultaneously release water, leading to subsequent flooding of downstream areas (Figure 4.5). The disaster claimed the lives of more than 480 people, displaced over 1.4 million, and caused losses amounting to $5.0 billion. We can utilize satellite-based observations to study and understand the scale of the flooding, by using indices that accentuate water and vegetation extents. We will also explore the use of synthetic aperture radar (SAR)-based satellite imagery to identify water extents in areas with high cloud cover.

We will be using Google Earth Engine (GEE) to perform this analysis. The entire set of scripts in GEE can be found here: https://code.earthengine.google.com/?accept_repo=users/saraths96/Tutorial (click on "Flood Analysis")

The reader is first encouraged to refer to the GEE documentation for details on any of the GEE methods and objects used in this tutorial, available at www.developers.google.com/earth-engine/guides.

First, let us open a new GEE script file by clicking the "NEW" option under the scripts tab and creating a new file (Figure 4.6).

This opens a new file in the GEE script editor window where we can enter the JavaScript code to run our analysis.

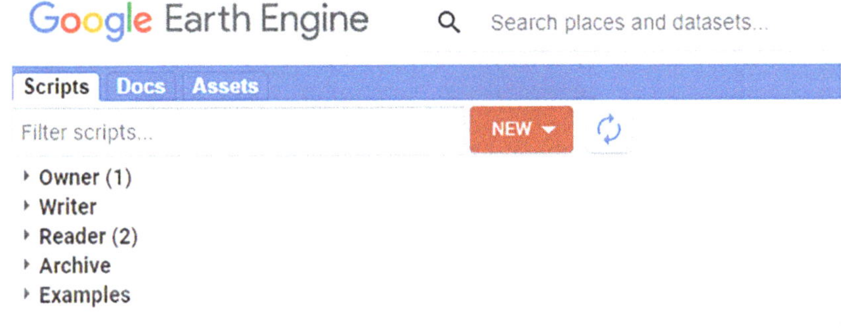

Figure 4.6 Google Earth Engine new scripts tab. [Credit: Google Earth Engine]

Part 1: Sentinel-1 SAR

The first analysis is going to use SAR imagery from the Sentinel-1 satellite. For now, let us just assume SAR is a satellite remote sensing dataset, based on microwave frequency, that can be useful for detecting water on the surface. SAR images also have the ability to penetrate cloud cover and allow us to view the extent of flood inundation even during periods of extreme cloud cover. However, the file size of SAR images can be large, thereby necessitating cloud computing for analysis. The step-by-step procedure for writing the GEE code for the analysis is as follows.

1. Let us begin by defining the region and date of interest. The region around Vembanad Lake, situated in the central part of Kerala, is chosen as it was one of the worst affected by the flooding.

   As a principle, it is always recommended to use native Earth engine objects and methods if they are available. For specifying the region of interest, we can use the **ee.Geometry .Rectangle** object to create a geometry object containing the geographic coordinates of the Vembanad Lake area. There are various other types of geometry objects that the reader can explore, such as **Point**, **MultiPoint**, **Linestring**, **Polygon**, and so on.

```
//Defining the region and date of interest
var roi_Vembanad = ee.Geometry.Rectangle([76.341854, 9.343205, 76.611320, 9.677872]);
var doi_preFlood = ['2018-02-10', '2018-02-11'];
var doi_postFlood = ['2018-08-21', '2018-08-22']
```

2. To view the results of our analysis, we must ensure that the GEE map located at the bottom part of the interface is centered on our area of study. The "**Map.centerObject**" method can be used for this purpose. The method accepts an object to center the map view on and a default numeric zoom value as the second argument, which ranges from 0 to 24.

   The region of interest (ROI) Earth engine object is also added to the map for visualization purposes using the "**Map.addLayer**" method.

```
//Setting the map center and adding the roi to the map
Map.centerObject(roi_Vembanad, 11)
Map.addLayer(roi_Vembanad, {color: 'lightgreen', opacity: 0.2}, 'Region of
Interest');
```

Upon executing the above lines of code using the "RUN" option (keyboard shortcut: Ctrl+Enter), we can see that the GEE map is centered on our area of interest and that the ROI is highlighted by a shaded rectangle (Figure 4.7).

3. Next, we need to search for the Sentinel-1 SAR imagery and access it from our GEE Code Editor. Enter "Sentinel-1" on the search bar found at the top edge of the GEE workspace and select the **Sentinel-1 SAR GRD** product (Figure 4.8).

A window containing the detailed description of the dataset is displayed. Various attributes of the dataset, such as the bands, image properties, and terms of use, are provided here. The reader is encouraged to explore this and associated articles for better understanding of the Sentinel-1 mission (Figure 4.9).

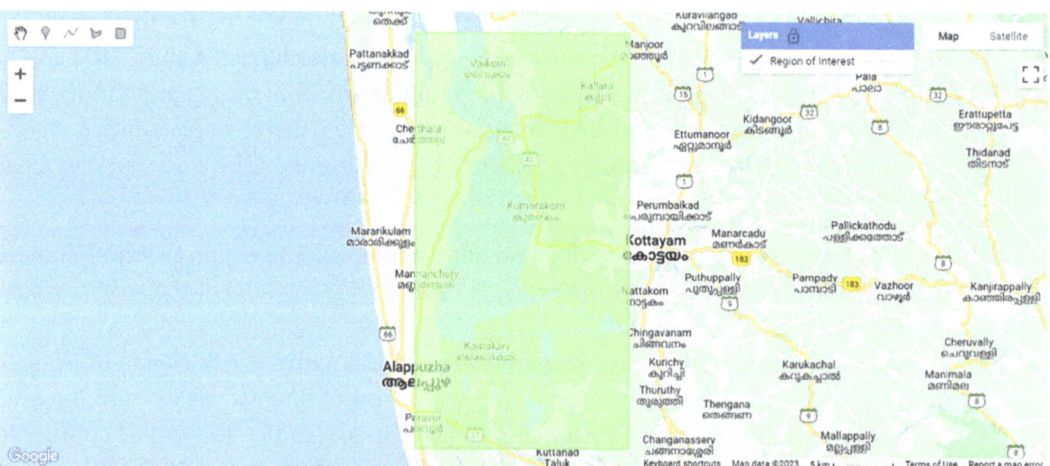

Figure 4.7 Region of interest (ROI) in GEE screen.

Figure 4.8 Searching for Sentinel-1 SAR data in GEE data repository.

Figure 4.9 Description of SAR dataset in GEE.

The Dataset can be imported into the code editor by either using the import option or by copying the Collection snippet code and pasting the same into the editor. For this tutorial, let us follow the latter, and copy the code provided in the collection snippet.

Raster datasets such as the Sentinel-1 imagery are often present as Earth Engine Image Collection objects and are identified by a unique Engine Asset ID, which is passed into the **ee.ImageCollection** constructor. In this instance, **'COPERNICUS/S1_GRD'** is the Asset ID for the Sentinel-1 SAR GRD dataset.

Various filters can be applied to ImageCollection objects to extract the exact scene we are looking for. In this case, we want the Sentinel-1 imagery for the Vembanad Lake region during the date of interest. We will also filter out the vertical transmit/vertical receive single co-polarization band (VV band), which will be used to extract the water extent. For this purpose, the **ee.Filter.listContains**, **ee.Filter.date**, and the **filterBounds** methods are used. GEE allows users to chain various filter methods together by separating them by a period "." as shown in the following code snippet.

```
//Importing the required Sentinel-1 image and applying required filters
var sentinel1 = ee.ImageCollection('COPERNICUS/S1_GRD')
  .filter(ee.Filter.listContains('transmitterReceiverPolarisation', 'VV'))
  .filter(ee.Filter.date(doi_preFlood[0], doi_preFlood[1]))
  .filterBounds(roi_Vembanad)
```

Observing this Image collection using a **print**(sentinel) statement, we notice that there are two images that make up the Sentinel-1 image collection for the Vembanad Lake ROI (Figure 4.10).

This is because the Vembanad Lake region that we have selected is not completely covered by a single Sentinel-1 image, but rather by two adjacent images.

4. SAR data such as that obtained from Sentinel-1 is prone to certain unwanted artifacts, which, if not corrected, can lead to incorrect results. One of the most common artifacts that is often filtered out is speckle. Speckle is a granular noise texture that degrades image quality and complicates image interpretation. It appears in SAR images owing to the interference of waves reflected from many scatterers. Techniques to filter out speckle can be broadly

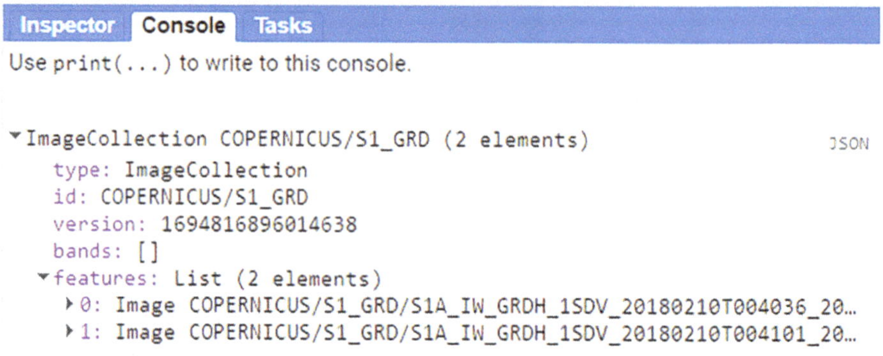

Inspector   **Console**   Tasks

Use print(...) to write to this console.

▼ ImageCollection COPERNICUS/S1_GRD (2 elements)   JSON
    type: ImageCollection
    id: COPERNICUS/S1_GRD
    version: 1694816896014638
    bands: []
  ▼ features: List (2 elements)
      ▶ 0: Image COPERNICUS/S1_GRD/S1A_IW_GRDH_1SDV_20180210T004036_20…
      ▶ 1: Image COPERNICUS/S1_GRD/S1A_IW_GRDH_1SDV_20180210T004101_20…

Figure 4.10 Image collection for Vembanad Lake.

classified into adaptive and nonadaptive filters. The reader may explore the different types of speckle filtering to get a more detailed understanding of the subject. In general, speckle removal is always a compromise between noise reduction and detail preservation. Let us apply a simple focal median filter here. This filter works by taking a pixel and replacing it with the median of the neighboring pixels.

We can achieve this by using the built-in **imageCollection.map** function and writing a user-defined function to carry out the filtering. The **map** function allows us to apply a function to every image in an ImageCollection object. It accepts a single parameter, the **ee.Image** object.

```
// Filtering speckle noise using Focal median filter
var filterSpeckles = function(img) {
  var vv = img.select('VV')
  var vv_smoothed = vv.focal_median(50,'circle','meters').rename('VV_Filtered')
  return img.addBands(vv_smoothed)
}
var S1_speckleCorrected = sentinel1.map(filterSpeckles)
```

Here, we are writing a function "filterSpeckles", which accepts an **ee.Image** object. The function selects the "VV" band from the image, and then applies the built-in **ee.Image.focal_median** method to it to generate a smoothed image devoid of noise. It renames the band as "VV_Filtered" and adds the newly generated band back to the image before returning it. The **ee.Image.select**, **ee.Image.rename**, and **ee.Image.addBands** methods are used here. Parameters such as kernel radius, kernel shape, and units are passed to the focal_median function to get the desired output. These parameters control the intensity and the type of kernel used in focal median filtering. Without going into extreme detail, a kernel is a small matrix or array of values that defines the pattern or mask used for the filtering operation.

5. Now that the SAR image from the Sentinel-1 dataset is corrected for speckle, we can apply a simple water classification technique called **backscatter thresholding** to identify the areas in the image that are water. This method works by extracting only those parts of the image that are less than a certain threshold of backscatter intensity, expressed in decibels (dB).

Typically, water pixels produce very much less backscatter than the surrounding terrain. A threshold of −16 dB is used in this example. The choice of this threshold value is arbitrary and depends on factors such as image characteristics and noise level. The reader may experiment with different threshold values while visualizing the results as an exercise. Automated thresholding algorithms such as the Otsu threshold are also often applied to automatically identify the backscatter threshold value.

```
//Classifying pixels as water based on its backscatter threshold values.
var classifyWater = function(img) {
  var vv_filtered = img.select('VV_Filtered')
  var water = vv_filtered.lt(-16).rename('Water')
  water = water.updateMask(water)
  return img.addBands(water)
}
var S1_water = S1_speckleCorrected.map(classifyWater)
```

Similar to the speckle filtering we did previously, a user-defined function "classifyWater" is written and applied to the speckle-corrected SAR image collection using the **ee.imageCollection.map** function. The function first selects the "VV_filtered" band and then creates a mask using the **ee.Image.lt** method. This method identifies all pixels in the selected band that are less than the provided threshold of −16 and assigns a value of 1 to them. All remaining pixels are assigned a value of 0. Then the **ee.Image.updateMask** method is applied, which removes all values from the mask that are 0. This essentially creates a mask that contains only those pixels that are identified as water pixels. The mask is then renamed as "Water" and added back to the image as a new band.

6. Let us now clip this mask to the Vembanad Lake region and visualize it on the map. Again, the **ee.imageCollection.map** function can be used for this purpose. If the user-defined function that is being passed to the map function is not elaborate, it is considered good practice to use an anonymous function (a function without a specific name) directly written within the map function to make the code more concise and readable.

```
// Clip the water classification layer to the region of interest and adding to map.
var clippedS1_water = S1_water.map(function(image) {
  return image.clip(roi_Vembanad);
});
Map.addLayer(clippedS1_water, {bands: 'Water', min: 0, max: 1, palette: ['white',
'blue']}, 'Water Classification');
```

Here, the user-defined function to clip the image to the ROI is written directly inside the map function, as it is relatively simple and is only used in this context. The **ee.Image.clip** method is used to clip the images within the ImageCollection to the ROI. The "Water" band from the clipped image collection is then added to the map as a new layer using the **Map.addLayer** method. The band is colored "blue" using the palette parameter of the addLayer method, and the layer is further named as "Water Classification" (Figure 4.11).

As can be seen, the water pixels are prominently displayed in blue, and we are able to get a good qualitative idea of the extent of the Vembanad Lake before the flood event.

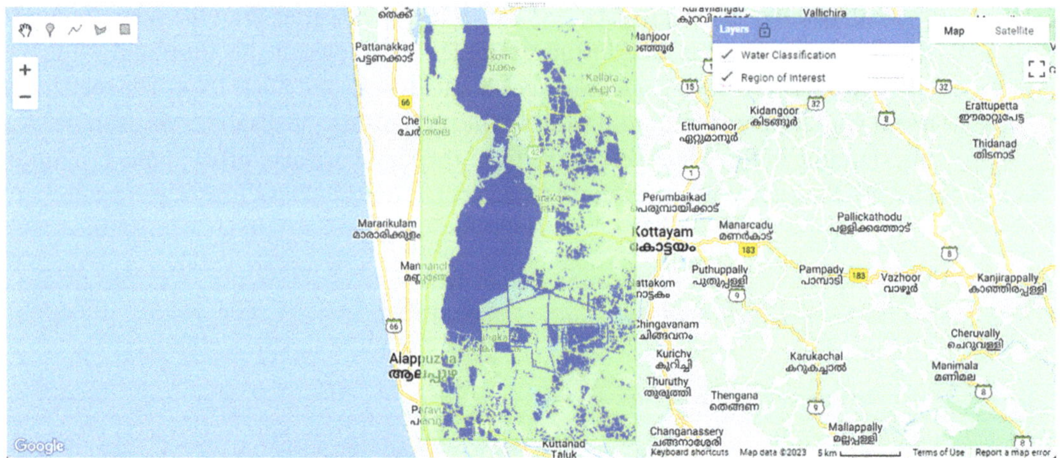

Figure 4.11 Water areas mapped in GEE for the region of interest. [Credit: Google Earth Engine]

7. Let us also compute the absolute area covered by all the water pixels in this scene. This can be achieved by computing the total number of water pixels and multiplying this by the area of an individual pixel. In the case of the Sentinel-1 GRD dataset, images captured in the "VV" band have a spatial resolution of 10 m. Hence, we must multiply the total number of water pixels by 100 to get the area in square meters. This can then be further divided by 10^6 to get the area in square kilometers. The **ee.Image.pixelArea** method automatically computes the area of a pixel for a given image. By default, it computes the area using the WGS84 coordinate reference system and a 1-degree scale.

To compute the total number of pixels in the scene, we make use of the reducer functionality in GEE. **Reducers** are a way in GEE to aggregate data over space, time, bands, or other data structures. At this point, the reader is encouraged to go through the documentation for Reducers in the GEE developer guide. In this tutorial, we use the **ee.Image.reduceRegion** method to apply a reducer on all individual pixels within the ROI.

```
// Computing total area covered by water
var waterArea = clippedS1_
water.select('Water').mosaic().multiply(ee.Image.pixelArea()).divide(1e6); //
Convert to sq.km
var totalWaterArea = waterArea.reduceRegion({
  reducer: ee.Reducer.sum(),
  geometry: roi_Vembanad,
  scale: 10,
  maxPixels: 1e9
});
// Get the total water area in sq.km
var waterAreaSqKm = totalWaterArea.get('Water');
print('Total Water Area:', waterAreaSqKm.getInfo().toFixed(2) + ' km²');
```

First, the "Water" mask band from the clipped Sentinel-1 image collection is selected. Because our scene contains two separate images covering the area of Vembanad Lake, we then use the **ee.ImageCollection.mosaic** method to create a single merged image from the two images. Next, all the pixel values in the water mask are multiplied with the pixel area using the **ee.Image.multiply** method with the **ee.Image.pixelArea** as its argument. Finally, the **ee.Image.divide** method is used for the purpose of converting the units to square kilometers.

Then, the total water area is computed by applying the **ee.Image.reduceRegion** method. The reducer is specified to be **ee.Reducer.sum**, since our objective here is to obtain the sum of all the water pixels. The scale and maximum pixel limit for the reduction computation are also specified. This parameter is essential for managing computational efficiency and preventing excessive resource usage.

Finally, the resultant Total Water Area is printed to the console, with some formatting applied to it to limit the number of decimal place values.

```
Inspector   Console   Tasks
Use print(...) to write to this console.


Total Water Area:                                              JSON
174.42 km²                                                     JSON
```

The Total Water Area in the Vembanad Lake region computed before the flood is obtained as **174.42 km^2**.

8. Now, let us re-run the GEE script by changing the date to a post-flood time period. The doi_postFlood variable that we defined in **Step.1** can be used in **Step.3** to achieve this

```
//Importing the required Sentinel-1 image and applying required filters
var sentinel1 = ee.ImageCollection('COPERNICUS/S1_GRD')
  .filter(ee.Filter.listContains('transmitterReceiverPolarisation', 'VV'))
  .filter(ee.Filter.date(doi_postFlood[0], doi_postFlood[1]))
  .filterBounds(roi_Vembanad)
```

The new run gives us the following results:

```
Inspector   Console   Tasks
Use print(...) to write to this console.


Total Water Area:                                              JSON
338.55 km²                                                     JSON
```

We can clearly see from the map (Figure 4.12) that the area covered by water is much higher in the post-flood scenario. It is easy to visualize that the Vembanad Lake was not able to hold the excessive amount of water it received leading to large scale flooding around it. Quantitatively, the area covered by water increased to **338.55 km^2**, which equates to an increase in water area of 94.1%.

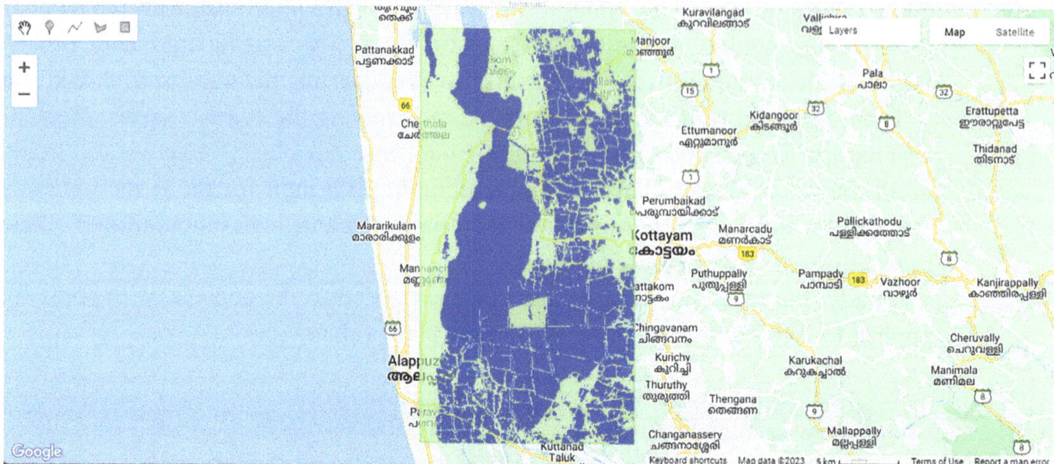

Figure 4.12 Updated water classification maps after changing date to a post flood time period. [Credit Google Earth Engine]

As can be seen, SAR images are a powerful tool to visualize and quantify the extent of flooding. Let us now do a similar analysis using an optical satellite.

Part 2: Sentinel-2

Just like the first part using Sentinel-1 SAR (microwave), let us see another example using the visible wavelength. Optical satellites like Sentinel-2 capture images in a wide spectrum of frequencies, known as bands, ranging from visible light to infrared. In this section, we will see how we can implement the same type of flood analysis using various band combinations in GEE. We will follow a very similar approach to what we did for the SAR imagery using Sentinel-1.

1. As before, let us begin by defining the ROI and the time of interest. Then the map is centered on the study region and the ROI object is added to the map. Finally, the Sentinel-2 Image Collection for the scene is obtained using the Asset ID ("COPERNICUS/S2_HARMONISED"), and filtered using the date and ROI bounds.

```
//Defining the area of interest for analysis: Vembanad Lake - Kerala and time
period for flood inundation study
var roi_Vembanad = ee.Geometry.Rectangle([76.341854, 9.343205, 76.611320,
9.677872]);
var doi_preFlood = ['2018-02-20', '2018-02-28'];
var doi_postFlood = ['2018-08-20', '2018-08-23'];
//Setting the map center and adding the roi to the map
Map.centerObject(roi_Vembanad,11)
Map.addLayer(roi_Vembanad, {color: 'lightgreen', opacity: 0.2}, 'Region of
Interest');
//Importing the required Sentinel-1 image and applying required filters
var sentinel2 = ee.ImageCollection("COPERNICUS/S2_HARMONIZED")
  .filter(ee.Filter.date(doi[0], doi[1]))
  .filter(ee.Filter.bounds(roi_Vembanad));
```

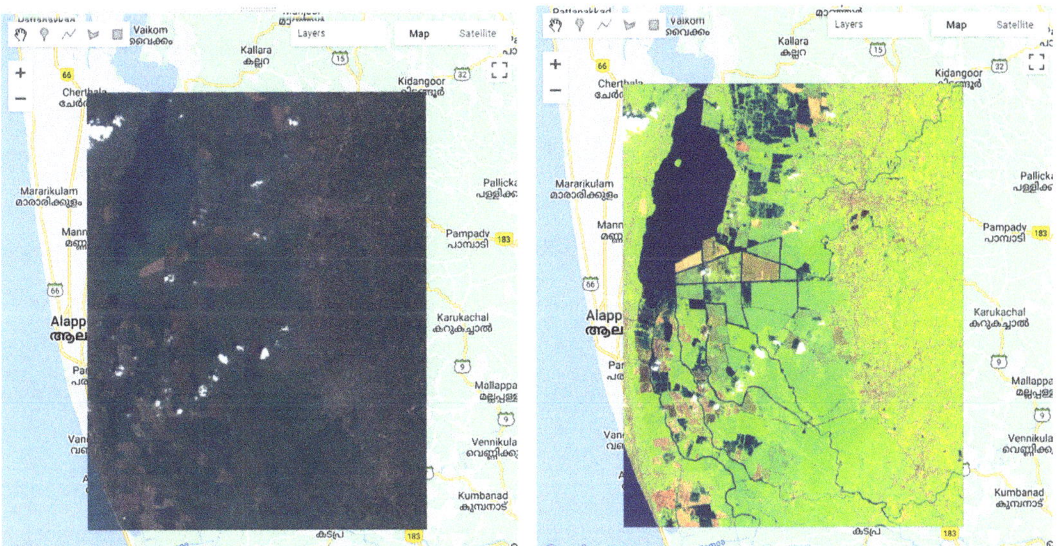

Figure 4.13 Sentinel-2 optical images from various bands (true color on left; and false color on right) to see where the water on the surface is likely to be. [Credit Google Earth Engine]

2. Let us quickly visualize the Sentinel-2 image in true color (Bands 4-3-2) and false color (Bands 11-8-2) composites to better visualize the region (Figure 4.13). GEE makes it easy to create such composites by directly specifying the band combination as arguments to the **Map .addLayer** method.

```
//Visualizing true color and false color composites for Sentinel-2
var rgbVis = {min: 0.0, max: 3000, bands: ['B4', 'B3', 'B2']};
var fcc = {min: 0.0, max: 3000, bands: ['B11', 'B8', 'B2']};
Map.addLayer(sentinel2.first().clip(roi_Vembanad), rgbVis, 'TCC');
Map.addLayer(sentinel2.first().clip(roi_Vembanad), fcc, 'FCC');
```

Once added to the map, GEE gives options to further fine-tune the layer through interactive controls right within the map area using the "Layers" dropdown menu (Figure 4.14).

Band combinations can be altered on the fly using this menu to enable quick visualization. Options to control the layer opacity and the overall brightness and contrast (gamma) are also provided.

As we can see, the area near Vembanad Lake is predominantly covered by agricultural fields, appearing in shades of red and green in the false color composite image.

3. We will be using two indexes to determine the extent of surface water and vegetation, namely the Modified Normalized Difference Water Index (MNDWI) and the Normalized Difference Vegetation Index (NDVI). These indices for water detection are covered again in Chapter 6 for surface water detection. Here, the reader should look at the formulations as examples on how to appropriately "code" it for automatic calculation in Google Earth Engine.

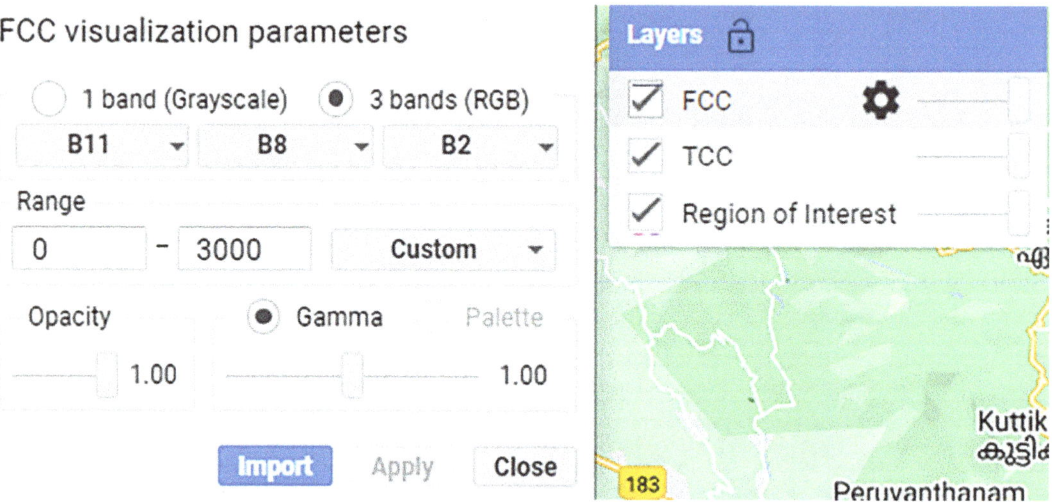

Figure 4.14. Options in Google Earth Engine to fine-tune layers. [Credit Google Earth Engine]

MNDWI is a particularly useful index in distinguishing water bodies from other land features. It works by utilizing the difference in the intensity with which water reflects various frequencies of the electromagnetic spectrum. Water is extremely reflective to green light (Band 3 – 560 nm) while absorbing most of the light in the short-wave infrared (SWIR) region (Band 11 – 1,600 nm). Hence a normalized difference of these two bands will result in an image that highlights and contrasts water bodies against features such as buildings and land (recall Figure 2.14). Similarly, NDVI works by utilizing the differential reflectance of healthy vegetation between red light (Band 4 – 665 nm) and near infrared light (Band 8 – 835 nm).

$$\text{MNDWI} = \frac{(\text{Green} - \text{SWIR})}{(\text{Green} + \text{SWIR})}$$

$$\text{NDVI} = \frac{(\text{NIR} - \text{Red})}{(\text{NIR} + \text{Red})}$$

```
//Computing ndvi and mndwi indexes for Sentinel-2 imagery
var ndvi = sentinel2.first().normalizedDifference(['B8', 'B4']).rename(['ndvi']);
var mndwi = sentinel2.first().normalizedDifference(['B3',
'B11']).rename(['mndwi']);
```

4. With the indices computed, we will now create water and vegetation masks using thresholding as we did before with the SAR imagery. The masks are then clipped to the ROI and added to the map (Figure 4.15).

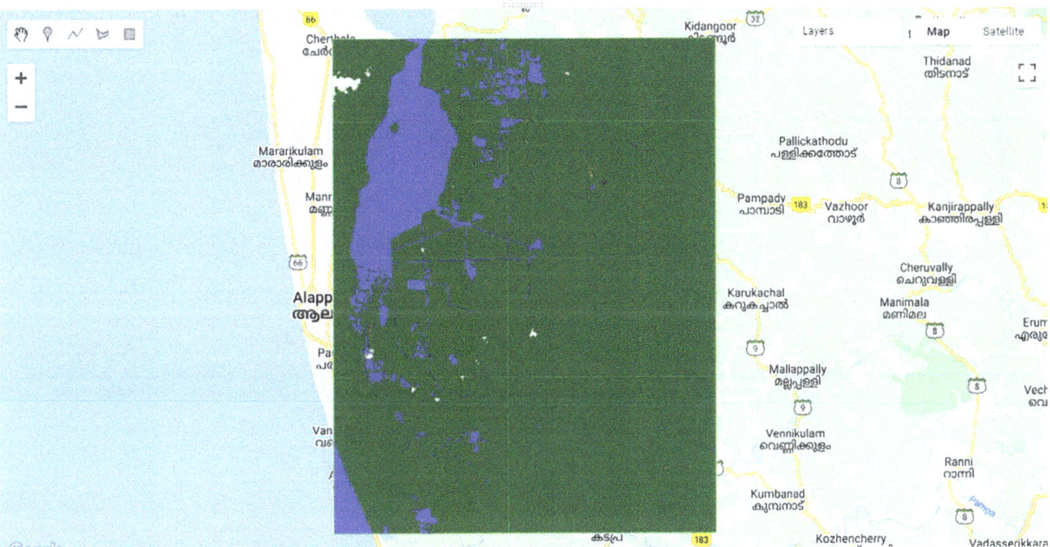

Figure 4.15 Water classified map in region of interest for optical imagery. [Credit: Google Earth Engine]

```
// Define thresholds for water and vegetation and creating water and vegetation
masks
var waterThreshold = 0.3;
var vegetationThreshold = 0.1;

var waterMask = mndwi.gt(waterThreshold).rename('WaterMask');
var vegetationMask = ndvi.gt(vegetationThreshold).rename('VegetationMask');
waterMask = waterMask.updateMask(waterMask);
vegetationMask = vegetationMask.updateMask(vegetationMask);

Map.addLayer(waterMask.clip(roi_Vembanad), {bands: 'WaterMask', min: 0, max: 0.5,
palette: ['white', 'blue']}, 'Water Classification');
Map.addLayer(vegetationMask.clip(roi_Vembanad), {bands: 'VegetationMask', min: 0,
max: 1,
palette: ['white', 'darkgreen']}, 'Vegetation Classification');
```

5. Next, let us compute the total water and vegetation areas using the reducer functionality and print the results to the console. We can write a simple function that computes area for a given feature mask and use the same for both the water and vegetation areas.

```
// Computing total area covered by water and vegetation
function computeArea(featureMask,roi)
{
var area = featureMask.multiply(ee.Image.pixelArea()).divide(1e6);
var totalArea = area.reduceRegion({
reducer: ee.Reducer.sum(),
geometry: roi,
scale: 10,
maxPixels: 1e9
});
```

```
return totalArea;
}

var waterArea = computeArea(waterMask,roi_Vembanad);
var vegetationArea = computeArea(vegetationMask,roi_Vembanad);

print('Total Water Area: ' + waterArea.get('WaterMask')
.getInfo().toFixed(2) + ' km²');
print('Total Vegetation Area: ' + vegetationArea.get('VegetationMask')
.getInfo().toFixed(2) + ' km²');
```

Inspector	**Console**	Tasks

Use `print(...)` to write to this console.

Total Water Area: 147.08 km²   JSON

Total Vegetation Area: 954.58 km²   JSON

6. Finally let us re-run the script using a post-flood date of interest and analyze the results. As before, we can simply change the filter date in Step 1 to achieve this (Figure 4.16, Figure 4.17).

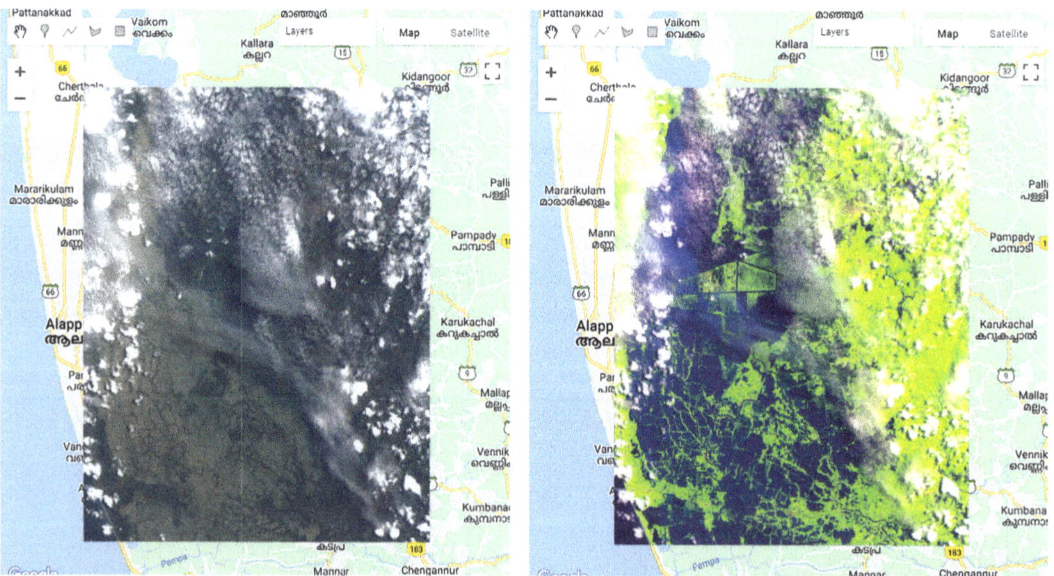

Figure 4.16 Images from a different date (post-flood date) in true color (left panel) and false color (right panel). [Credit: Google Earth Engine]

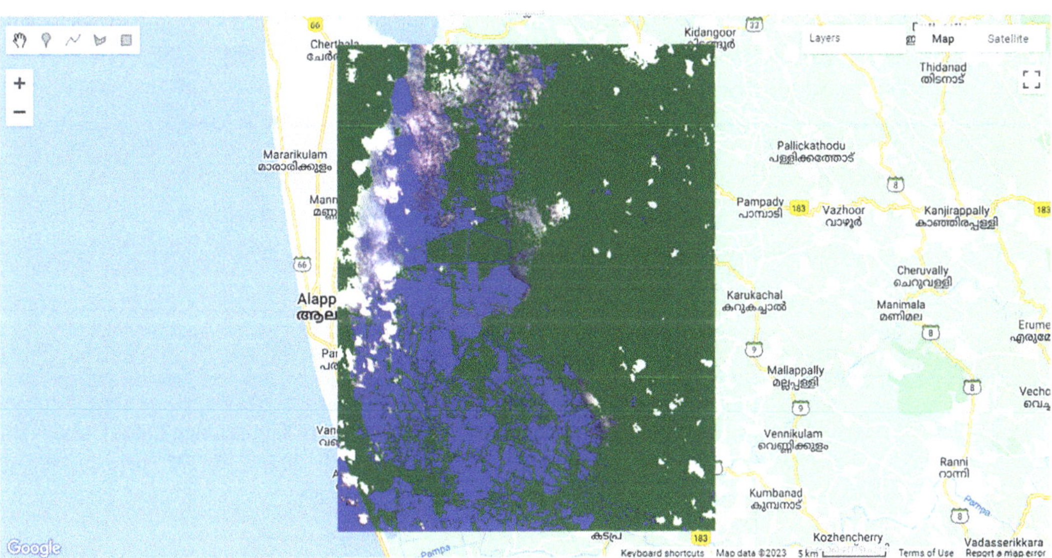

Figure 4.17 Water classified map in region of interest for a post-flood date. [Credit: Google Earth Engine]

We can clearly see the difference between the results in the pre-flood and post-flood images, with the post-flood water area increasing by **78.3%** and the vegetation cover decreasing by **29.4%**.

However, it should be noted that parts of the Sentinel-2 image that are under cloud cover give us incorrect information. Often, cloud correction techniques are implemented to overcome this issue.

EXERCISES: CHAPTER 4 (CONTRIBUTED BY SARATH SURESH, UNIVERSITY OF WASHINGTON)

Use Sentinel-2 and Sentinel-1 satellites for studying the effects of the 2022 flood in the Sylhet region of Bangladesh. Compute the total water and vegetation area for the pre-flood and post-flood scenarios.

Use the following region geometry coordinates and time periods:

For Sentinel-2
 Sylhet geometry coordinates: [92.102, 24.696, 92.256, 24.876]
 Pre-flood date of interest: 2022-05-08
 Post-flood date of interest: 2022-06-27

For Sentinel-1
 Sylhet geometry coordinates: [91.399, 24.347, 92.151, 25.146]
 Pre-flood date of interest: 2022-05-01
 Post-flood date of interest: 2022-06-28

REFERENCE

Lewis, A., S. Oliver, L. Lymburner, et al. (2017). The Australian Geoscience Data Cube – foundations and lessons learned. *Remote Sensing of Environment*, vol. 202, 276–292. https://doi.org/10.1016/j.rse.2017.03.015.

SUGGESTED READING

Amani, M. A. Ghorbanian, S. A. Ahmadi, et al. (2020). Google Earth Engine cloud computing platform for remote sensing big data applications: a comprehensive review. *IEEE Journal of Selected Topics in Applied Earth Observations and Remote Sensing*, vol. 13, 5326–5350. https://doi.org/10.1109/JSTARS.2020.3021052

5   Satellite Remote Sensing of Precipitation

5.1   Chapter Overview

In the previous chapter, we covered cloud computing. In this chapter, we will cover the remote sensing of precipitation to understand how precipitation is estimated. Precipitation is considered one of the most important components of the water cycle that drives the availability of water and its management. For example, it leads to runoff and streamflow, irrigates a field of crops and provides the water for crop growth, and fills up lakes, reservoirs, and ponds that are a key source for water management. The understanding of precipitation remote sensing will pave the way for learning more complex water management applications that are being increasingly carried out around the world using satellite water data. We will first cover the history of precipitation remote sensing that began with active sensing and ground radar. Next, we will cover satellite-based sensing, where the challenges and complexities are different. The pros and cons of various electromagnetic wavelengths will be covered. Finally, we will cover the topic of multi-sensor precipitation estimation based on the synergistic use of multiple satellite sensors spanning different wavelengths of the electromagnetic spectrum.

5.2   Introduction

For water management, the important components of the water cycle are precipitation, surface water storage, discharge, and groundwater. These flux components drive many of the water management activities such as irrigation and crop management, flood forecasting and flood management, reservoir monitoring, and hydropower management. To understand water management from space for such activities, it is important to first understand the remote sensing of such water cycle components.

Precipitation is inclusive of snowfall and rainfall. This chapter is primarily concerned with rainfall (liquid water precipitation), and the term precipitation will mainly be used to indicate rainfall. Snow is an equally important water cycle component for water management for which the occurrence and quantity can be remotely sensed, but it is not covered in this chapter.

5.3   Precipitation Remote Sensing: How It All Began

To understand precipitation remote sensing from space, it is important to have a basic overview of how it all began with the ground-based platform. Remote sensing of precipitation is nothing new. Ever since radar (radio detection and ranging) was developed before the Second World War to detect incoming enemy aircraft, engineers and scientists started noticing that the same idea could be used to detect and track clouds and their "potential" for rain.

5.3.1 Ground-Based Remote Sensing as a Precursor to Tracking Precipitation from Space

Today, precipitation can be remotely sensed from space using passive or active sensing, or a combination of both methods. However, the first ground-based remote sensing using a radar was active and used microwave (MW) wavelengths. Hereafter, when the term radar is used, let us always think "active" and microwave electromagnetic (MW E-M) energy at the same time. Let us review this ground-based radar technique for tracking rainfall (again – we are using rainfall and precipitation to mean the same here).

Radars create their own electromagnetic energy that is transmitted from the sensor toward the target. Radar waves are largely unaffected by the atmosphere during propagation, which means that MW wavelengths have large atmospheric windows or plenty of transmission bands (see Figure 3.7b). This MW E-M energy then interacts with the target, producing a backscatter of E-M energy that is then recorded by the radar's receiver. Analogously, we can think of this like a streetlight in the rain at night. The light that scatters or reflects back helps us to detect the falling drops (Figure 5.1).

Figure 5.1 Streetlights at night during rain. The scattered light makes the rainfall quite visible. [Source: Wayne Speedy @ Pexels.com]

Figure 5.2 A microwave pulse emitted from a radar towards a target that can be a cloud (simplified here). [Source: National Weather Service, NOAA]

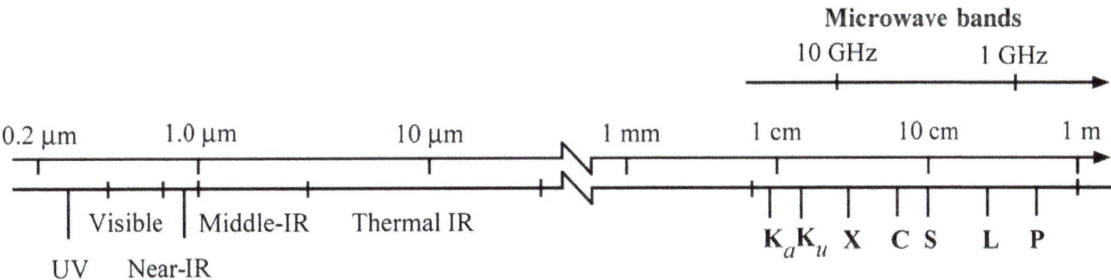

Figure 5.3 A close-up view of the microwave region of the E-M spectrum. (K_a, K_u are the same as Ka, Ku.)

Here, the target is cloud and the cloud droplets inside them. We know that clouds have the potential to produce rain, although not all clouds do so. So, if we can track the cloud, its location, size, and even speed, and can also infer the amount of cloud droplets it may have, then we have a way to indirectly estimate (using inverse methods) how much rain the cloud may produce.

In a more technical way, Figure 5.2 summarizes how a radar would actually work. Imagine in Figure 5.2 the blue circle to be our target – the cloud.

As mentioned earlier, the wavelengths for a radar are in the MW range, as shown in Figure 5.3. Because the original radar development was shrouded in secrecy due to national security during the Second World War, engineers used code for different bands within the MW wavelength range. These codes are the Ka, Ku, X, C, S, L, and P bands, in order of increasing wavelength, and their naming persists today in remote sensing literature.

Going back to Figure 5.2, the radar transmits an MW signal in one of the specific wavelength bands towards a target and detects the backscattered radiation or the reflected radiation that returns to the radar. The strength of the backscattered signal indicates the target property, which in this case is the cloud's likelihood for precipitation. Also, the time delay between the transmitted and reflected signals can help determine the distance (or range) to the target. Radar, by virtue of being an active MW sensor, has all-weather and all-day capability to track clouds and estimate rainfall.

We should remember that the ground-based platform affords some unique advantages that are not available in space. These are access to unlimited power supply to run the radar at any wavelength of choice. It also allows easy maintenance and control of scanning and the pulse of the MW energy generated. However, ground-based precipitation radars are not without their fair share of limitations. These limitations are important to understand. Some of them can be mitigated using the vantage of space that allows looking down on the cloud vertically (called near-nadir) and over oceans or difficult terrain where ground radars cannot be installed.

5.3.2 Radar Equation

The key thing to consider for understanding how rainfall is estimated is the radar equation. Figure 5.4 shows a more detailed view of the E-M pulse, which propagates and widens as it travels further. The pulse length and the beamwidth determine the pulse volume. The pulse volume can be quite large at long ranges, which means that consecutive pulses can receive backscattered radiation from a larger number of targets the farther away they are from the radar.

Power radiated, per unit area (W m^{-2}), by a radar antenna with beamwidth as in Figure 5.4 is

$$\frac{P_t g}{4\pi r^2} \tag{5.1}$$

Here P_t is the total radar power transmitted, r is range, and g is antenna gain.

Now if A_σ is the cross-sectional area of the target (i.e. the area that the radar pulse propagates through orthogonally), then the power received by the target can be defined as:

$$\frac{P_t g A_\sigma}{4\pi r^2} \tag{5.2}$$

However, only a fraction of this power received by the target returns as reflection (or backscatter) to the radar's receiver with effective antenna size Aa. This fraction is:

$$\frac{P_t g A_\sigma}{4\pi r^2} \frac{Aa}{4\pi r^2} \tag{5.3}$$

The effective antenna size is $Aa = \dfrac{g\lambda^2}{4\pi}$ (for spherical antennas).

Now reorganizing everything, the power returned P_r (which is what the radar's receiver can measure and attribute to the target) is defined as:

$$P_r = \frac{P_t g^2 \lambda^2 A_\sigma}{(4\pi)^3 r^4} \tag{5.4}$$

From Equation 5.4 it is clear that the power returned $\left(P_r\right)$, which is what the radar's receiver can record, is a function of power transmitted $\left(P_t\right)$, the range, and the target's cross-section. Since we can control the power transmitted, and we already know the range, we are essentially left with cross-sectional area of the target as the unknown variable.

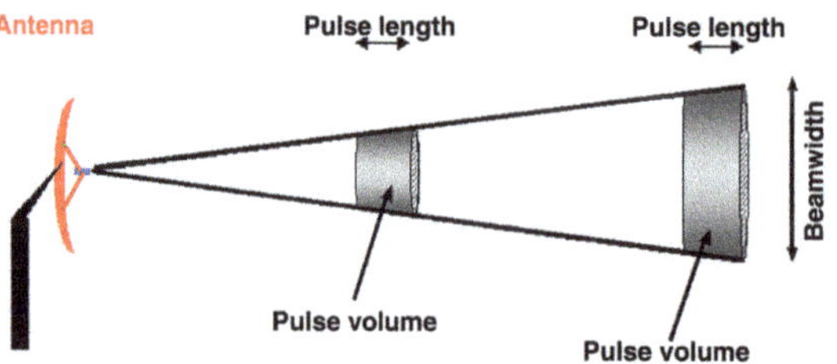

Figure 5.4 Radar pulse widening as it propagates. [Source: COMET Classroom, University Corporation for Atmospheric Research]

Through some manipulation, we can show that this is in fact related to cloud droplet size distribution (Equation 5.5). Here the droplet size D is typically in mm and the volume is in m^3, and the most common unit for Z is $mm^6\,m^{-3}$, which is expressed using the logarithmic scale. In fact, we can define a new parameter here called reflectivity Z, which is nothing but the sum of the sixth power of diameter of *all* the cloud droplets contained in the cloud (or the scattering cross-section).

$$Z = \sum_{vol} D^6 \tag{5.5}$$

Through some further manipulation, we can show that the power returned, which is the recorded "interaction" with the target, is proportional to Z, reflectivity (Equation 5.6). So, if radar is measuring P_r, it is essentially measuring Z of the target from which P_r is returning. The bigger question is: how do we figure out the diameter of *all* the cloud droplets in the cloud? Luckily, and thanks to steadfast work by pioneers in the field in the early 1950s through 1970s, we now can say with some confidence that the Z is related to the rainfall rate R mostly in the form of a power-law equation:

$$Z = aR^b \tag{5.6}$$

where

Z = "reflectivity" $(mm^6\,m^{-3})$
R = rainfall rate $(mm\,hr^{-1})$
a and b are empirically derived constants.

Also, in the modern world, we can use high-speed cameras to measure literally the actual number of droplets and their diameters. Such a camera is called a video disdrometer.

Please note that the above is an oversimplified description of how MW radars on the ground are applied to estimate precipitation rate. There have been many decades of research and comprehensive theory behind radar precipitation estimation. For water management using satellite remote sensing, we have provided the basic background needed to transition to the topic of satellite remote sensing of precipitation. Interested readers can refer to the classic book *Radar Meteorology* (Battan, 1959).

The Z–R relationship is at the core of using radar for precipitation remote sensing. To recap, the radar measures power returned from the target. This power returned is known to be related to the cross-section of the target, which in turn is related to the reflectivity. The reflectivity, Z, in turn is related to the sum of the sixth power of diameter of all cloud droplets. Using this, researchers have found that Z is related to rainfall in the form of a power-law relationship and hence relatively easy to invert in the retrieval problem (see Chapter 3). The Z–R relationship needs to be calibrated against in-situ rainfall data for a region and a rainfall regime. Table 5.1 shows some Z–R relationships for radar that researchers have come up with.

The United States, like most industrialized nations, has invested in a network of ground radars called NEXRAD (also known as WSR-88D) for precipitation and weather tracking. Figure 5.5a shows the location of these radars. The radars use the S band of the MW spectrum (see Figure 5.3 for pinpointing the S-band wavelength) and therefore require a large antenna. Typically, a radar makes a 360-degree sweep at a low angle to the ground and repeats the sweep at gradually increasing angles to complete what we call a volume scan of the target. Remember that the cloud as a target is a 3D structure. Figure 5.5b shows what a typical map of radar reflectivity Z (in units of dBZ) can look like. These reflectivity values can now be converted into rainfall values in mm hr^{-1} for each pixel of the image, based on the Z–R relationship that is deemed most appropriate for the region, the season, and the storm system.

Table 5.1 Examples of a few *Z–R* relationships for various rainfall regimes around the world.

Relationship	Optimum for:	Also recommended for:
Marshall-Palmer ($Z = 200R^{1.6}$)	General stratiform precipitation	
East – Cool Stratiform ($Z = 130R^{2.0}$)	Winter stratiform precipitation – east of the US continental divide	Orographic rain – east
West – Cool Stratiform ($Z = 75R^{2.0}$)	Winter stratiform – west of the US continental divide	Orographic rain – west
WSR-88D Convective ($Z = 300R^{1.4}$)	Summer deep convection	Other nontropical convection
Rosenfeld Tropical ($Z = 250R^{1.2}$)	Tropical convective systems	

(a)

Figure 5.5a Location of US government ground-based weather radars for precipitation estimation. [Image credit: National Weather Service]

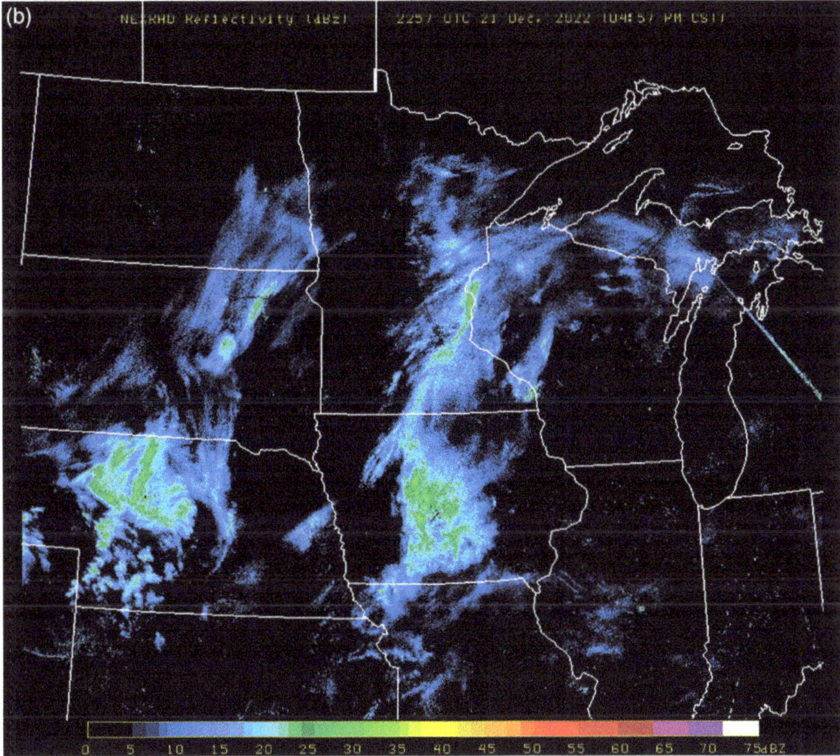

Figure 5.5b An example map of radar reflectivity over the mid-western United States. [Image credit: National Weather Service]

5.3.3 Limitations of Radar Rainfall Remote Sensing

As mentioned earlier, radar rainfall remote sensing from a ground-based platform has its fair share of limitations, despite the several advantages that still make this the best source of rainfall data using remote sensing. It is helpful to understand some of these, as many of them can be mitigated by using the vantage point of space.

The first limitation that is quite easy to grasp is that posed due to terrain and sampling geometry (Figure 5.6a). Being on the ground, a radar may have to shoot its beam at an angle to look beyond the tree or terrain "canopy". This can imply that some of the low-lying clouds at close range may be under-sampled. Similarly, at long ranges, the beam can still be blocked by mountains and consequently see only part of a large cloud system (see left side of Figure 5.6a).

The next limitation is due to anomalous propagation of the radar beam when the power returned is mistakenly thought to be from a cloud target. Anomalous propagation, also known as super-refraction, is where the radar beam is bent more than usual back toward Earth due to an unusual temperature and humidity profile of the atmosphere. Normally, the air is thinner and cooler as the radar beam climbs in elevation. But if it were the opposite (this is known as an "atmospheric inversion"), the beam could experience severe bending towards Earth and at one point hit the ground (and reflect back to the receiver). Without awareness of this anomalous propagation, the receiver would record this power returned as reflectivity from clouds and interpret this as potential rainfall using the Z–R relationship (Figure 5.6b). With super-refraction or anomalous propagation, not only can the radar not see the more distant "real" precipitation, it can also misidentify the return from the ground reflection as real precipitation. Super-refraction is more likely when a strong low-level inversion is in place, especially when the low-level air is moist and the air above the inversion is very dry. Fortunately, there are corrections or corrective measures that radar can take to minimize the impact of anomalous propagation.

Other limitations of radar precipitation remote sensing can be due to the vertical profile of reflectivity (or VPR) and cloud microphysics. Precipitation is three-dimensional in space. The vertical distribution of precipitation (and thus reflectivity) is typically nonuniform (Figure 5.6c). As the elevation of the radar beam increases with the distance from the radar location (through beam propagation and the Earth's curvature), one sweep samples from different heights. The effects of the nonuniform VPR and the different sampling heights need to be

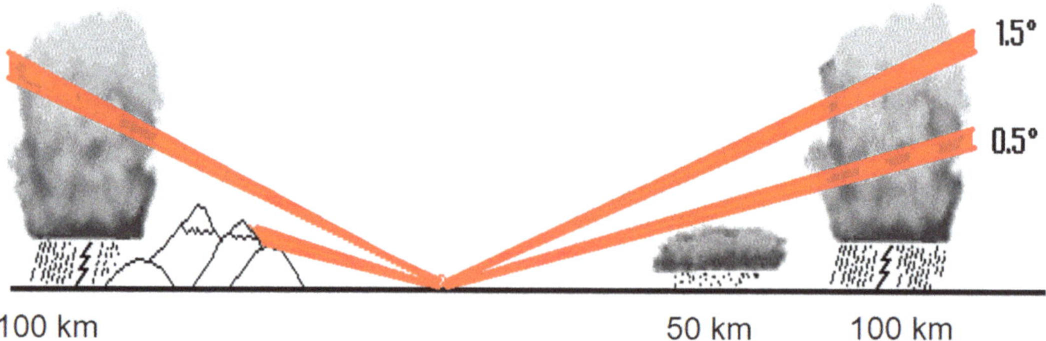

Figure 5.6a Challenges posed by terrain in ground-based radar remote sensing of precipitation. The mountains block beams and result in partial sampling of a cloud system. Many times, the very shallow clouds can be completely missed when the radar is forced to sweep at a higher angle to avoid terrain blocking.

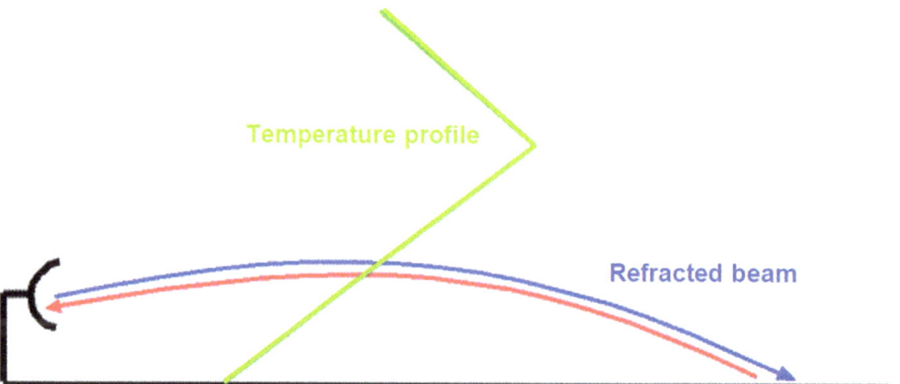

Figure 5.6b Anomalous propagation or super-refraction. Here the temperature increases (and air density decreases) at elevation increases up to a certain level, causing the beam to bend away until it hits the ground.

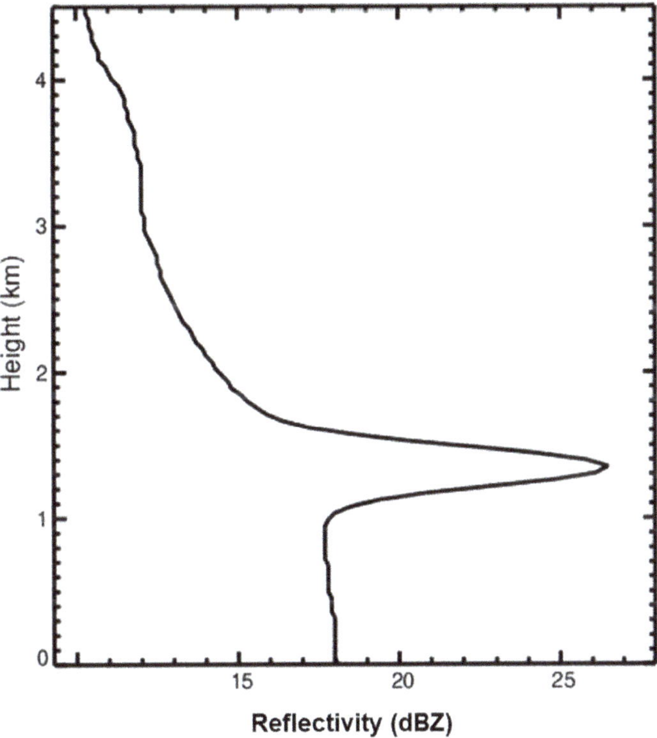

Figure 5.6c Vertical profile of reflectivity, showing the nonuniform reflectivity as a function of elevation.

accounted for if we are interested in the precipitation near the ground or at defined heights. On cloud microphysics, if the exact distribution of drop size distribution is not well accounted for, the reflectivity may produce nonunique rainfall estimates. For example, one 3-mm cloud droplet and 729 1-mm droplets theoretically have the same reflectivity in terms of the sum of sixth power of the diameters. So, the power returned can be the same even though the rain rate will be different, owing to the vastly different drop size distribution in each case (Figure 5.6d).

5.4   Satellite Remote Sensing of Precipitation

Among the limitations of ground-based radar, two problems are mitigated quite easily if the remote sensing is performed using the vantage of space, looking vertically down using a radar. These are: (i) anomalous propagation/super-refraction (beams do not bend in the atmosphere when travelling at 90 degrees towards the ground target – known as normal propagation); (ii) cloud overshoot and partial beam coverage. However, satellite precipitation remote sensing began with passive techniques, and active remote sensing was introduced later in the 1990s. This is primarily because passive sensors are cheaper and easier to launch and maintain in space. Today, most precipitation sensors in space are based on passive sensing in the IR or MW

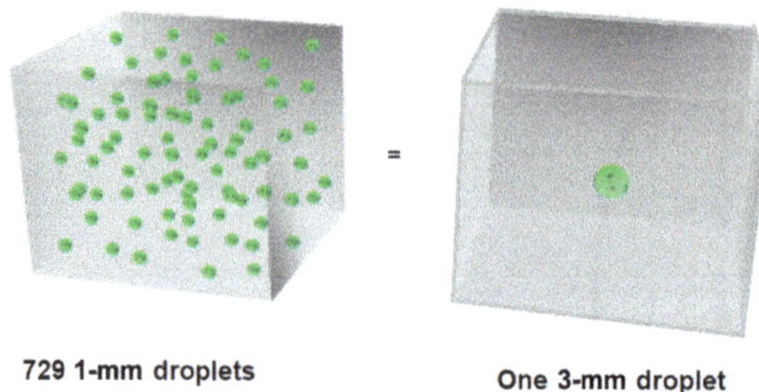

Figure 5.6d Uncertainty in cloud microphysics (or drop size distribution) can lead to nonunique estimation of precipitation.

wavelengths. In the following, we will first build an understanding of the passive approach to estimating rainfall. This will then help us understand the current state-of-the-art of satellite precipitation products, where multiple sensors that are both active and passive sensing are used to generate the best possible quality precipitation data.

5.4.1 Passive Remote Sensing

Recall from Planck's equation that for low temperatures of black bodies in ranges we are exposed to on planet Earth (273–320 K), most of the E-M energy is emitted at the longer wavelengths of IR and MW (Figure 5.7).

There is radiation from thermal (terrestrial) bodies even at long wavelengths that extend into the MW region. MW radiation is emitted just like IR at low "earthly" temperatures, and it is generally much weaker in intensity than short-wavelength radiation from high-temperature surfaces such as the Sun. However, this natural emission (let us call it "passive" energy) is still detectable by sensitive instruments in space. The MW range we are talking about is 0.15–30 cm (1–200 GHz). Being passive, the energy source is weak, and therefore the sensor needs a large field of view and wide wavelength ranges (low spectral resolution).

Clouds emit MW and IR radiation. In other words, the target has an E-M signature in these two wavelength ranges. One thing that helps the precipitation remote sensing using passive MW signature is that, in these longer-than-visible wavelengths (0.15–30 cm), Planck's curve is almost a straight line (Figure 5.8). In other words, the MW radiance emitted by clouds, known as bright temperature T_b (that the satellite sensor records), is linearly correlated to kinetic temperature $T_{kinetic}$ (through an approximation known as Rayleigh–Jeans approximation):

$$T_b = \varepsilon T_{kinetic} \tag{5.7}$$

Here ε in Equation 5.7 is the MW emissivity of the target. So passive MW brightness temperatures can be used to monitor temperature as well as properties related to emissivity. In the MW region, Earth's surface can have large variations in emissivity (Figure 5.9). We can take advantage of this large variability, and the consistent emissivity of clouds, to apply the golden rules of remote sensing, especially over oceans where the background ocean appears "colder" than the cloud in terms of brightness temperature.

Let us take a closer look at how MW passive remote sensing would work for estimating rainfall. Rain, when present, is the major source of attenuation when viewing the surface from

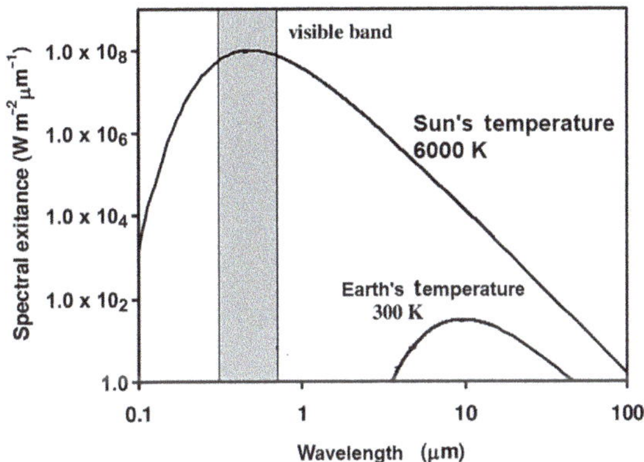

Figure 5.7 Planck's equation, showing that for black bodies at planetary temperatures (~300 K), most of the energy is radiated in the longer wavelengths in the IR and MW ranges (>5 micrometers). Here spectral exitance is the same as spectral radiance shown in Figure 3.2 earlier.

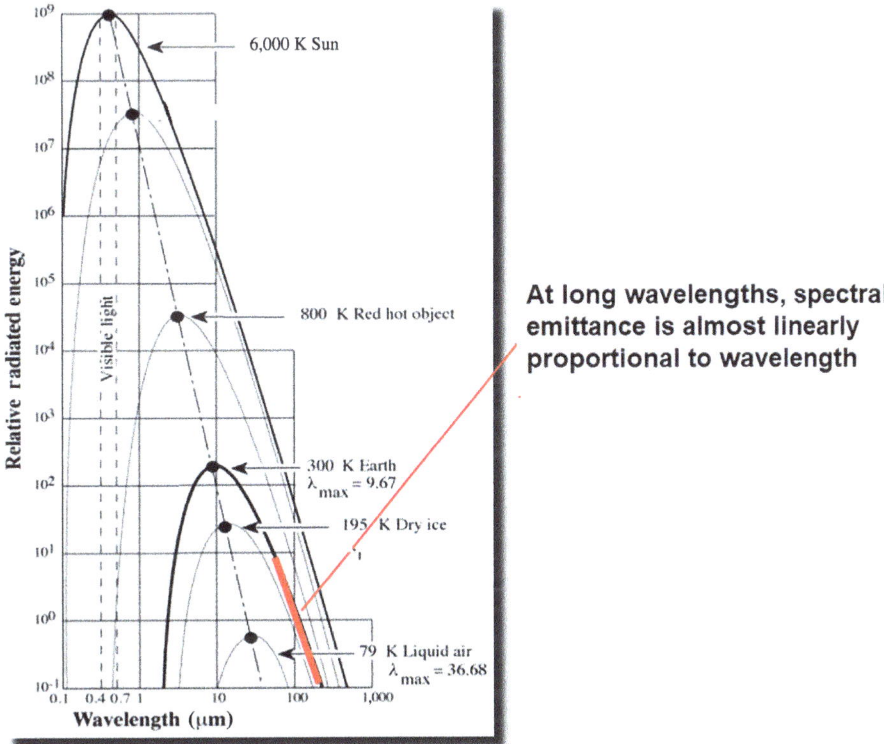

Figure 5.8 Planck's equation at long wavelengths, where the spectral emission is effectively a linear function of wavelength.

space using the atmospheric window (transmission) channels of the MW range. The emissivity of raindrops is almost 1 (at 19 GHz), while the emissivity of the ocean is 0.4 (at 19 GHz), which consequently helps us distinguish clouds from ocean using the golden rules of remote sensing.

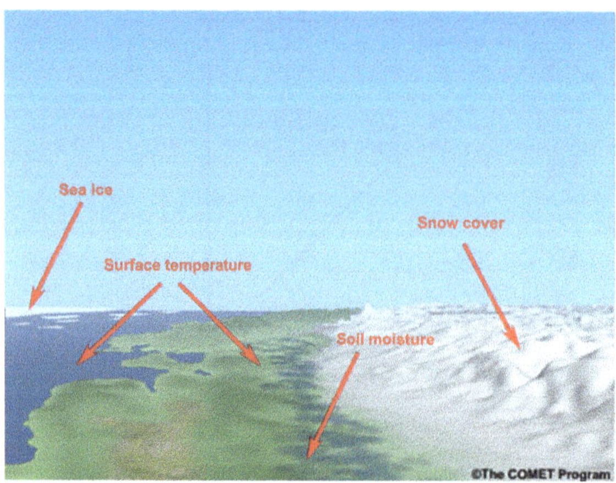

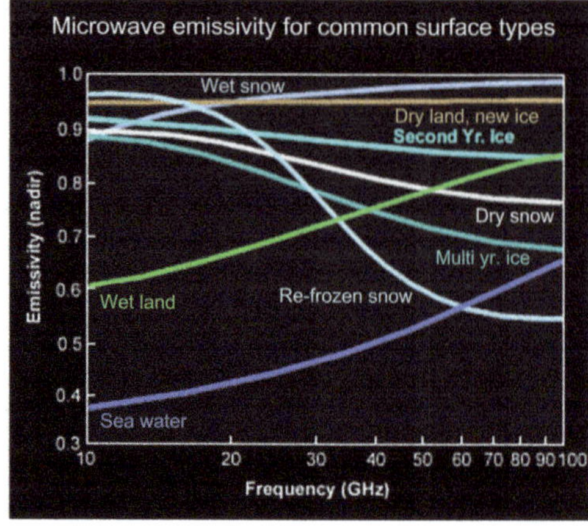

Figure 5.9 Microwave emissivity of various targets and surfaces/backgrounds from the vantage of space. [Sources: left panel – the COMET Program of University Center for Atmospheric Research; right panel – adapted from Weng et al., 2001.]

This large contrast between rain and ocean has been used to estimate precipitation over oceans. The emission-based method (19 GHz) is found to be highly accurate in identifying light as well as heavy precipitation over ocean. This method based on 19 GHz frequency is, however, inadequate for precipitation over land, because the emissivity of land varies widely (Figure 5.9). In water management, we are mostly concerned with precipitation over land, which generally works fine using ground radars. So, another MW wavelength, known as the scattering channel, comes to the rescue here. It turns out that the ice particles produced aloft in rain-bearing clouds over land scatter microwave radiation with negligible absorption or emission. If we try to detect the scattering, which is mostly at 85.5 GHz, we can improve our discriminatory power overland to detect and estimate clouds of their rainfall rate. As an example, we can define a metric called the Scattering Index, which is a statistical combination of brightness temperature at key atmospheric transmission windows of 19 GHz, 22 GHz, and the scattering channel of 85.5 GHz. When this index exceeds a threshold, it is usually considered to be a good indicator of rainfall. Subsequently, by using calibration with in-situ rainfall data, one can develop relationships between this index and rainfall rate similar to the Z–R relationship.

Today, there are numerous algorithms that researchers have proposed for estimation of rainfall using passive MW or IR radiances. The goal of this book is not to review the state-of-the-art of such algorithms, but rather to help readers develop a basic understanding and data literacy skill for water management using satellite precipitation datasets available and used globally today. Here, we will intentionally avoid showing the numerical algorithms (of which there are too many), as the goal is to help readers conceptually understand what goes behind in their preparation rather than how to reproduce such data. This will help readers understand the pros and cons of specific satellite precipitation data products for water management.

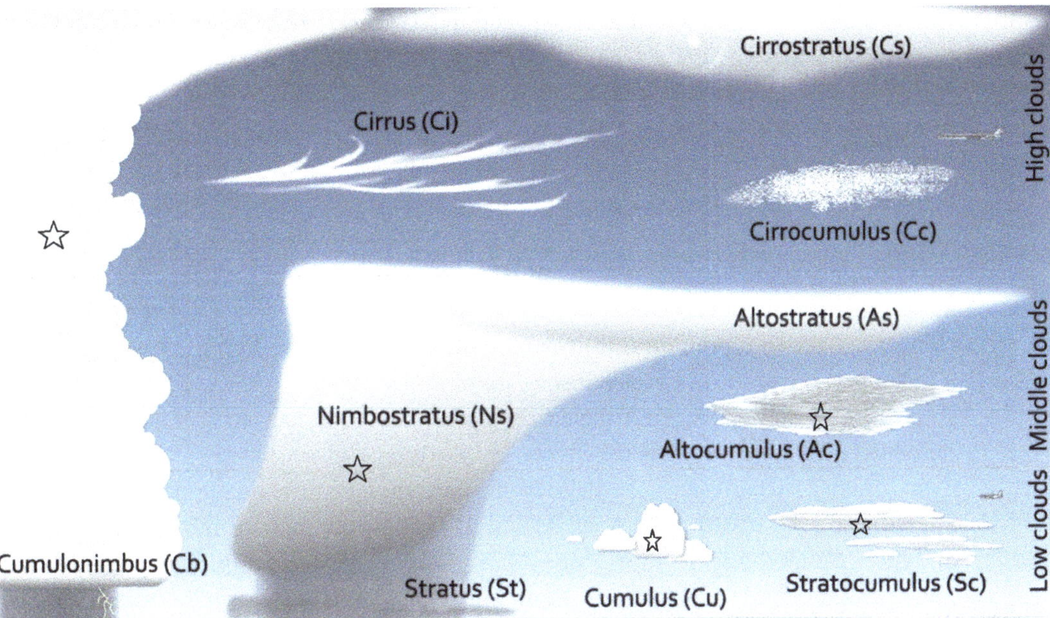

Figure 5.10 Various types of clouds that bear or do not bear rain. The star sign indicates clouds that typically bear rain, while the others do not. [Source: National Weather Service, NOAA]

With this in mind, let us now explore how the thermal IR signature of clouds can be used to estimate rainfall. The heritage of using IR emission from clouds to estimate rainfall is much longer than for the MW sensors because of the global and near-continuous availability of IR radiation data from geostationary platforms. One of the first pieces of research work to show how IR brightness temperature from clouds could be used to estimate rainfall came out in the late 1970s (Griffith et al., 1978). The physical basis is that clouds that are "cold" in the IR emission range produce more rain than warm, shallow clouds. In other words, the IR method assumes that the "colder" the cloud, the taller, larger, and more rain-bearing it should be. Using that as a premise, there have been several algorithms proposed. However, the physical connection between cloud IR observations and rainfall rate is weak. This is mainly because IR sensing reveals to us only the cloud top temperature, which is not representative of the underlying layers. This is best understood by looking at various types of rain- and non-rain-bearing clouds, and their vertical structure, in the context of temperature of the cloud top (Figure 5.10).

It is clear from Figure 5.10 that there are many clouds that are at very high altitude and therefore colder but are not rain-bearing (see those not marked with a star). So an IR radiometer sensor would have no way of knowing that those clouds are in fact just floating at very high altitude rather than being tall and large. Similarly, many clouds are not as cold and are known as "warm clouds" because of their lower altitude, but can bear precipitation of significance for water management. The premise of cloud brightness temperature as a strong proxy for precipitation on the ground will not work here. We should point out two other complicating factors for MW, and for both MW and IR in general for passive remote sensing of rainfall: (1) not all clouds, especially warm ones, have ice crystals for scattering; (2) the background or the ground surface can be just as cold as the cloud during winter with snow cover, in mountainous regions, or in high-latitude regions.

What emerges from this short overview of passive remote sensing of precipitation is that neither IR nor MW is perfect, and there are pros and cons for each wavelength. However, MW and IR have complementary strengths that could be combined (or merged) to create better precipitation products than what is possible using individual wavelengths. For example, MW has a much more physical connection to cloud and rainfall property. On the other hand, satellites recording MW radiation do not provide high temporal resolution as they are orbiting. IR has a much weaker physical connection to cloud and rainfall property (by being opaque to cloud structure). However, IR radiation can be detected at high temporal and also spatial resolution on board geosynchronous platforms.

So, today's precipitation products from space that we have available for water management are based on "teamwork" between MW and IR sensors, passive and active (where available), and multiple methods. In short, today's rainfall products are multi-sensor and multi-method. Such products are a far cry from what was available in the early 2000s and have opened up numerous water management application opportunities, as we shall see later. We will show one example of such teamwork between MW and IR sensors using various clever methods that continue to make satellite-based precipitation data more usable for water management applications. But first, we need to quickly overview two satellite rainfall missions that were at the heart of quantum improvement in precipitation remote sensing from space today and have spurred water management from space.

5.4.2 Tropical Rainfall Measuring Mission

The Tropical Rainfall Measuring Mission (TRMM, Figure 5.11) was a joint space mission between NASA and Japan's National Space Development Agency, designed to monitor and study tropical and subtropical precipitation and the associated release of energy. TRMM was the first Earth Science mission dedicated to studying tropical and subtropical rainfall: precipitation that falls within 35 degrees north and 35 degrees south of the equator. Tropical rainfall comprises more than two-thirds of the world's total. The mission uses five instruments: the Precipitation Radar (PR), TRMM Microwave Imager (TMI), Visible Infrared Scanner (VIRS), Clouds & Earths Radiant Energy System (CERES), and Lightning Imaging Sensor (LSI). Note that the radar here allows the active remote sensing of precipitation, which has been the norm in the US and other places for remote sensing of precipitation using ground platforms. The TMI and PR are the main instruments used for precipitation. Some of the precipitation data products produced with the help of TRMM data are 3B42 and 3B43, which are available at 0.25° spatial resolution (approximately 25 km at the equator), covering 50° N to 50° S for 1998–2015. TRMM carried the first ever precipitation radar in space. By virtue of its accuracy, the TRMM PR afforded better calibration of the passive techniques using the MW and IR wavelength.

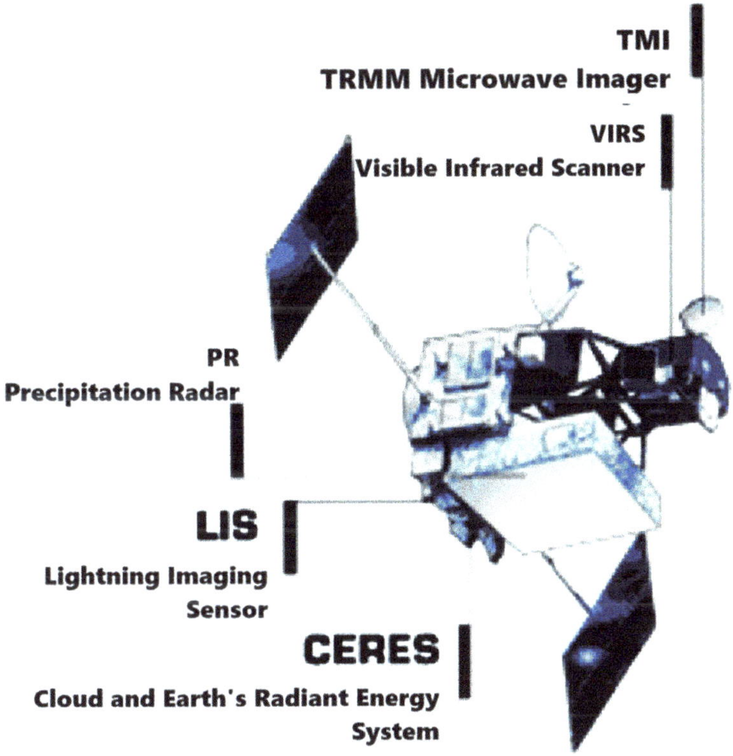

Figure 5.11 The key remote sensing instruments on board the Tropical Rainfall Measuring Mission (TRMM). [Source: NASA]

5.4.3 The Global Precipitation Measurement Mission

The Global Precipitation Measurement (GPM) mission (Figure 5.12) is an international network of satellites that provide next-generation global observations of rain and snow. Building upon the success of the TRMM, the GPM concept centers on the deployment of a "core observatory" satellite carrying an advanced radar/radiometer system to measure precipitation from space and serve as a reference standard to unify precipitation measurements from a constellation of research and operational satellites (Figure 5.13). The constellation of satellites represents passive MW sensors to help GPM provide a spatially more complete measurement of the global precipitation. Through improved measurements of precipitation globally, the GPM mission's goal is to advance our understanding of Earth's water and energy cycles, improve forecasting of extreme events that cause natural hazards and disasters, and extend current capabilities in using accurate and timely information of precipitation to directly benefit society. Most importantly, GPM, by virtue of providing global and high-temporal-resolution precipitation products developed using active radar (PR) and passive remote sensing (MW and IR from geosynchronous orbits), has opened the door for many water management applications from space.

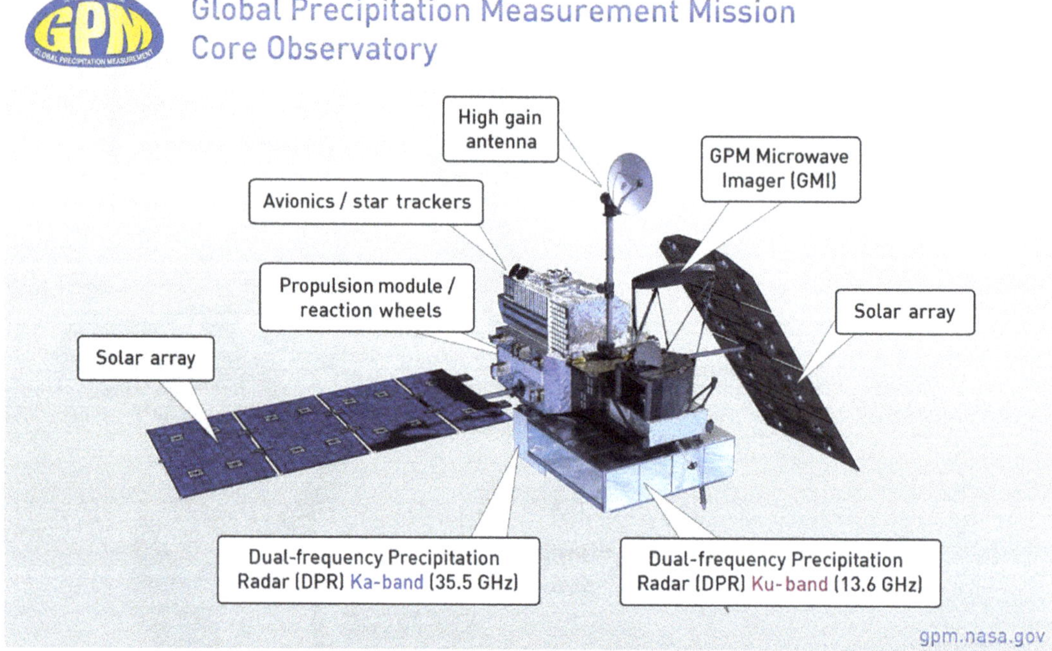

Figure 5.12 The key instruments on board the Global Precipitation Measurement (GPM) mission's core satellite observatory (platform). [Source: NASA; gpm.nasa.gov]

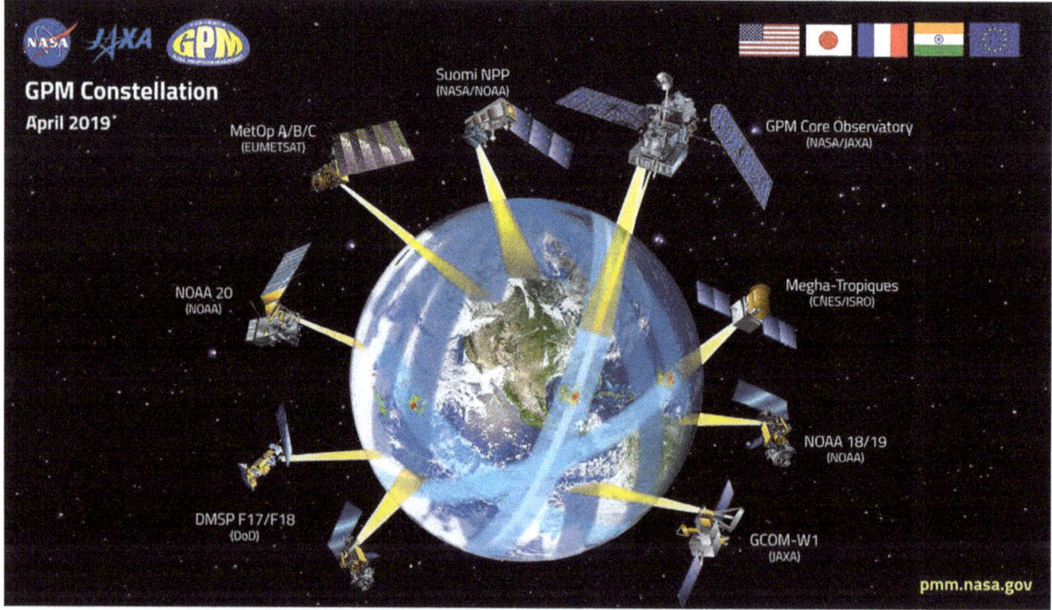

Figure 5.13 The GPM constellation of satellite platforms, showcasing the multi-national and multi-mission collaboration unified by the GPM core observatory to create globally consistent precipitation data. [Source: NASA]

GPM comprises a consortium of international space agencies around the world such as NASA, the Japanese Space Agency JAXA, Centre National d'Études Spatiales (CNES), the Indian Space Research Organization (ISRO), the National Oceanic and Atmospheric Administration (NOAA), the European Organization for the Exploitation of Meteorological Satellites (EUMETSAT), and others.

5.4.4 Multi-Sensor and Multi-Method Precipitation Products of Today for Water Management

As mentioned earlier, today's global precipitation products from space that have enabled water management applications are based on multiple wavelengths (usually MW and IR), platforms (orbiting and geosynchronous), and types (active and passive). TRMM and GPM have helped revolutionize such application of multi-sensor precipitation development products. We shall review here the basic concept of how the "teamwork" happens between multiple sensors to create something that is better than individual sensors or platforms. We should remember that there are numerous methods and data products based on such teamwork for global precipitation estimation. The field continues to evolve, and it is fair to assume that precipitation products will keep improving in accuracy, scale, and availability for water management applications. The purpose here is to give the readers a basic idea of how global precipitation products are produced today. We will do this using one example out of many that are out there for multi-sensor, multi-technique estimation methods. This will help readers become informed users of such data for water management. For more information on how the state-of-the-art of satellite precipitation remote sensing is evolving using multiple sensors and methods in a changing climate, a good reference is Levizzani and Catani (2019).

The first thing to remind ourselves is that, courtesy of missions like GPM, we now have a constellation of many orbiting MW sensors and one (single) precipitation radar. These sensors provide intermittent spatial coverage (i.e. every few hours) of the same region but have a strong physical connection to the rainfall process. On the other hand, we have near-24/7 coverage of IR imagery from geostationary platforms, which has weaker connection to rainfall (Figure 5.14). In the lower panel of Figure 5.14, we can therefore ask ourselves: for an assumed 6-hour window, how do we create a spatially and temporally continuous estimate of precipitation by leveraging the best of both the MW and IR sensors?

The question essentially boils down to this – *how do we fill the gaps between MW estimates, using IR data to provide estimates that will be just as good as MW but without directly using MW data*? See Figure 5.15 for a visual rendition of this question. There are multiple ways to tackle this question, and many multi-sensor data algorithms have been proposed. One of the more common ones is based on some type of calibration so that the more reliable information content in the higher-accuracy sensor can be "transferred" over to the less accurate sensor. In other words, we can try to calibrate the passive MW source to the active MW source first (note there is currently only one precipitation radar) and then calibrate the passive IR source to the passive MW sensor data that is already calibrated to the active MW sensor. This way, information

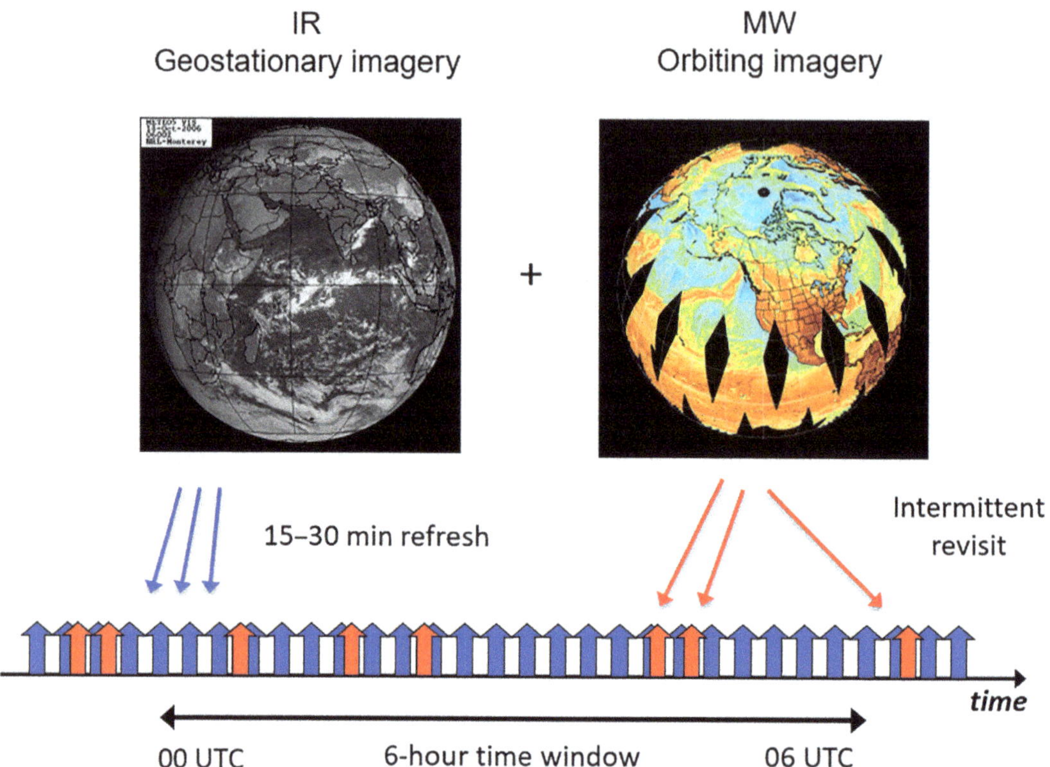

Figure 5.14 The MW passive sensors have a strong physical relationship to the rainfall process but, as a whole, provide intermittent spatial coverage of the same region (shown in red). On the other hand, IR sensors on geostationary platforms provide continuous coverage but have a weaker relationship to rainfall (shown in blue). [Image courtesy of Dr. Joe Turk, NASA Jet Propulsion Laboratory, Caltech]

skill from the most accurate source, the active MW sensor, gets transferred downstream in a cascading manner to the next less accurate sensor (passive MW on orbiting platforms and then passive IR on geostationary platforms). One such algorithm, called the TRMM Multi-satellite Precipitation Algorithm (TMPA) and developed by Huffman et al. (2010), uses this concept of calibration to transfer skill from highest-accuracy sensor to the lower ones for estimating precipitation.

Figure 5.16 (a to c) shows one such calibration summarized above based on histogram-matching. Let us assume that the passive MW estimates are already calibrated accurately to the active MW rainfall estimates from the precipitation radar and start from there (Figure 5.16a). The first panel shows an area that is covered by both passive MW and IR sensors (but with different temporal frequency). If we focus on MW and IR data that overlap, we can construct histograms of rainfall (from MW, which we trust more than IR) and brightness temperatures (from IR). This is shown in Figure 5.16b. Finally, by combining the two histograms, we can create an "adjustment curve" or calibration curve that will allow us to estimate "MW-like" rainfall based on IR brightness temperatures through this histogram matching. We can use this adjustment

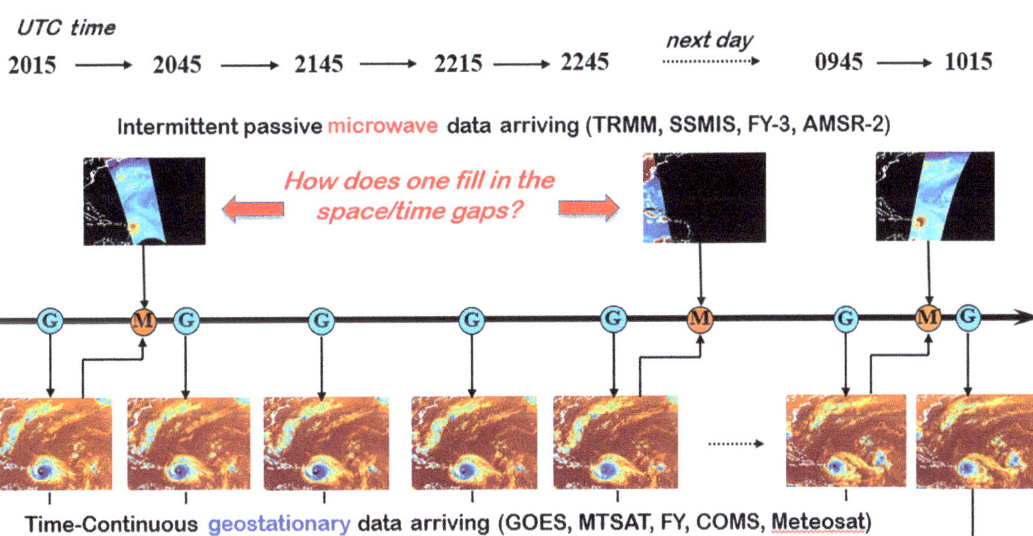

Figure 5.15 How do we fill the gaps between MW estimates, using IR data to provide estimates that will be just as good as MW but without directly using MW data? [Image courtesy of Dr. Joe Turk, NASA Jet Propulsion Laboratory, Caltech]

curve to convert the IR brightness temperatures, which are available to us near-24/7 globally, to MW-like rainfall estimates at times when actual MW sensor data is unavailable and fill in the gaps in time (Figure 5.17).

We have just overviewed the making of one of the many multi-sensor precipitation data products that is available globally at fairly high temporal resolution to enable numerous water management applications. One such product, which is the flagship product of the GPM mission, is the IMERG precipitation data product available at very short latency (i.e. we get the data in hand within a few hours of the time of observation). We will use this IMERG to demonstrate several of our GEE tutorials and exercises.

5.5 Like Most Things, Satellite Precipitation Data Is Not "Perfect"

Before we start applying multi-sensor precipitation data from space to carry out water management applications and understand their potential, it is useful for us to recognize that they are not perfect, like most things in life. In other words, global precipitation data products such as IMERG have errors and uncertainties depending on how we use them. We could get into

Figure 5.16 An example of histogram matching to create "adjustment curves" to obtain "MW-like" rainfall estimates using IR brightness temperatures that are available almost continuously. This adjustment curve is created by first identifying common overlapping areas with coincident MW-rainfall and IR-brightness temperature estimates (panel a). Histograms are created of MW-rainfall and IR brightness temperatures (panel b). Next, by combining the histograms, the adjustment curve is created (panel c). [Image courtesy of Dr. Joe Turk, NASA Jet Propulsion Laboratory, Caltech]

(a)

(b)

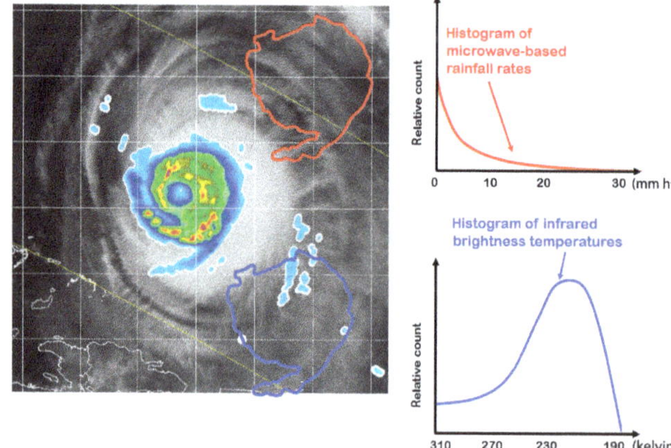

(c)

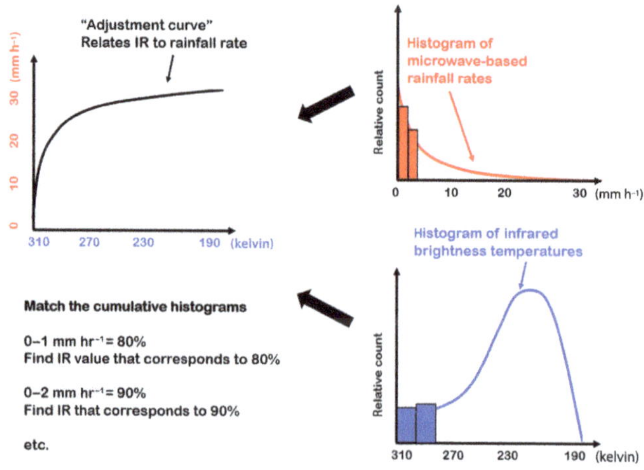

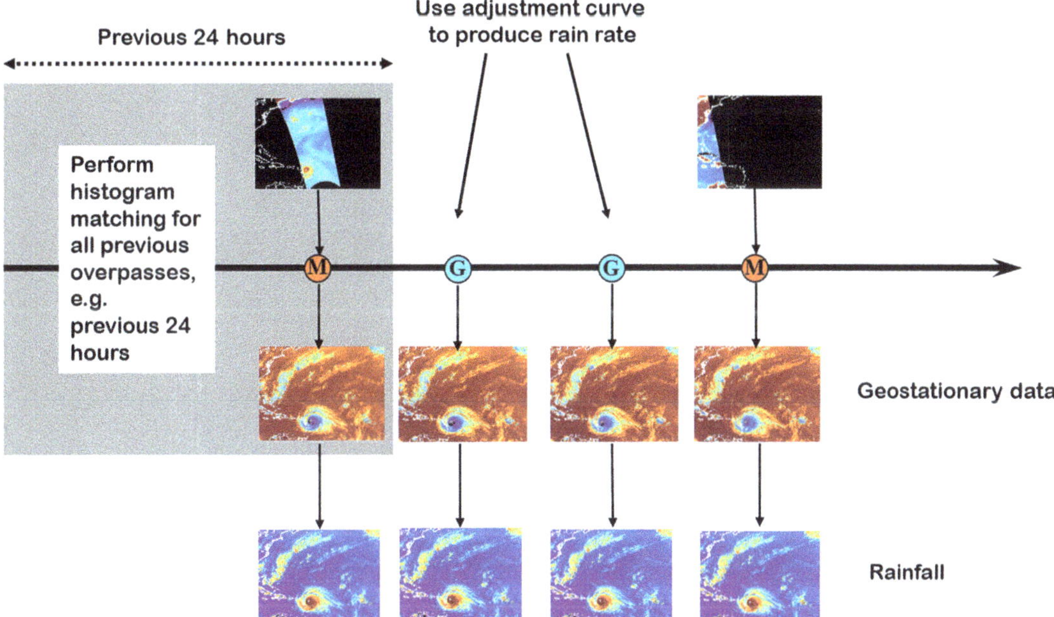

Figure 5.17 Application of the "adjustment curve" to transfer the accuracy of MW rainfall estimates (assume that they are already calibrated to active MW precipitation radar) to IR brightness temperature, so that IR-based rainfall estimates are "MW-like". [Image courtesy of Dr. Joe Turk, NASA Jet Propulsion Laboratory, Caltech]

an endless argument about sources of data that we normally consider accurate and claim that they too are imperfect or do not represent the true rainfall, which can never be measured. For simplicity's sake, let us believe that ground-based precipitation data from ground radars or in-situ gauges, which has been the mainstay of traditional water management in the past, is the reference for comparison.

Although error and uncertainty do not mean the same thing, in this chapter we will use the terms to refer to the same notion of imperfection. A satellite precipitation data product may not always match the rainfall estimate based on ground gauges that has been the traditional benchmark for water management applications. Ground gauges are essentially point-based, whereas satellite products represent a temporal and spatial aggregation. Ground gauges are direct methods of estimating rainfall and therefore do not suffer from the limitations of having to invert the problem to estimate rainfall (see "The Retrieval Problem" in Chapter 3).

If we assume ground-based precipitation from a gauge or even from a ground radar network to be more accurate, we will notice different types of mismatches with satellite precipitation data. It is helpful to understand these different mismatches and why they may be occurring because of the nature of clouds, ground, rainfall systems, and the underlying method used for remote sensing. There are numerous methods to characterize these mismatches. For water management, we will use the simplest way of classifying mismatch in the form of "bias", as follows.

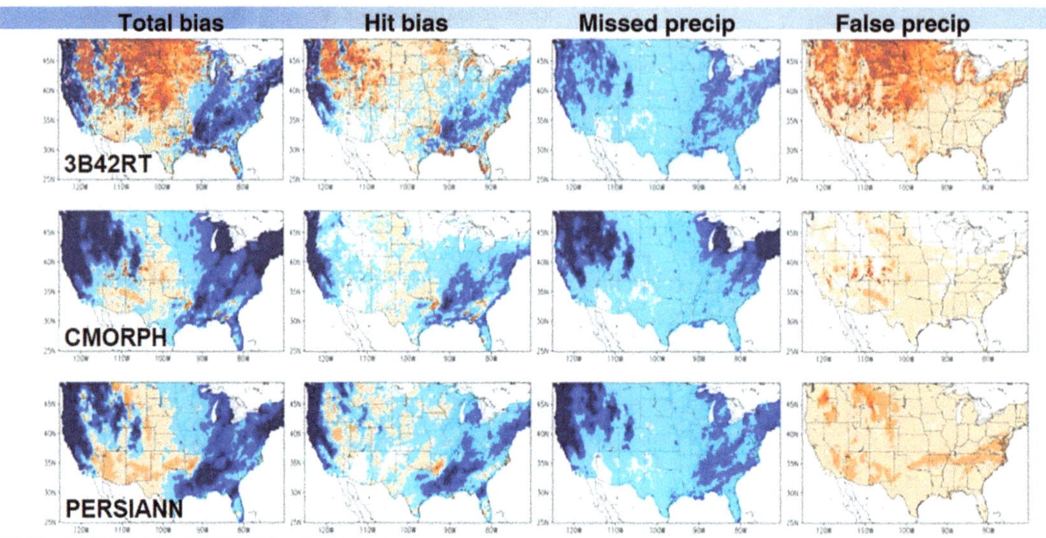

Figure 5.18 Conceptual understanding of error components of three global precipitation data products (3B42RT – upper panel; CMORPH – middle panel; and PERSIANN – lower panel) in terms of metrics of (from left to right) total bias, hit bias, missed precipitation, and false precipitation. The reference precipitation data is from ground radar, and error metrics are shown for the winter season of 2006. The color represents the magnitude from negative (blue) to positive (brown). For more information on the satellite precipitation products, refer to the International Precipitation Working Group at http://ipwg.isac .cnr.it/ [Credit: Image modified from Tian et al., 2009].

First, we will call "total bias" the difference between satellite rainfall and ground rainfall estimates over the same domain and same temporal window. This could be over an area of 100 square kilometers (10 km ×10 km) or an entire river basin. This total bias is made up of hit bias, false precipitation, and missed precipitation. This means that a satellite-based product will not always agree with ground and direct sources that reported that it actually rained at a given location and time. The difference between satellite precipitation and reference for such cases is known as missed precipitation (and is a negative number to signify the "missed" aspect of the satellite precipitation data). And, vice versa, a satellite product may report rainfall when in reality it did not rain (the difference is known as false precipitation). When satellite rainfall data does agree in terms of occurrence, the numerical value will generally be different from the ground one (the arithmetic difference known as hit bias). These mismatches happen because multi-sensor precipitation estimation is still an indirect method of inverting rainfall from radiances or radar reflectivity. Referring to Figure 5.18 and the section on passive remote sensing will be helpful here.

The sum of these three components produces total bias. Figure 5.18 shows a visual map of these error components for the continental United States for a particular month, using ground radar rainfall as reference, based on analysis reported by Tian et al. (2009). The bias components are shown for different types of satellite data products. If we overlaid the maps of hit bias, false precipitation, and missed precipitation numerically, we would generate a map of total bias on the leftmost side of Figure 5.18. This figure also shows that not all data products are the same when it comes to errors, and that these mismatches matter in water management applications. For example, for seasonal crop management applications such as yield prediction or forecasting using week to monthly averaged precipitation data, the hit bias may be a more important component to be concerned with so that crop growth is not overestimated. For flood forecasting applications at much finer time periods (daily or sub-daily), very high false precipitation may lead to overprediction of the magnitude of the flood peak. Vice versa, a very high (negative) value of missed precipitation may result in underestimation of the severity of the flood pulse that needs to be captured via hydrologic modeling. For drought management over multiple years for a large region, the user may really need to just look into total bias and pick a product that has the minimum of that.

In water management, the important thing is to drive the use of satellite precipitation data according to the end goal or key objectives of the application. For example, is the manager more interested in getting trends right to make a decision, or do they want to get an accurate time series of what has happened numerically? In this chapter and in this book, it is impossible to cover every possible objective and end goal of water management from space. Rather, what is more important is for readers to build literacy on how to compute these error metrics and why they might be occurring (owing to remote sensing), so that users can decide for themselves the right precipitation data and the type of water management decision making that can be afforded. Using real-world case studies, hands-on exercises, and tutorials, readers will hopefully build the basic literacy needed to be informed users of satellite data for water management.

Case Study 5.1: **Drought Monitoring Using Satellite Precipitation Data**

One of the many applications of satellite precipitation data for water management is in the domain of drought monitoring and famine warning. Droughts of meteorological nature require a long time series of precipitation data to understand how recent precipitation is changing relative to long-term climatologic patterns. For global monitoring, this is often based on multi-sensor precipitation data (Figure 5.19). For example, NOAA monitors global and US drought conditions based on the "National Integrated Drought Information System" at www.drought.gov/. A key metric for meteorological drought is often Standard Precipitation Index (SPI), which for global regions is calculated using a multi-sensor precipitation product by NOAA called CMORPH (which is very similar to 3B42RT, using methods to combine the complementary skill of passive IR and passive MW sensor data).

The drought.gov site specifically mentions the following on the calculation of SPI:

This technique uses precipitation estimates that have been derived from low orbiter satellite microwave observations and interpolated in space and time from geostationary satellite IR data. CMORPH data has a low latency, but does not detect snow effectively and should not be used in areas with snow during the time period selected, in this case the last month.

Case Study 5.1 (cont.)

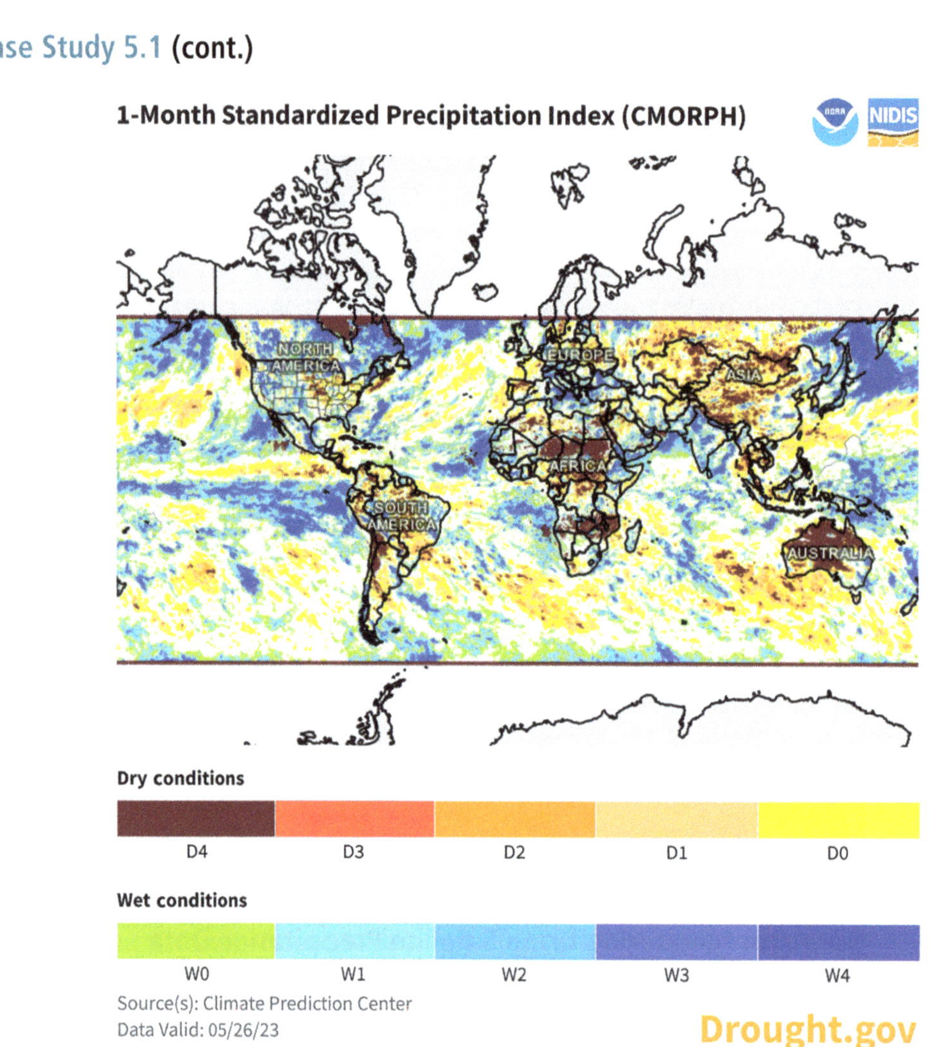

Figure 5.19 Use of multi-sensor precipitation data product for global drought monitoring of meteorological nature (drought-prone regions are shown in colors of brown, red, orange, or yellow). [Source: Climate Prediction Center, NOAA]

Case Study 5.2: Streamflow Monitoring in Transboundary River Basins Using Satellite Precipitation Data

Recall from Chapter 1 that almost half the world's land surface is actually made up of river basins that are international (multiple countries). These river basins, known as transboundary basins, are challenging for water management decision making for reasons ranging from lack of in-situ water monitoring to lack of sharing of data freely, and in near real-time, across political

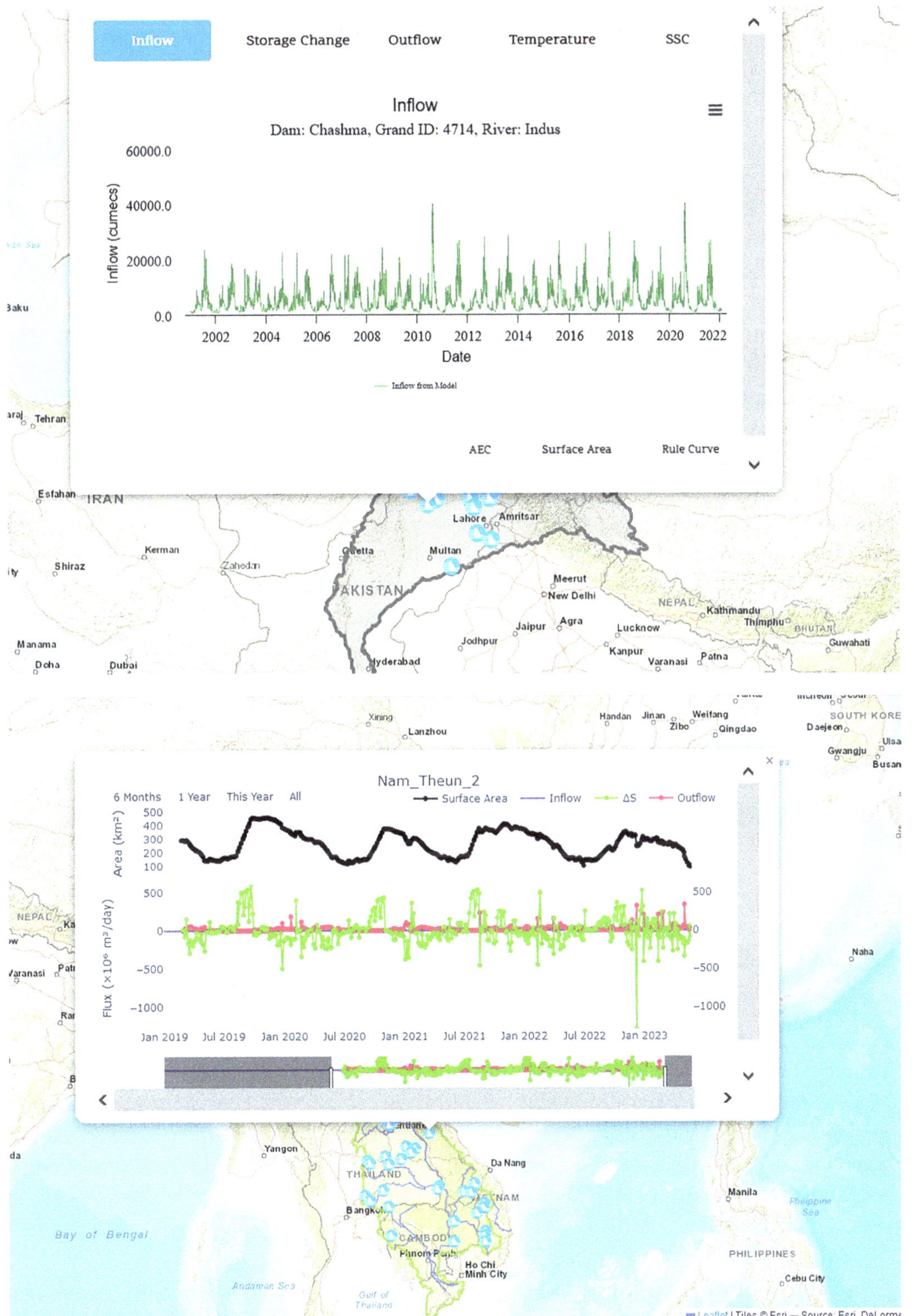

Figure 5.20 Using IMERG satellite precipitation data for near real-time monitoring of streamflow in transboundary basins of Indus (upper panel) and Mekong (lower panel) rivers.

Case Study 5.2 (cont.)

boundaries. Naturally, satellite precipitation data is therefore the most logical alternative source. Using this, we can set up hydrologic models to predict the streamflow in such transboundary basins and make water management decisions. Of course, this is not the same as directly measuring streamflow. Nor is it the same as remotely sensing a river's hydraulic conditions to estimate streamflow, which will be covered in Chapter 7. The uncertainties in precipitation remote sensing will propagate and compound with modeling uncertainty of the hydrologic model to produce streamflow estimates with uncertainty. Nevertheless, in the current state-of-the-art, given the wide availability of hydrologic models and satellite precipitation data, estimating streamflow in this manner has been found to be fairly useful for water management in many international river basins of today where the flow is transboundary.

For example, today we can monitor in near real-time, using the low-latency multi-satellite IMERG precipitation product from the GPM mission, the streamflow conditions in transboundary basins of the Mekong (http://depts.washington.edu/saswe/mekong) or Indus (see http://depts.washington.edu/saswe/indus), and in the southern Indian state of Kerala (http://depts.washington.edu/saswe/kerala). If the streamflow prediction is based on a hydrologic model, it represents the naturalized flow. This naturalized flow can then be adjusted for any upstream regulation by reservoirs, using various reservoir monitoring techniques that also use satellite data of other types. Figure 5.20 shows a screenshot of the front-end web interface for Mekong and Indus streamflow and reservoir monitoring systems. Chapter 1 already alluded to how various satellite data products can be pieced together to monitor how our reservoirs are moving water around (or have moved it in past decades) using rules made by humans and not just nature.

Case Study 5.3: Irrigation Advisory Using Satellite Precipitation Data in Bangladesh

In many places in South Asia, an overwhelming majority of farmers continue to waste water and over-irrigate owing to old irrigation practices, archaic irrigation financing, or just lack of awareness of how much water the plants need. Bangladesh is no exception. Fortunately, now we can monitor crop water need using satellites, we can also track evapotranspiration of crops using satellites. By analyzing the two, and also keeping an eye on weather forecast, particularly precipitation, it is possible to alert farmers on their cellphones to situations when irrigation can be avoided, or when they really need to keep the irrigation going. For precipitation data driving such advisories, we use the IMERG multi-sensor data product from GPM. In Chapter 1, we showed in Figure 1.11 the basic idea of how it works.

In Figure 5.21, the exact role of satellite precipitation data is shown. Currently such a system is now serving farmers of Bangladesh and helping the country conserve groundwater, fuel consumption and improve farmer livelihoods. The advisory on irrigation is disseminated to millions of farmers nationally via the Bangladesh Agro-meteorological Information

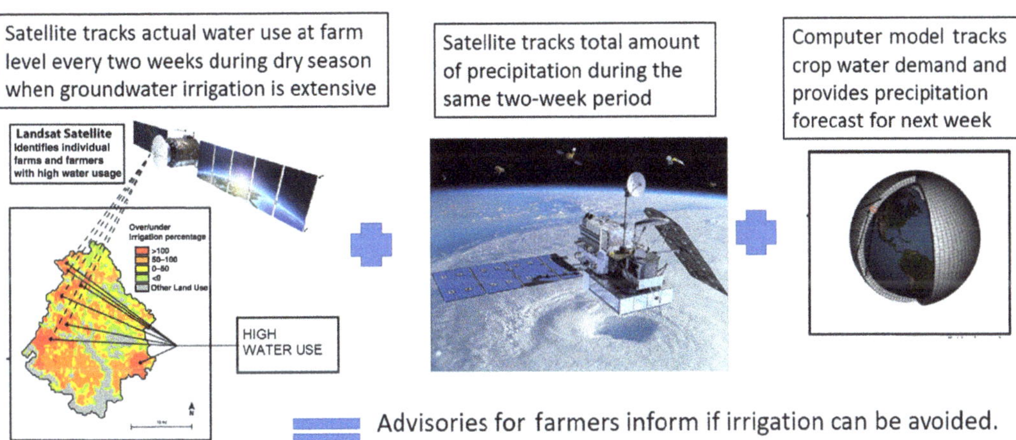

Figure 5.21 The use of IMERG satellite precipitation data from GPM in the routine generation of irrigation advisory for farmers of Bangladesh. The system is called the Integrated Rice Advisory System (IRAS; see Figure 5.22) and also uses other satellites such as Landsat. [Sources: middle and right images from NASA and NOAA, respectively]

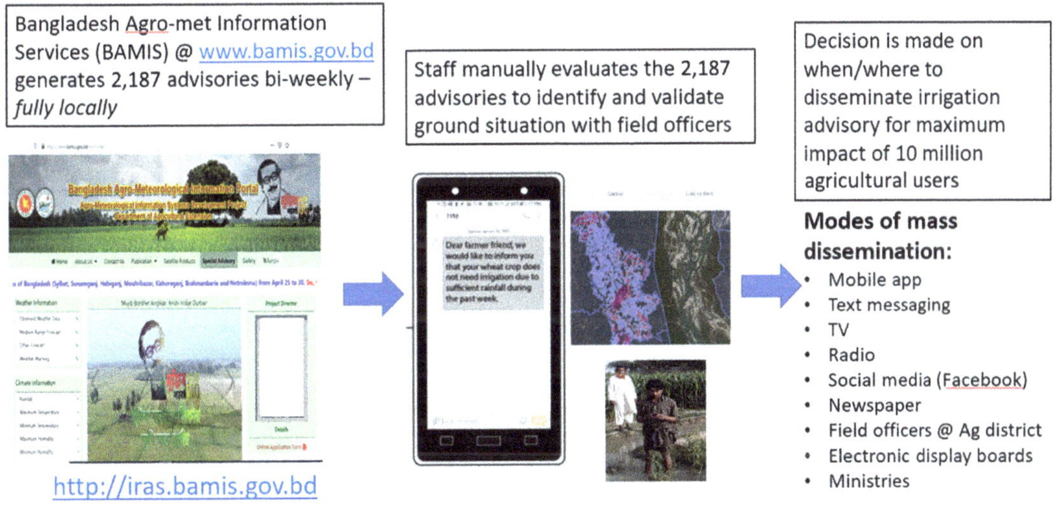

Figure 5.22 The generation of IRAS irrigation advisory using satellite data in the Bangladesh Agro-meteorological Information System of Bangladesh Government (www.bamis.gov.bd) and its mass (national) dissemination to millions of farmers to help with agricultural water management. [Acknowledgement: *Memorandum of Understanding between University of Washington Civil and Environmental Engineering and Ministry of Agriculture*]

System (www.bamis.gov.bd) The system based on IMERG precipitation data is called IRAS (Integrated Rice Advisory System) as it is currently optimized rice and operates for farmers from January-June of each year when groundwater irrigation is extensive (see http://iras .bamis.gov.bd).

5.6   Conclusion

In this chapter, we have covered the topic of precipitation remote sensing. We first overviewed remote sensing using ground radars and covered some of the challenges that this platform faces. Next, using the vantage of space, we covered satellite-based precipitation estimation and discussed the pros and cons of using passive infrared (IR) from geostationary platforms and microwave (MW) from orbiting platforms. Finally, we covered multi-sensor precipitation, as the norm today for creating precipitation data products is based on synergistic use of multiple platforms and sensors. As uncertainty and error are always associated with precipitation estimates, this chapter provided a simple way of understanding the various dimensions of uncertainty and error and what they potentially mean for water management applications. In the next chapter, we will move to the target of surface water. We will cover how satellite remote sensing can be applied to track water on land and its variation in spatial extent and storage.

TUTORIAL ON SATELLITE PRECIPITATION DATA[1]

In this tutorial, we will be looking at the use of satellite data for estimating the amount of precipitation that occurred in a region during a specific time period.

As already seen in Chapter 4, in the month of August 2018, the Indian state of Kerala experienced one of its worst ever floods. Unexpectedly high precipitation triggered by a propagating low-pressure system from the Bay of Bengal forced 35 dams to simultaneously release water, leading to subsequent flooding of downstream areas. The long-term average annual rainfall for the state of Kerala observed over the period of 1901–2017 is about 2,400 mm with a standard deviation of 400 mm. However, from June 1, 2018, to August 19, 2018, a span of only three months, the state received 2,346.6 mm of rainfall, which is 42% above the expected precipitation for the period. Through this tutorial, let us see how we can leverage satellite-borne sensors to estimate the amount of rainfall that Kerala received.

We will be using **Google Earth Engine (GEE)** (cloud-computing) and **Python** (non-cloud/local) to perform the same analysis.

[Note: This is the first instance in this book of providing a tutorial and exercise to handle satellite data in a local (i.e. non-cloud-computing) framework, using a widely popular programming language called Python. The Python-based tutorials and exercises provided hereafter use one particular style known as Jupyter Notebook. We have tried to maintain consistency in this style in the rest of the book. However, we should note for the reader and instructor that Jupyter Notebook is not the only way to apply the Python codes provided here. One other way, which is quasi-local, is Google Colab, provided by Google. This is a free platform where users can run their Python codes with minimum prerequisites in terms of software and hardware in most cases. With some modifications, the Python codes provided here as Jupyter notebooks can be potentially repurposed for Google Colab. For more information on Google Colab, readers and instructors can refer to https://colab.research.google.com/]

---

[1] Contributed by Sarath Suresh, University of Washington.

Readers can find the entire set of GEE scripts for this tutorial at: https://code.earthengine .google.com/?accept_repo=users/saraths96/Tutorial (click on "Precipitation")

The reader is encouraged to refer to the GEE documentation for details on any of the GEE methods and objects used in this tutorial, available at www.developers.google.com/ earth-engine/guides.

To perform this analysis, we will be using the **Integrated Multi-Satellite Retrievals for GPM** (IMERG) gridded precipitation dataset. The dataset is created as part of a joint NASA and Japan Aerospace Exploration Agency (JAXA) initiative called the Global Precipitation Measurement (GPM) mission. The mission utilizes a constellation of satellites equipped with microwave radars. The constellation contains a core satellite called the GPM Core Observatory and several other satellites provided by a consortium of international partners (see Figure 5.13).

First, let us open a new GEE script file by clicking the "NEW" option under the scripts tab and creating a new file (Figure 5.23).

This opens a new file in the GEE script editor window where we can enter the JavaScript code to run our analysis.

Let us first visualize the precipitation for a single day and then create a time series of daily precipitation values over Kerala for the month of August. The time series data will be plotted within GEE using the chart functionality.

The procedure to perform the analysis using GEE is as follows:

1. Let us begin by first defining the region of interest (ROI). Since we are interested in analyzing the precipitation over the state of Kerala, we will need to obtain a geometry object for the state. The Global Administrative Unit Layers (GAUL) dataset is a collection of all administrative units at varying levels with state level geometries at Level 1. Search from "GAUL" in the search bar located at the top part of the GEE interface and select the **FAO GAUL 500m: Global Administrative Unit Layers 2015, First-Level Administrative Units** dataset

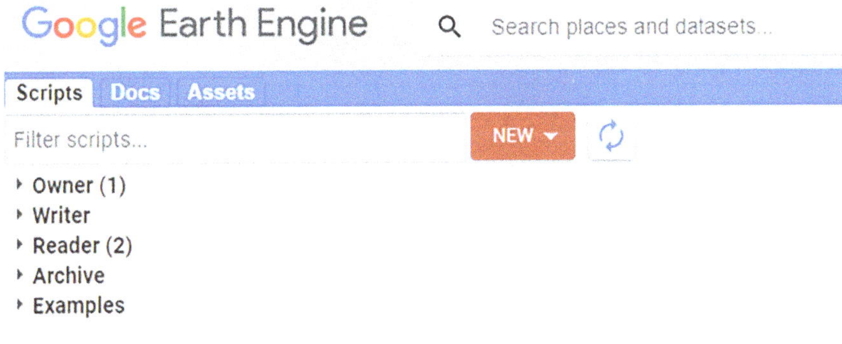

Figure 5.23 New option under scripts tab in Google Earth Engine (GEE). [Credit: Google Earth Engine]

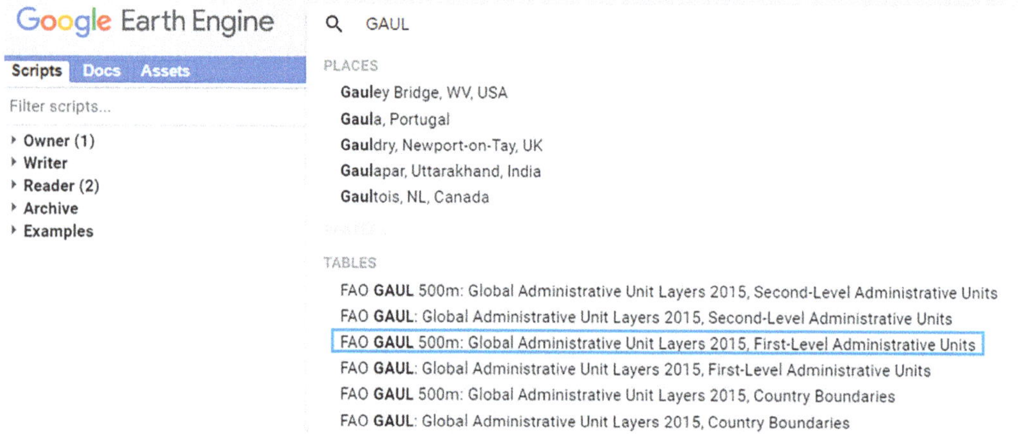

Figure 5.24 Searching for Global Administrative Unit Layers in GEE. [Credit: Google Earth Engine]

(Figure 5.24). Copy the collection code and paste it into the code editor window by creating a new variable. We will then filter out the Kerala geometry by using the **ADM1_NAME** attribute of the dataset, which is useful for filtering out features using the administrative unit name (Figure 5.25). Then the map area is centered on the extracted feature so that we can begin the analysis (Figure 5.26).

```
//Extracting Kerala geometry from GAUL
var kerala_geom =ee.FeatureCollection('FAO/GAUL_SIMPLIFIED_500m/2015/level1')
                .filter(ee.Filter.eq('ADM1_NAME','Kerala'));
Map.centerObject(kerala_geom, 7)
Map.addLayer(kerala_geom, {color: 'green'})
```

2. Next, let us define the time period of the study and access the necessary GPM dataset. Again, search for GPM on the search bar and select the **GPM: Global Precipitation Measurement (GPM) v6** dataset. This dataset provides a gridded global precipitation map at a spatial resolution of 0.1° (11.13 km) at half-hourly intervals in units of mm hr^{-1} (Figure 5.27).

   The GPM v6 dataset contains many bands. For this analysis, we will be working with the "precipitationCal" band. As seen in the Chapter 4 tutorial for flood analysis, we can extract the exact scene we are looking for by chaining filtering functions on the Image Collection. Note that since the GPM data we have is at half-hourly intervals, we have a total of 48 images for a day. Hence, in order to get the total daily precipitation, we sum up all the half-hourly images and divide by 2.

```
//Defining doi and accessing GPM data
var doi = ee.Date('2018-08-15').getRange('day');
var imerg_data = ee.ImageCollection("NASA/GPM_L3/IMERG_V06");
var imerg_kerala = imerg_data.filterBounds(kerala_geom)
                  .filter(ee.Filter.date(doi));
var precip_daily = imerg_kerala.select('precipitationCal').sum().divide(2)
.clip(kerala_geom);
```

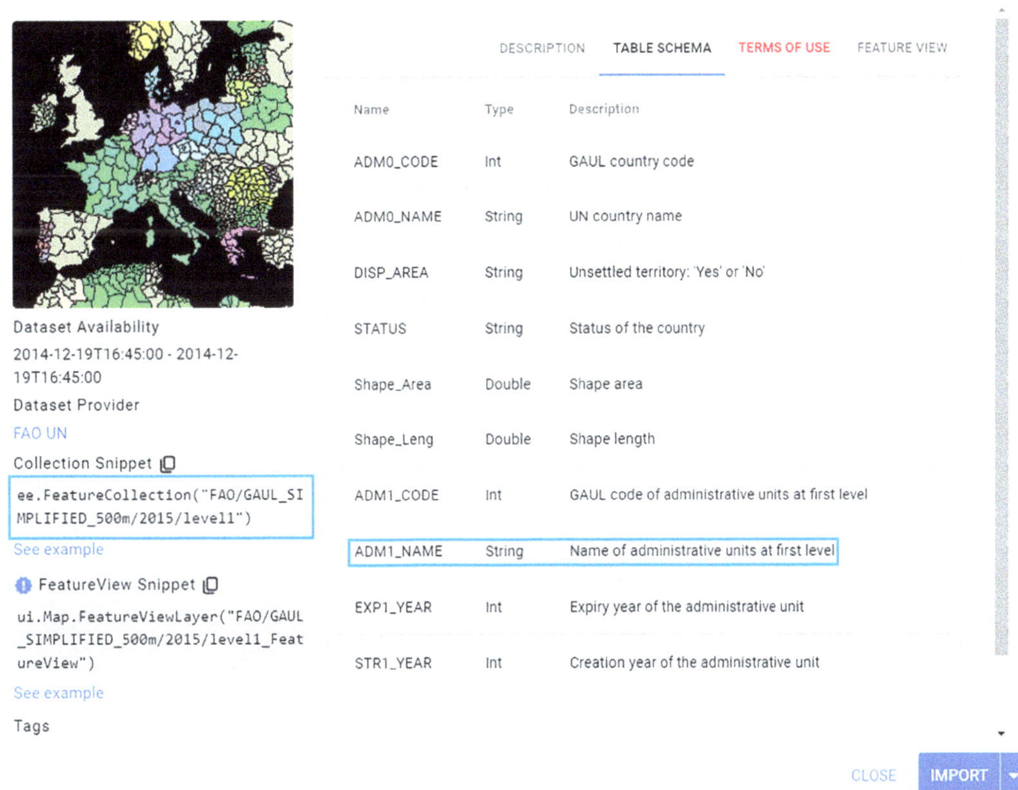

Figure 5.25 Screenshot that appears after copying from Figure 5.24. [Credit: Google Earth Engine]

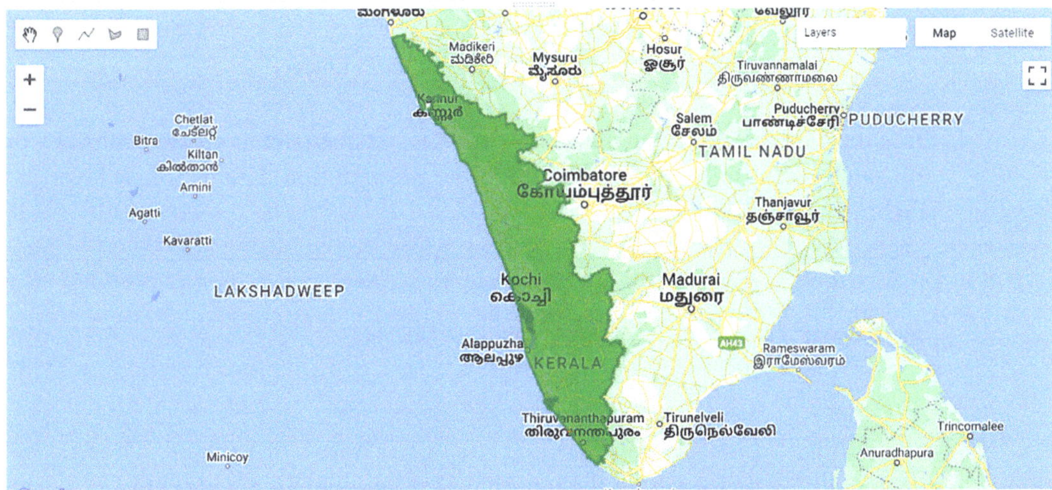

Figure 5.26 Mapped area for selected administrative layer units. [Credit: Google Earth Engine]

Figure 5.27 Searching for Global Precipitation Measurement (GPM) v6 dataset in GEE. [Credit: Google Earth Engine]

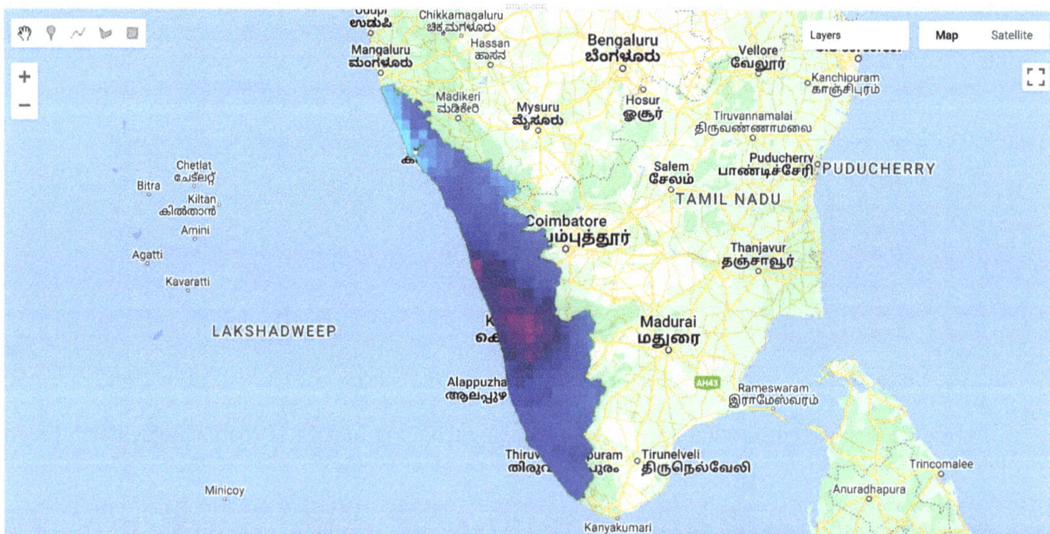

Figure 5.28 Creating precipitation map [Credit: Google Earth Engine]

3. Once the daily precipitation image collection is obtained, let us create a mask by discarding any values that are less than 0, and add the precipitation data as a new layer in the Map (Figure 5.28).

```
//Creating precipitation mask and adding to Map
var mask = precip_daily.gt(0);
var precip_daily = precip_daily.updateMask(mask);
var precipitationVis = {bands: 'precipitationCal', min: 0, max: 180, palette:
['cyan','blue','darkblue','purple']};
Map.addLayer(precip_daily, precipitationVis, 'Precipitation (mm)');
```

Using the inspector tab on the top right of the screen, we are also able to quickly inspect the pixel values in the Map layer that we have added. Switch to the inspector tab and click

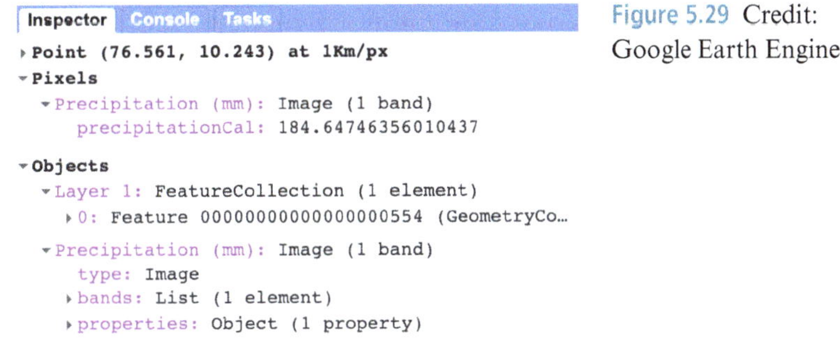

Figure 5.29 Credit: Google Earth Engine

```javascript
// Calculating the mean precipitation over Kerala
var maxDailyPrecipitation = precip_daily.reduceRegion({
  reducer: ee.Reducer.mean(),
  geometry: kerala_geom,
  scale: 11132,
  maxPixels: 1e9
});

var maxPrecipValue = maxDailyPrecipitation.get('precipitationCal');
print('Mean Precipitation (mm) over Kerala:', maxPrecipValue.getInfo().toFixed(2));
```


Figure 5.30 Credit: Google Earth Engine

on any part of the newly plotted Precipitation map layer. The inspector displays details on the objects that make up the map layer and its value at the selected point (Figure 5.29).

4. Let us also compute the mean precipitation over Kerala for that particular day. We can use the **ee.ImageCollection.reduceRegion** method with the reducer as the **ee.Reducer.mean**. The scale is set to be 11,132, as the GPM dataset we use here has a spatial resolution of 11,132 m per pixel (Figure 5.30).

```javascript
// Calculating the mean precipitation over Kerala
var maxDailyPrecipitation = precip_daily.reduceRegion({
  reducer: ee.Reducer.mean(),
  geometry: kerala_geom,
  scale: 11132,
  maxPixels: 1e9
});

var maxPrecipValue = maxDailyPrecipitation.get('precipitationCal');
print('Mean Precipitation over Kerala for the date 2018-08-15 :',
maxPrecipValue.getInfo().toFixed(2) + ' mm');
```

5. Now, we will compute the mean precipitation for Kerala over a longer time period of 2 months from July to August 2018. Since the image collection for a single day has 48 images, we need to implement a logic to create a new image collection containing a sequence of images for each day of the month with the total accumulated precipitation.

   To achieve this, we will create a new Image Collection by passing an **ee.List.sequence** object into an **ee.ImageCollection** constructor. First, we define the start date and end date as 2018-07-01 and 2018-08-31. Then we compute the total number of days between the start and end dates using the **ee.Number.difference** function. Next, a list object is created using the **ee.List.sequence** constructor that ranges from 0 to the total number of days. Then a new image collection "precipByDays" is created using this list object. For this purpose, the **ee.ImageCollection.map** function is used. An anonymous function is passed as the argument for the map function where for every day in the list sequence, an initial and end date are computed using the **ee.Date.advance** method. Then the GPM dataset is filtered for the initial and end dates, the total precipitation is computed, and the "system:time_start" attribute of the image is set to the initial date. This attribute is important if we want to create a time series plot of the precipitation using the **ui.Chart** functionality.

```javascript
//setting start-date and end-date for daily time series
var startDate = ee.Date.fromYMD(2018,7,1);
var endDate = ee.Date.fromYMD(2018,8,31);

//Creating daily image collection from the GPM dataset
var imerg_dataset = ee.ImageCollection('NASA/GPM_L3/IMERG_V06')
var nDays = ee.Number(endDate.difference(startDate,'day'));

var precipByDays = ee.ImageCollection(
  ee.List.sequence(0, nDays).map(function (n) {
    var ini = startDate.advance(n,'day');
    var end = ini.advance(1,'day');
    return imerg_dataset.filterDate(ini,end)
               .select('precipitationCal').sum().divide(2)
               .set('system:time_start', ini);
}));
```

The reader is encouraged to go through the above code line by line to understand how the daily precipitation image collection is being created.

Once this image collection is created, we can easily create a time series plot using the **ui.Chart.image.seriesByRegion** method. The **ui.Chart** functionality in GEE helps us to create a chart widget within the console tab. We will pass the daily precipitation image collection, the ROI geometry, the type of reducer to apply (in this case, the reducer will be mean), the specific band that is to be plotted, and the scale.

Please refer to the documentation for GEE Charting functionality, available at: https://developers.google.com/earth-engine/guides/charts_overview, for more details.

```javascript
//Creating time series chart in GEE for daily mean precipitation
var dailyPrecipChart =
    ui.Chart.image
    .seriesByRegion({
```

```
imageCollection: precipByDays,
band: 'precipitationCal',
regions: kerala_geom,
reducer: ee.Reducer.mean(),
scale: 11132
});
```

GEE also makes it easy to style the chart widget using the **ui.Chart.setOptions** method. A few styling options are shown in the following code snippet (Figure 5.31).

```
// Setting chart style properties
var chartStyle = {
  title: 'Daily Mean Precipitation over Kerala (July & August 2018)',
  hAxis: {
    title: 'Date',
    titleTextStyle: {italic: false, bold: true},
    gridlines: {color: 'FFFFFF'}
  },
  vAxis: {
    title: 'Precipitation (mm)',
    titleTextStyle: {italic: false, bold: true},
    gridlines: {color: 'FFFFFF'},
    format: 'short',
    baselineColor: 'FFFFFF'
  },
  series: {
    0: {lineWidth: 2, color: 'blue', pointSize: 5}
  },
  chartArea: {backgroundColor: 'EBEBEB'},
  legend: {position: 'none'}
};

dailyPrecipChart.setOptions(chartStyle)
print(dailyPrecipChart);
```

We can also easily export the time series data in a tabular format by clicking on the pop-up icon on the top right corner of the chart widget (Figure 5.32).

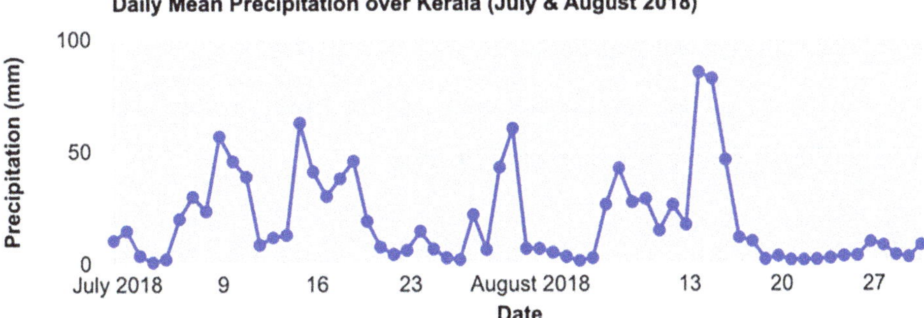

Figure 5.31 Plotting time series of precipitation. [Credit: Google Earth Engine]

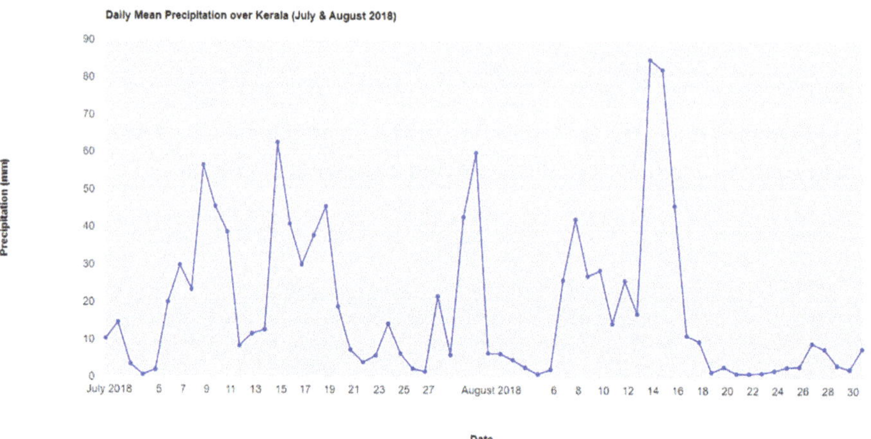

Figure 5.32 Credit: Google Earth Engine

Part 2: Python

We will also take a brief look at how to create the same daily precipitation time series for a given region using a general-purpose programming language such as Python in a local computing environment. Readers can find the complete Python script at: www.cambridge.org/hossain (copy and paste this in your browser)

Prerequisites

Before beginning this part of the tutorial for analyzing GPM precipitation using Python, the reader is advised to set up the following:

1. An interactive Python notebook (Jupyter Notebook) file to work in. The notebook may be named as "precipitationAnalysis.ipynb".

   The Python environment that the notebook is running on should include the following libraries: rioxarray, pandas, geopandas, matplotlib, os.

2. Download the GPM IMERG precipitation files and the region of interest shapefiles from the link provided above. Place the "Precipitation" folder downloaded from this link in the same folder as the Jupyter Notebook.

The procedure to obtain the daily precipitation time series using Python is as follows.

1. Let us start by importing all the required libraries:

```python
# Importing the necessary libraries
import rioxarray as rxr
import os
import pandas as pd
import geopandas as gpd
import matplotlib.pyplot as plt
```

We will be using the "rioxarray" library for loading the GPM IMERG data to perform the analysis. "rioxarray" is a Python library that extends the functionality of the popular xarray library to work seamlessly with geospatial data, particularly gridded raster data.

2. Next, we will define the date range for the analysis, and the various folder and file paths for the GPM data and the ROI data.

```python
# Defining the start and end dates
startDate = pd.to_datetime('2018-07-01')
endDate = pd.to_datetime('2018-08-31')
# Defining the folder path where the GPM & roi data is stored and the GPM scale
factor
GPM_folderPAth = './Precipitation/IMERG_data'
GPM_scaleFactor = 0.1 #From GPM documentation
roi_filePath = './Precipitation/roi/kerala.geojson'
```

3. An empty list is created to store the panda data frames that will be created for the daily precipitation records. The ROI shapefile is also loaded as a geopanda dataframe.

```python
# Creating an empty list to store daily precipitation panda dataframes.
precip_df_list = []
# Load the roi shapefile as a GeoPanda DataFrame
kerala_gdf = gpd.read_file(roi_filePath)
```

4. Then a function "**computeMeanPrecip**" is defined that does the following:
   - Opens the GPM data using rioxarray.
   - Clips the raster data to the ROI geometry.
   - Multiplies the raster values in the clipped data with the scale factor.
   - Calculates the mean precipitation value from the clipped data and returns it.

```python
# Function to compute mean precipitation.
def computeMeanPrecip(precip_global_fp, roi, scaleFactor):
    # Opening the GPM global precip file using rioxarray.
    precip_global = rxr.open_rasterio(f'{precip_global_fp}', masked = True)
    # Clipping dataset to roi and multiplying with GPM scale factor
    precip_clipped = precip_global.rio.clip(roi.geometry)
    precip_clipped = precip_clipped*scaleFactor
    # Computing mean precipitaiton and returning the mean value.
    meanPrecip = precip_clipped.mean()
    return(meanPrecip.values)
```

5. Next, we iterate over all the dates between the start and end date and call the function defined above for all the individual daily precipitation files:

```python
# Running the computeMeanPrecip function for every day in the date range.
for date in pd.date_range(startDate, endDate):
    try:
        formatted_date = date.strftime('%Y-%m-%d')
```

```python
            # creating GPM filepaths for each date
            GPM_filePath = os.path.join(GPM_folderPAth, formatted_date + '_IMERG.tif')
            # creating panda dataframe with mean precipitation for each day and
            appending to list
            precip_df = pd.DataFrame(
                    {'Date': [date],
                     'MeanPrecipitation':
                 [computeMeanPrecip(GPM_filePath, kerala_gdf, GPM_scaleFactor)]
                    })
            precip_df_list.append(precip_df)
    except Exception as e:
        print(f'Precipitation analysis for {formatted_date} did not complete due
        to error: {e}')
```

6. The panda dataframes within the list are then concatenated to create the final panda dataframe containing mean daily precipitation values for each day.

```python
# Concatenating all daily panda dataframes into single dataframe
precip_df = pd.concat(precip_df_list, ignore_index=True)
# Convert the 'Date' column to a datetime type and set as index
precip_df['Date'] = pd.to_datetime(precip_df['Date'])
precip_df.set_index('Date', inplace=True)
```

7. Finally, let us plot the daily precipitation values for visualization.

```python
# Plotting the daily precipitation time series using matplotlib.pyplot
fig, ax = plt.subplots(figsize = (15,4))
ax.plot(precip_df.index, precip_df['MeanPrecipitation'],color = 'blue',
        linewidth = '2',marker = 'o', label = 'Precipitation (mm)')
plt.title('Precipitation over Kerala (July to August 2018)')
plt.ylabel('Mean Precipitation (mm)')
plt.xlabel('Date')
plt.legend()
```

EXERCISES: CHAPTER 5

Q5.1   For ground-based radar remote sensing of precipitation, name the key bands (wavelengths) used. Why are they named like this? What is the band used in the US network of radars?

Q5.2   What is reflectivity, and how is it related to rainfall rate?

Q5.3   Describe briefly the ground weather radar network of the United States, Canada, and Japan (use the internet; summarize the band, number, and geographic coverage).

Q.5.4   What are typical challenges or limitations faced by radar precipitation estimation?

Q5.5   What is anomalous propagation, and why does it happen?

Q5.6   What are the advantages and disadvantages of using active microwave (MW) remote sensing from space and using a ground platform? What specific challenges of ground-based radar could be mitigated using the vantage of space?

Q5.7   Describe briefly the TRMM and GPM missions and their uniqueness.

Q5.8   What are the pros and cons of using orbiting MW and geosynchronous IR radiation data for precipitation estimation? What will be the impact of placing MW sensors in geosynchronous orbit?

Q5.9   Summarize briefly the challenges of satellite precipitation estimation using passive technique over land, compared with over ocean. How are the challenges typically resolved?

Q5.10   By overviewing the various types of clouds, summarize the types that make rainfall remote sensing difficult for the IR passive technique.

Q5.11   What is histogram matching, and how can it be used to create an adjustment curve to improve IR-based rainfall estimates when MW rainfall estimates are unavailable?

Q.5.12   **Data-based exercise (contributed by Sarath Suresh, University of Washington):** Compare monthly average precipitations over the regions of Pakistan and Türkiye for the months of December 2020 to February 2021. Plot the mean monthly precipitation of the two regions. Use either the Google Earth Engine (cloud) or Python option (local). Hints are provided below to get started.

1. Google Earth Engine
   - The GPM: Monthly Global Precipitation Measurement dataset may be used for monthly precipitation data.
   - Türkiye and Pakistan geometries can be obtained from the GAUL Level0 (Country Boundaries) dataset by filtering for the "ADM0_NAME" attribute
   - The following algorithm may be followed to complete the exercise:
     - Obtain geometries
     - Define start date and end date as 2020-12-1 and 2021-02-28
     - Obtain GPM monthly data for both regions.
     - GPM data is to be multiplied by 720 to get total monthly precipitation. (30 days × 24 hours)
     - Create a new ImageCollection for both regions containing monthly mean values.
     - Merge the two ImageCollections using **ee.ImageCollection.merge** method
     - Plot the merged data onto a chart.
       Note: The region geometries may need to be merged into a FeatureCollection for plotting on the same chart. Use the **ee.Feature** and **ee.FeatureCollection** methods to achieve this.

2. Python
   - Use the "turkey.geojson" and "pakistan.geojson" files provided in the "roi" folder for the ROI geometry.
   - The required IMERG global files for the time period are provided in the "IMERG_data" folder.
   - Tip: You may use the **pandas dataframe.resample** function to easily compute monthly means from a dataframe containing daily precipitation values.

REFERENCES

Battan, L. J. (1959). *Radar Meteorology*. University of Chicago Press.

Griffith, C. G. W. L. Woodley, P. G. Grube, et al. (1978). Rain estimation from geosynchronous satellite imagery – visible and infrared, studies. *Monthly Weather Review*, vol. 106, 1153–1171. https://doi.org/10.1175/1520-0493(1978)106<1153:REFGSI>2.0.CO;2

Huffman, G. J., R. F. Adler, D. T. Bolvin, and E. J. Nelkin (2010). The TRMM Multi-Satellite Precipitation Analysis (TMPA). In Gebremichael, M. and Hossain, F. (eds.) *Satellite Rainfall Applications for Surface Hydrology*, 3–22. Springer. https://doi.org/10.1007/978-90-481-2915-7_1

Levizzani, V. and E. Cattani (2019). Satellite remote sensing of precipitation and the terrestrial water cycle in a changing climate. *Remote Sensing*, vol. 11, 2301. https://doi.org/10.3390/rs11192301

Tian, Y., C. D. Peters-Lidard, J. B. Eylander, et al. (2009). Component analysis of errors in satellite-based precipitation estimates. *Journal of Geophysical Research*, vol. 114, D24101. https://doi.org/10.1029/2009JD011949

Weng, F., B. Yan, and N. C. Grody (2001). A microwave land emissivity model. *Journal of Geophysical Research*, vol. 106, 20115–20123. https://doi.org/10.1029/2001JD900019

SUGGESTED READING

Joyce, R. J., J. E. Janowiak, P. A. Arkin, and P. Xie (2004). CMORPH: a method that produces global precipitation estimates from passive microwave and infrared data at high spatial and temporal resolution. *Journal of Hydrometeorology*, vol. 5, 487–503. https://doi.org/10.1175/1525-7541(2004)005<0487:CAMTPG>2.0.CO;2

Levizzani, V., C. Kidd, D. B. Kirschbaum, et al. (eds.) (2020). *Satellite Precipitation Measurement*. Springer. https://doi.org/10.1007/978-3-030-24568-9

Sorooshian, S., A. AghaKouchak, P. Arkin, et al. (2011). Advancing the remote sensing of precipitation. *Bulletin of the American Meteorological Society*, vol. 92, 1271–1272. http://dx.doi.org/10.1175/bams-d-11-00116.1 Retrieved from https://escholarship.org/uc/item/0hz1d89z

6   Satellite Remote Sensing of Surface Water in Lakes and Reservoirs

6.1   Chapter Overview

In Chapter 5, we learned how precipitation can be estimated by remote sensing using ground-based platforms or space platforms such as satellites. In this chapter, we will focus on surface water – notably the water that is in lakes and reservoirs, rather than rivers and groundwater. This is the water that remains directly on land and represents a significant reservoir for the water cycle. The storage of such water drives many water management applications, as we shall see later, such as reservoir and flood management (Chapter 8), or irrigation (Chapter 9). Here, we will overview the various remote sensing techniques that can be used to detect whether land is covered with water and if so, to what extent. Later in the chapter, we will learn how two successive satellite overpasses can help us estimate the storage change a water body may have experienced. This storage change can be a crucial component for various water management applications as it helps us understand how much water lakes or reservoirs are storing, losing to diversion or evaporation, or releasing.

6.2   Introduction

Surface water can be generated from many sources such as precipitation, runoff, streamflow, and snowmelt. Surface water is a very important variable for most water management applications because it provides the manager with a direct answer to the most fundamental question for making decisions – *how much water do we have that is readily available or affecting us directly?* In most cases, we can answer this question qualitatively by just finding out the spatial extent of this surface water. Knowledge of the area of surface water can help decisions related to inundation mapping, flood risk assessment, reservoir management, and crop and irrigation operations. Fortunately, satellites have been used quite effectively for many decades to track the surface water extent. A great example of this is the Landsat mission that was first launched in 1972 and continues to provide near-global imaging of surface water over land during cloud-free and clear sky conditions.

Water management decisions may also require information on the quantity of water in terms of volume or change of volume over time, which is known as discharge. Today we can actually track volume changes in water bodies by piecing together satellite observations from different platforms on areal extent and elevation (from altimeters). By piecing together various parameters of a river channel, researchers have made significant progress in developing remote-sensing-based methods to estimate discharge.

Satellite-based discharge estimation methods continue to mature that will gradually improve their user-readiness for water management applications. One satellite mission that is specifically optimized to provide simultaneous, all-weather, and all-day capability of tracking surface water volume changes and discharge is the Surface Water and Ocean Topography (SWOT) mission. Launched on December 16, 2022, SWOT is a new type of satellite mission optimized for surface water observation of changes in volumetric terms, to help water managers understand how much surface water one really has at a given location and time.

In this book, we will talk about SWOT only as a mission, given that it is still relatively new at the time of writing. But first, let us start from the very beginning, when the extent of surface water began to be tracked in the early 1970s. Figure 6.1 shows that almost all the wavelengths we have covered so far in the E-M spectrum can be used for tracking surface water extent. We already know from Chapter 3 that near infrared (NIR), when used wisely with other visible bands, can help us distinguish surface water from the rest of the background. Similarly, thermal infrared (TIR) can be useful too, in detecting the shoreline of water bodies by tracking the difference in temperature at the land–water interface (the water surface temperature will be cooler or warmer than land). This is also a passive technique, and like all passive techniques in the visible or IR range, there must be clear sky conditions. Microwave (MW) passive sensing

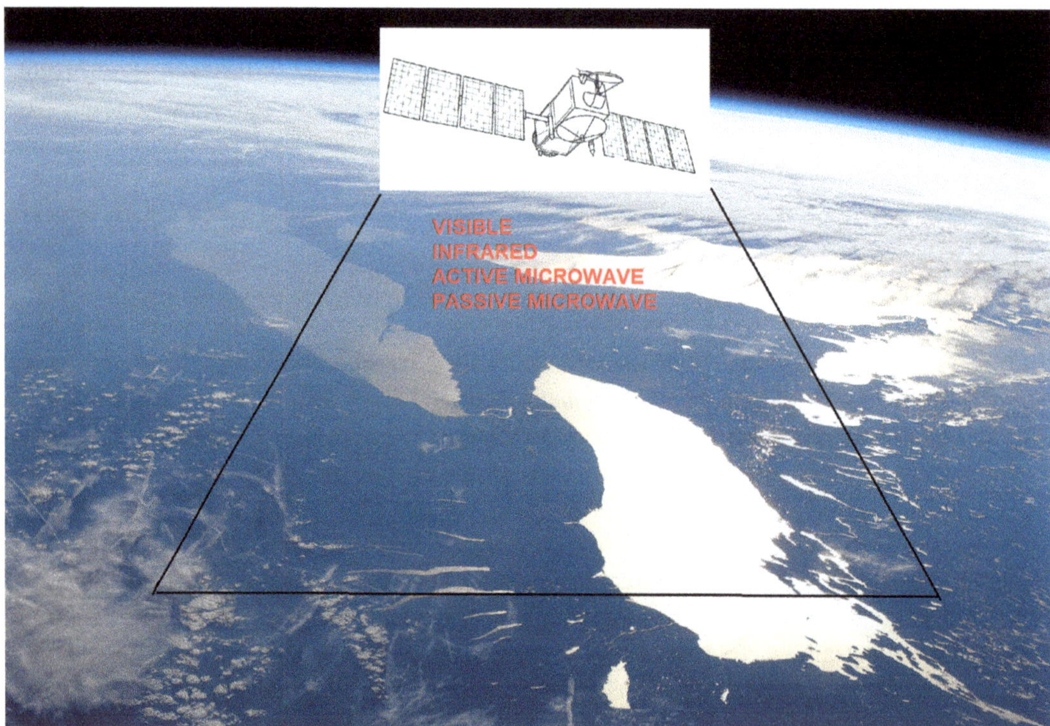

Figure 6.1 The wavelengths of the E-M spectrum that can be used to detect surface water. [Source: background image of land and satellite schematic taken from NASA]

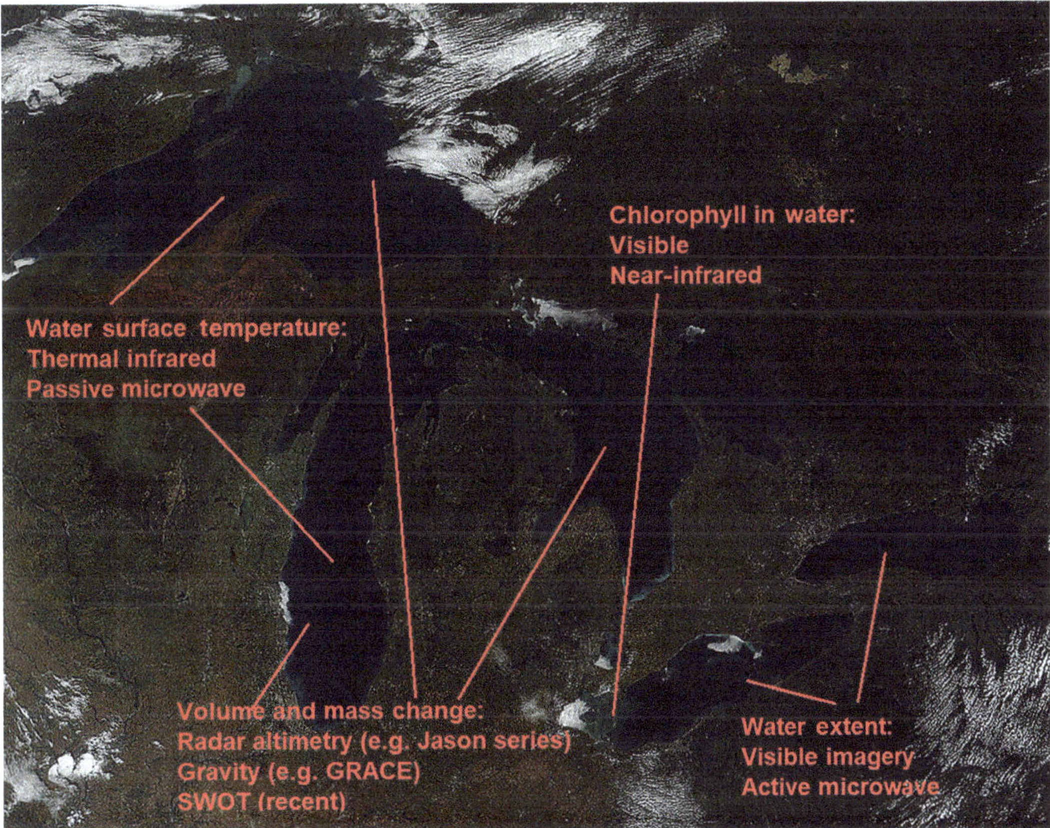

Figure 6.2 Wavelengths suited for specific aspects of surface water (quantity, quality). Some of the satellite mission names are also listed. [Source: base map of Great Lakes from NASA]

can estimate water surface temperature during cloudy conditions, but unlike TIR, we lose spatial resolution owing to the wavelengths being longer than in the visible or IR bands (roughly kilometer scale for MW, compared with tens of meters for IR and visible from orbiting satellites). Recall from Planck's equation in Chapter 3 that MW emission is a weaker energy source than TIR at planetary temperatures and therefore requires a larger field of view (and hence coarser spatial resolution). A closer breakdown of different remote sensing techniques for surface water detection is shown in Figure 6.2, where we also see the non-imaging techniques.

6.2.1 Estimating Surface Water Extent in the Visible and NIR Range

Using the golden rules of remote sensing, we can use a combination of visible and NIR wavelength during daytime to accurately detect from space the occurrence of water on the surface. We did conceptually overview this in Chapter 3 to decide the specific wavelength to choose. Numerous metrics have been proposed for this as well as many methods. In this chapter, we will overview just a few, to help readers develop an idea of how a surface water data product is typically created using reflectance in these wavelengths. But first let us review the basic

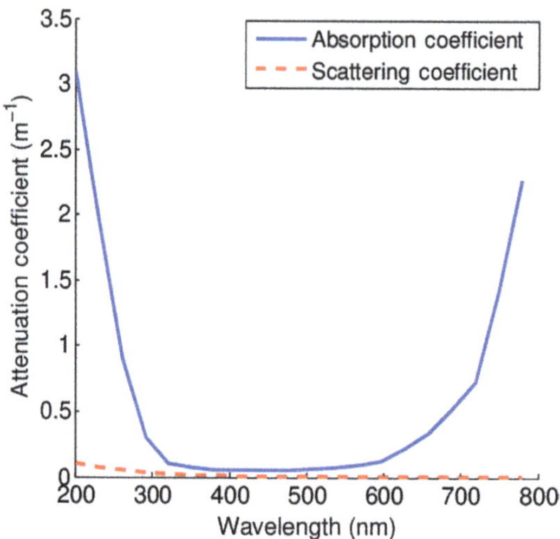

Figure 6.3 The absorption and scattering coefficients for pure sea water. The relationship across the visible and NIR wavelengths are applicable for pure freshwater as well in concept. In the NIR range, the properties diverge for absorption and scattering. [Source: the data used in this graph was obtained from Smith and Baker (1981), later reproduced in Mobley (1994)]

principle from where we left off in Chapter 3 for remote sensing of pure water using visible and NIR E-M reflection from the Sun.

Figure 6.3 shows that the least amount of absorption and scattering of incident light (from the Sun) takes place in the blue wavelength region (400–500 nm or 0.4–0.5 μm; here nm stands for nanometer, 10^{-9} meter). The minimum amount of absorption or maximum amount of reflection is located in the range of 460–480 nm. Nearly all incident E-M energy from the Sun in the near and middle infrared (740 nm–3,000 nm) entering a pure body of water is absorbed (Figure 6.3). Thus this scattering or reflection of violet and blue, and absorption of wavelengths 520–700 nm, results in pure water appearing blue.

One of the most common ways to estimate surface water extent is to use water indices, which are calculated from two or more wavelength bands, to identify the differences between water and non-water areas. Many indices have been developed to extract surface water areas or flood inundation extent. For example, the Normalized Difference Vegetation Index (NDVI), which is actually a vegetation index, has been used to detect water and flood in some studies. More useful indices for this purpose are those that can better highlight water bodies, such as the Normalized Difference Water Index (NDWI; McFeeters, 1996), and modified NDWI (MNDWI; Xu, 2006). McFeeters' NDWI can be regarded as the first generation of water index to track where the surface water is by using visible and NIR reflection in daytime. Later, Xu (2006) found that the short-wave infrared (SWIR) band is able to reflect some subtle characteristics of water, and so replaced the NIR band in NDWI with the SWIR band and proposed the MNDWI. It is now widely accepted that MNDWI is more stable and reliable than NDWI, because the SWIR band is less sensitive to concentrations of sediments and other optical active constituents in the water than the NIR band is (see Figure 6.4). Reader should keep in mind that a key limitation of the index-based approaches, such as MNDWI, is that they cannot

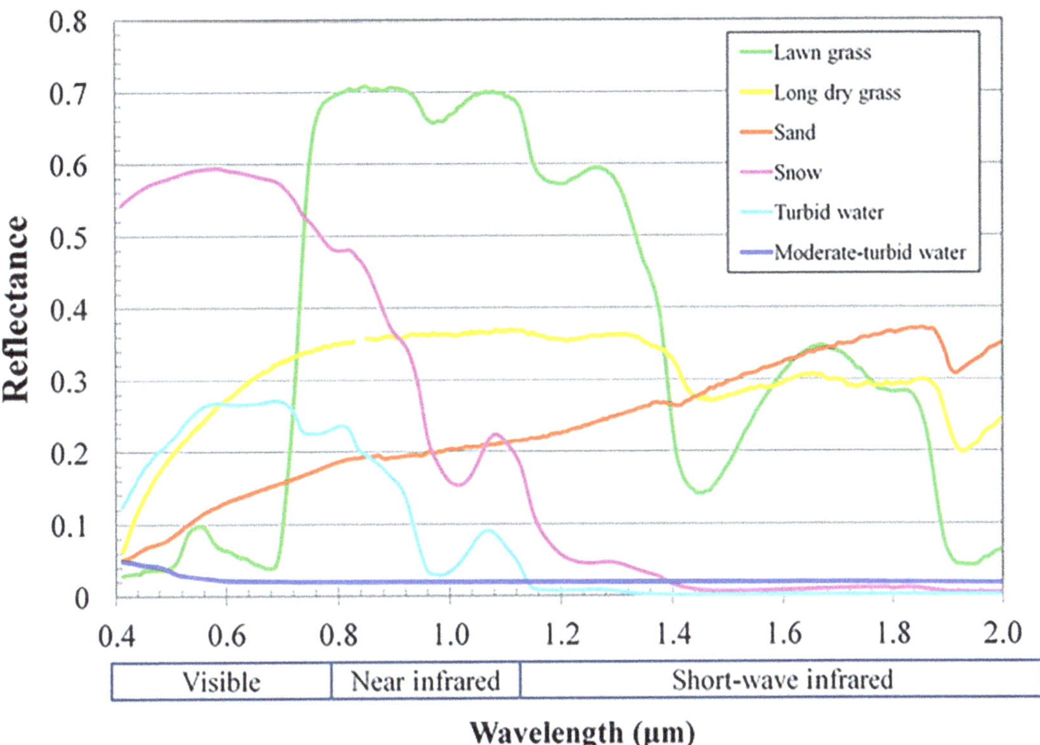

Figure 6.4 Reflectance of different types of land cover and water. [Credit: US Geological Survey digital spectral library]

discriminate between water and snow, because, although snow has a generally higher reflectance than the water in all the visible and infrared channels, the normalized difference between the green band and SWIR band for snow is as high as that of water. The formulas for calculating NDWI and MNDWI are as follows:

$$\text{NDWI} = \frac{(\text{Green} - \text{NIR})}{(\text{Green} + \text{NIR})} \tag{6.1}$$

$$\text{MNDWI} = \frac{(\text{Green} - \text{SWIR})}{(\text{Green} + \text{SWIR})} \tag{6.2}$$

In the above, the righthand-side terms are reflectances at a given wavelength that the passive sensor can measure during daytime. G, green wavelength; NIR, near IR; SWIR, short IR.

Thresholding is one of the most critical issues in using water indices such as NDWI and MNDWI to identify the extent of a surface water body. Based on the reflectance characteristics of water, NDWI and MNDWI values for water are usually greater than 0. Therefore, a threshold of 0 is often applied to extract water from index images (Xu, 2006). After counting the pixels from satellite data that exceed the threshold, the total surface area that is water can be calculated. Many have suggested that adjustment of the threshold value could usually achieve better extraction results. Fortunately, many studies have been devoted to working on this issue

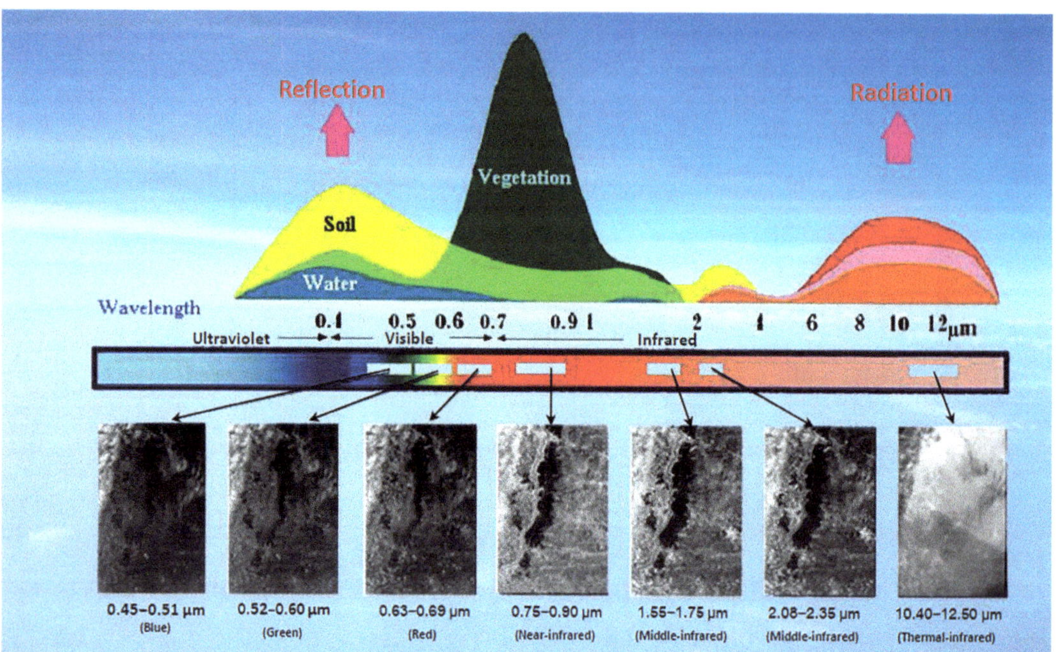

Figure 6.5 Summary of how a water body would "look" for each wavelength in the visible to NIR range when trying to identify its spatial extent.

and today, the identification of surface water extent using visible and NIR wavelengths from passive sensors during clear sky conditions is used widely in water management. Figure 6.5 recaps for a lake how the various wavelength bands in the visible and NIR/TIR ranges would typically "look".

6.2.2 Tracking Surface Water Extent in the MW Wavelength Using Angle-Looking Synthetic Aperture Radar

All the above techniques based on NDWI or MNDWI for tracking extent of surface water are passive and will work only during daytime and when skies are clear. Fortunately, there is a technique called angle-looking synthetic aperture radar (SAR) that is based on active MW and can come to the rescue to complement visible and NIR wavelengths from passive sensors. Until recently (before 2015), SAR was not as publicly accessible with reasonable short latency for use in water management. There were a few SAR platforms optimized for water extent or flood inundation mapping at long latencies or on special request. The Sentinel-1 SAR mission by the European Space Agency (ESA) changed all that and accelerated operational management of flooding and surface water extent identification by providing low latency and free data. Readers can learn more about the Sentinel-1 mission here: www.esa.int/Applications/Observing_the_Earth/Copernicus/Sentinel-1

Figure 6.6 summarizes the concept of how an angle-looking SAR works for water detection. Basically, if the target (water surface) has a smooth surface, then any MW energy incident on it

Figure 6.6 The basic concept of how an angle-looking SAR works to detect water surfaces. Specular reflection is reflection away at the same angle to the normal direction as the incident rays and thus no return back to the SAR's receiver. Such specular reflection can be expected from smooth surfaces such as water surfaces with little or no roughness. [Source: base map and satellite image taken from NASA]

at an angle will be reflected away, causing the receiver to see mostly dark (no or low return of energy or backscatter). On the other hand, land or a wavy water surface, which would typically be a rough surface, would return more backscatter reflection to the receiver. This reflection away for a smooth surface is called specular reflection, which is the core principle of angle-looking SAR for detecting water on surfaces. Such a SAR technique offers high-resolution mapping with all-weather and all-day capability, as it is an active technique in the MW E-M wavelength. The challenge of this technique, however, is that many water surfaces may be rough, owing to waves or protruding vegetation, causing higher backscatter than a smooth surface.

Applying Visible, NIR and MW Techniques for Monitoring of Lake Area

In this section, we will take a satellite sensor-centric approach to explain how current satellites in the visible, NIR, and MW ranges can be used to estimate lake surface area. First, Table 6.1 summarizes some of the many thresholding (NDWI, MNDWI) and SAR-based techniques for tracking surface water. Next, we will overview each satellite and explain how its data can be used to detect surface water.

Table 6.1 Summary of satellite sensors and techniques that can be used for lake area estimation

Satellite sensor	Revisit time	Technique	Band	Threshold
Landsat 8	16 days	Dynamic Surface Water Extent (DSWE)	Multiple (visible and NIR)	N/A
		Modified Normalized Difference Water Index (MNDWI)	Green band, short-wave infrared	0.3
Sentinel-1	10 days	Backscattering thresholding	MW	<-13 dB
Sentinel-2	5 days	DSWE	Multiple (visible and NIR)	N/A

Landsat 8 (L8)

Atmospherically corrected L8 data using the Land Surface Reflectance Code (LaSRC; Vermote et al., 2018) can be used to classify water. Table 6.1 shows classification techniques based on L8 data. The first is the Dynamic Surface Water Extent (DSWE; Jones, 2019). DSWE has the ability to extract the water surface where the pixel is partially covered with vegetation and water. In addition to Landsat imagery, DSWE uses a digital elevation model, slope, and hill and cloud shade. These parameters are calculated using the Fmask function. The output of the DSWE consists of several possible classes: not water; water – high confidence; water – moderate confidence; potential – wetland/partial surface water conservative; and masked out owing to the cloud, cloud shadow, or snow. The second technique used to extract the water surface area is the MNDWI. As mentioned earlier, Xu (2006) developed the definition using the NIR band with SWIR band to detect the water features better or more adaptively.

Sentinel-2 (S2)

Optical imagery from the Sentinel-2 (S2) sensor has a spatial resolution of 10 m, which is an improvement over the Landsat 8 spatial resolution of 30 m and can also provide water mapping. The DSWE technique can also be applied to S2 images. As the DSWE algorithm was designed specifically for L8 images, scaling of S2 reflectance data is required to make DSWE work for S2 data. Surface reflectance transformation functions between S2 and L8 can be used to transform the S2 bands to L8 bands.

Sentinel-1 (S1)

Sentinel-1 (S1) is an angle-looking C-band SAR that sends radar signals which can penetrate clouds. Water classification using the S1 imagery can be achieved with the help of the backscattering thresholding technique. The water-like surfaces appear dark in the imagery because of their smooth surface. However, one of the drawbacks of the SAR is speckle noise, which degrades the quality of the image and causes information loss. Over the years, various techniques have been used to reduce the speckle noise. With pre-processed image, a typical backscatter threshold of -13 dB or lower is quite effective in identifying surface water area (Ahmad et al., 2020; Khan et al., 2023).

Case Study 6.1: Satellites over the Amazon Capture the Choking of the "House of God" by the Belo Monte Dam

It's time for a case study to appreciate the water management value of using satellites to detect surface water extent. The Xingu River is revered as the "house of God" by the Indigenous people living along its Volte Grande, or Big Bend, in the Brazilian Amazon (Figure 6.7). The river is essential to their culture and religion, and a crucial source of fish, transportation and water for trees and plants. A few years ago (early 2012), the Big Bend was a broad river valley interwoven with river channels teaming with fish, turtles, and other wildlife (Figure 6.7). Today, as much as 80 per cent of the water flow is gone. That is because in late 2015, the massive Belo Monte Dam project began redirecting water from the Xingu River upstream from the Big Bend, channeling it through a canal to a giant new reservoir. The reservoir now powers one of the largest hydropower dams in the world, designed with enough capacity to power around 20 million households, although it has been producing far less.

Most of the river's flow now bypasses the Big Bend, and the Indigenous peoples who live there are watching their livelihoods and way of life become endangered. Some of the most devastating effects are during the rainy season, when wildlife and trees rely heavily on having high water. The consortium of utilities and mining companies that runs the dam has pushed back on government orders to allow more water to reach the Big Bend, claiming it would cut their generation and profits. The group has argued in the past that there was no scientific proof that the change in water flow harmed fish or turtles.

There is now proof of the Belo Monte Dam project's impact on the Big Bend, though – from above. Satellite data shows how dramatically the dam has altered the hydrology of the river there (Figure 6.7). We can use satellite data from Landsat mission and one of the indices mentioned earlier to track the change in water surface area (using NDWI) and vegetation cover (using NDVI) before, during, and after the completion of the Belo Monte Dam (Figures 6.8 and 6.9)

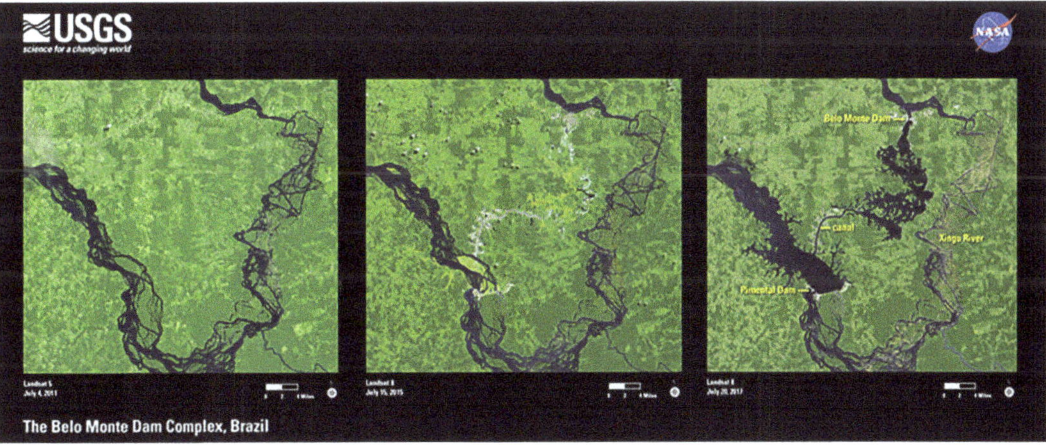

Figure 6.7 Satellite imagery from Landsat showing how the Belo Monte Dam has created artificial lakes (middle and rightmost panels) and choked the natural river that existed before the dam (shown on the leftmost panel). [Credit: US Geological Survey and NASA]

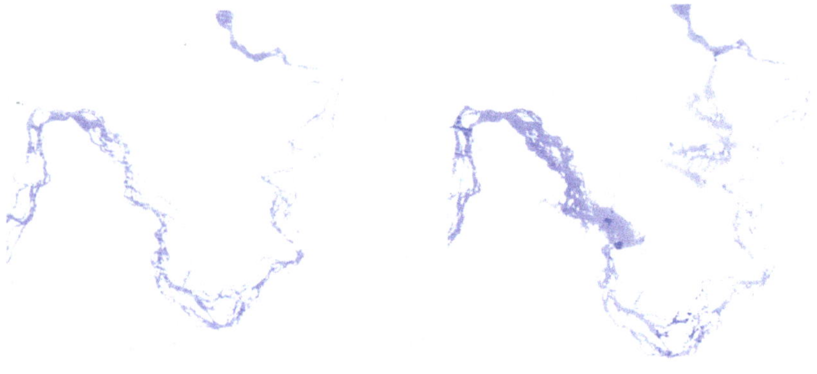

Average(2-yr)	Before 2013	After 2016
Water	0.363%	2.1%

Figure 6.8 The extent of surface water over a fixed domain before (2013) and after filling of Belo Monte reservoirs (2016), based on NDWI.

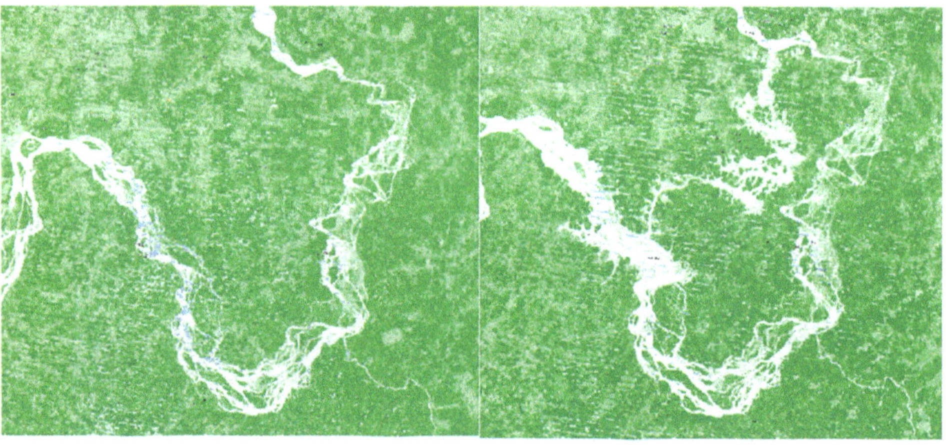

Average(2-yr)	Before 2013	After 2016
Vegetation	38.53%	31.28%

Figure 6.9 The extent of vegetation cover over a fixed domain before (2013) and after filling of Belo Monte reservoirs (2016), based on NDVI.

6.3   Microwave Radar Altimeter for Tracking Water Surface Elevations (Height)

As mentioned earlier, water management is concerned not only with water extent but also with the volumetric amount or change, or the volume change over time. The volumetric dimension of water is what truly helps to answer the question of how much water is available and then allows judicious decision-making to manage water. Now that we have learned how the area of a water surface can be estimated, we will explore how satellites can be used to estimate elevation of the water surface using radar altimeters that are active sensors and can work as all-weather and all-day platforms.

For nearly three decades, satellite nadir altimeters have provided essential information to understand primarily ocean and also inland water dynamics (see Figure 6.1). A variety of parameters can be inferred via altimeter measurements, including sea surface height, sea surface wind speeds, significant wave heights, and topography of land, sea ice, and ice sheets. The satellite altimeter is now an established space-based sensor providing sea surface height (SSH), near-surface winds, and significant wave height globally.

Originally, altimeters had the main objective of monitoring the oceans that account for three-quarters of the Earth's surface, where it is impossible to track globally with in-situ methods of using buoys. However, it was not long before the community realized that the same altimeter could be used to track the surface height of water bodies inland (Figure 6.10). Today, we have notable altimeter missions such as Topex/Poseidon (launched 1992) followed by the Jason series (1, 2, 3; first launched in 1992) and then Sentinel-3 (launched 2016) and Sentinel-6 Mike Freilich (launched 2020). We now have a constellation of typically three to four altimeters orbiting Earth at any given time. These altimeters, however, have a narrow swath, and hence we do not have global coverage. Fortunately, SWOT, as a "wide-swath altimetry" mission, currently provides this global coverage to expand water management applications. Such satellite-based information on water levels and storage changes is important for many surface water applications where rivers are extensively managed by reservoirs, and more so in transboundary and ungauged regions of the world.

6.3.1   How Does Altimetry Work?

Altimetry satellites basically determine the distance from the satellite to a target surface by measuring the satellite-to-surface round-trip time of a radar pulse (Figure 6.11). Technically, if the elevation of the datum (an assumed baseline or reference surface) is known, then the altimeter can provide the elevation of the water surface relative to that datum. This is different from height and should be noted. With some manipulations, one can arrive at water surface height, which is often the more practical parameter used in water management.

A lot of other information can be extracted from altimetry. Besides surface height, by looking at the return signal's amplitude and its time history (known as "waveform") we can also measure wave height and wind speed over the oceans, and, more generally, backscatter coefficient and surface roughness for most surfaces off which the signal is reflected.

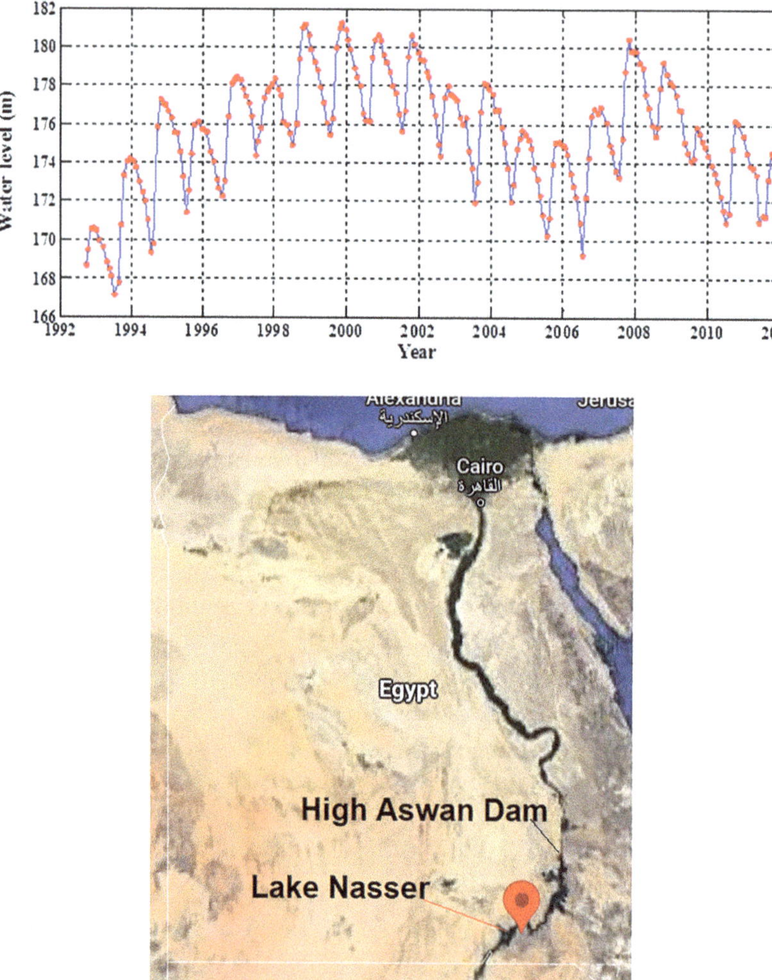

Figure 6.10 Example of time series of reservoir levels estimated by satellite radar altimeter for Lake Nasser, formed by the High Aswan Dam. The point where the altimeter passes over Lake Nasser is shown in red (long. 32.57° E, lat. 22.80° N). The seasonal (within-year) variability as well as the gradual hedging (storing) and loss of storage of the reservoir can be detected by the altimeter. [Credit: upper panel – Hydroweb by LEGOS – www.legos.omp.eu/en/hydroweb-2/; lower panel – Google Earth]

Any interference with the radar response signal also needs to be taken into account. Water vapor and electrons in the atmosphere, sea state, and a range of other parameters can affect the signal round-trip time, thus distorting range measurements. We can correct for these interference effects on the altimeter signal by measuring them with supporting instruments, or at several different frequencies, or by modelling them. Surface water elevation (or height as SSH) measurements require extremely precise knowledge of the satellite's orbital position: SSH = orbit altitude – range – corrections. Here SSH can be interpreted as the same thing as water surface height (elevation, to be precise, since it is relative to a datum) for a large water body over land.

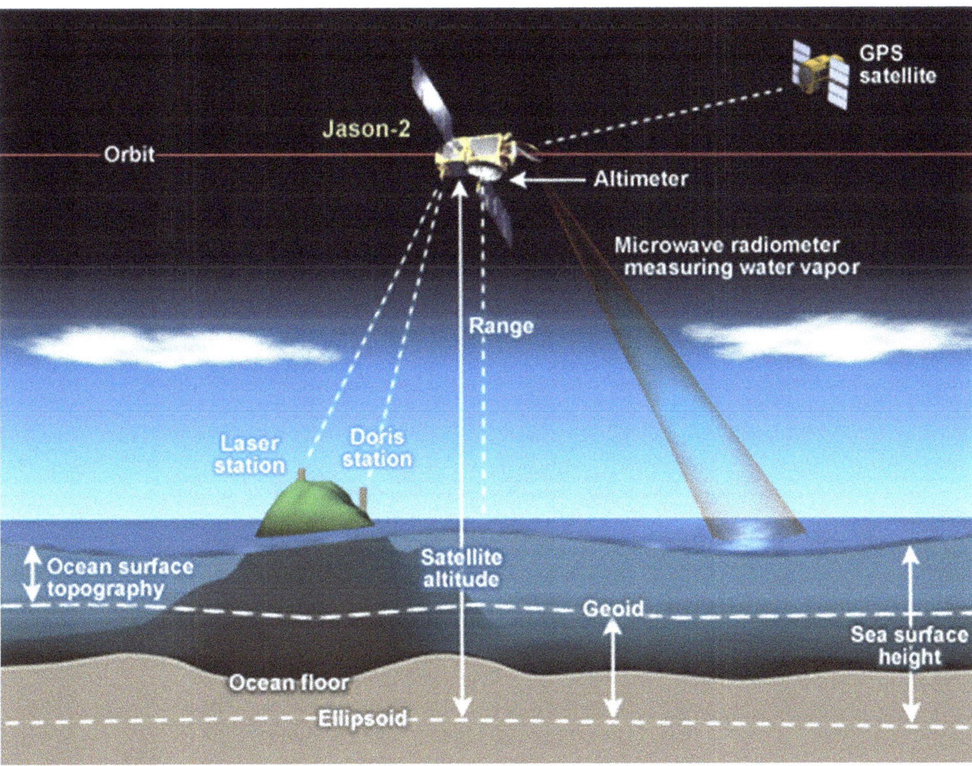

Figure 6.11 How radar altimetry works. [Image courtesy of NOAA/NESDIS/STAR Laboratory for Satellite Altimetry]

The Ku band (13.6 GHz) of the MW spectrum is the typical band used for altimeters. The Ku band is the most commonly used frequency (used for the Topex/Poseidon and Jason series). Other bands used are C, S, and Ka. A passive MW radiometer such as C-band radiometer is on board an altimeter for performing on the fly atmospheric corrections, as a MW radiometer can also detect atmospheric vapor in the absorption band.

Corrections needed to altimeter measurements are shown in Figure 6.12. For water management applications, it is often important to understand the datum of altimeter-based water surface elevation data. For global consistency, altimeters use a reference datum called an ellipsoid. But in reality, water management applications need water elevation data relevant to the local gravity datum, often known as mean sea surface (Figure 6.13). It is the elevation due to the gravity datum that "drives" water movement from a higher to a lower potential. This gravity datum is called a geoid and is the irregular surface Earth would have if it were completely made up of water. It is therefore important to convert altimeter elevation data relative to the ellipsoid to the local datum used by water managers. The accurate way to establish the geoid at a place is to use a gravimeter and figure out the geoid surface. Unfortunately, in many places, the geoid–ellipsoid relationship is not known. Thus, often, it is common to deal with anomalies in height (differences).

Satellite altimeters are mostly near-polar orbiting. Readers familiar with Google Earth can find Google Earth files for tracks for recent altimeters at: www.aviso.altimetry.fr/en/data/tools/pass-lo cator.html. An example of a typical single altimeter track from the Jason mission is shown in

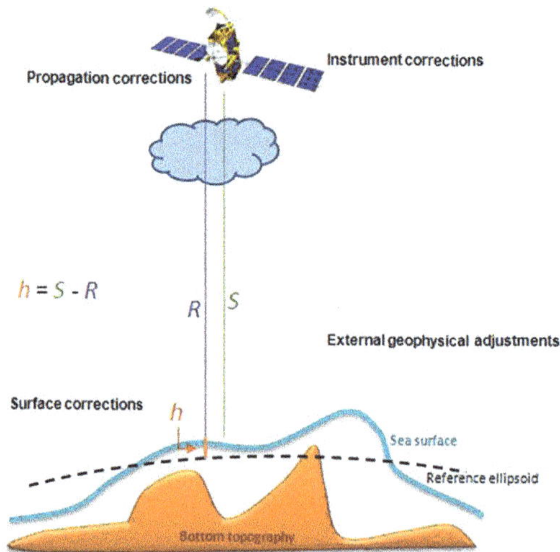

Figure 6.12 The various types of corrections (shown conceptually) needed for an altimeter to derive the elevation of the water surface relative to a datum. The datum here is the reference ellipsoid. h is the height; S is what the altimeter can measure at a given location where the range R will be known based on precise GPS location and knowledge of the ellipsoid datum. Even though the example is shown for the ocean, the idea is similar for large water bodies over land, including rivers where the changes of anomalies relative to previous measurement can be estimated fairly well. [Source image from NASA]

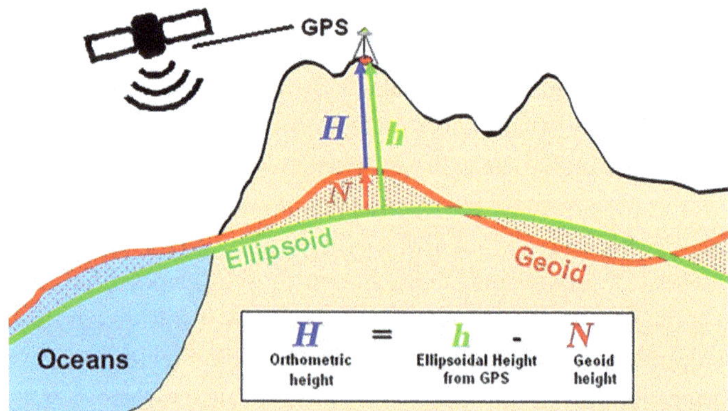

Figure 6.13 Datums for altimeter, and their relationship between each other and how the elevation is calculated relative to a datum using GPS. [Image based on Federal Geographic Data Committee at www.fgdc.gov]

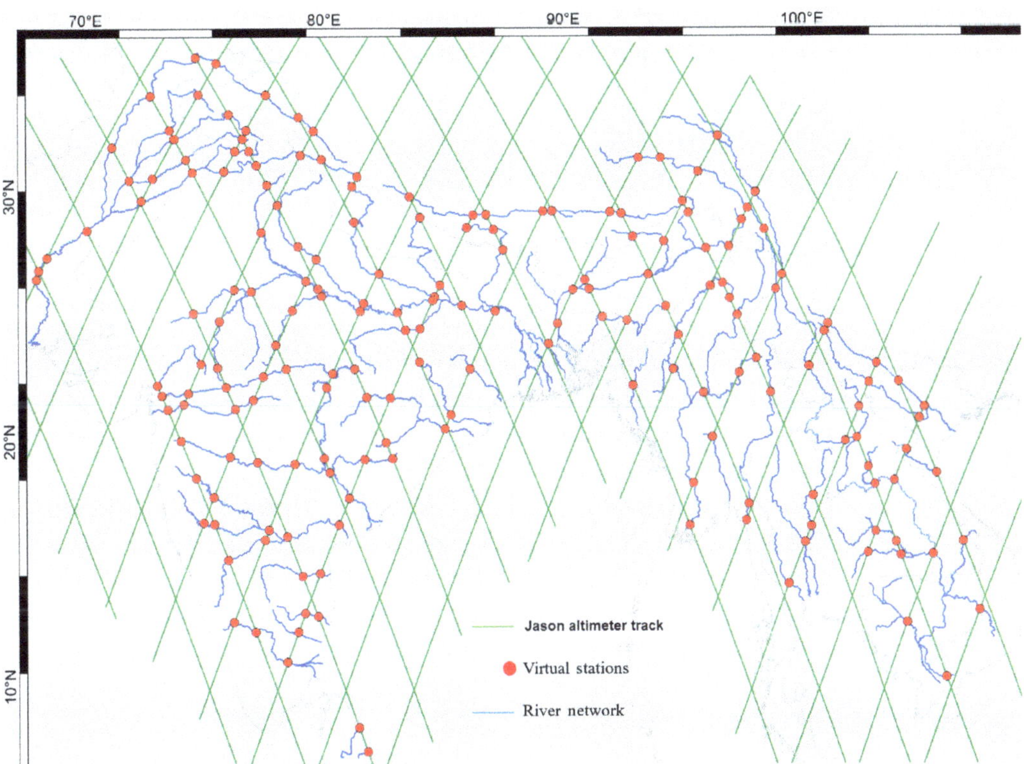

Figure 6.14 Jason-2 track coverage over the South and Southeast Asian region comprising Pakistan, India, Nepal, Bhutan, Bangladesh, Myanmar, and Thailand. The red dots indicate locations where an altimeter crosses a river – known as a virtual station. The tracks were first visualized in Google Earth using the track kml file which can be obtained from: www.aviso.altimetry.fr/en/data/tools/pass-locator .html [Source: Biswas et al., 2019]

Figure 6.14, where we can see that there are still many areas over land with rivers and lakes that are missed. At the same time, there are many locations where altimeters cross rivers, which can be considered as "virtual stations" (or space gauges). A lot of this gap can be "filled" with additional altimeters, but a large spatial sampling gap will remain, owing to the narrow swath of radar altimeters. Fortunately, this issue has been mitigated by the SWOT mission, as we shall see later.

Altimeters have a sampling frequency of typically 20 or 40 times per second. What this means is that for every second an altimeter is in flight, it pings the Earth's surface 20 or 40 times a second and gets as many return signals in the same time period. Typically, altimeters, like most orbiting satellites, fly at 6 km s^{-1}. Readers can find out why or how by referring to the Chapter 3 orbital mechanics problem on satellite altitude and circumnavigation time. This means that if an altimeter is sampling at 20 Hz (cycles per second), then each sample of height is typically spaced at 300 m. In other words, the use of altimeter data for rivers and lakes should be for those that are more than 300 m long along the direction of the altimeter track. For water management, it is important to bear in mind this limitation or requirement, along with the temporal sampling which for an individual altimeter can be 10 days (for Jason-3, Sentinel-6 MF) or up to 35 days (for an altimeter like Envisat) or even 90+ days for laser altimeters like ICESat-2.

Case Studies 6.2: **Two Practical Applications of Altimeter for Water Management**

In the following section, we review two major success stories for altimeters in the field of water management (Eldardiry et al., 2022).[1] To help readers understand the water management implications for practical decision making, we review each success story in the "Problem" and "Solution" format.

Monitoring Wetland Dynamics

Problem

Wetlands are crucial elements in the hydrological water cycle that can significantly affect terrestrial water storage. Ecosystem services provided by wetlands include fisheries support, water quality improvement, carbon sequestration, coastal protection from storm waves, and many more (Mitsch et al., 2015), and rank among the highest value in ecosystem service assessments (Costanza et al., 1997) for all landscapes. Knowledge of the extent, seasonality, and other temporal and spatial characteristics of wetlands is of vital importance to promote their wise use and to halt the alarming decline in global areal extent of this crucial ecosystem resource. This is particularly important for managing shared wetlands and river basins that cross international borders (Ramsar Convention Secretariat, 2010).

Solution and Proof of Concept

Because a wetland's physical extent changes seasonally, satellite altimeters, where available, adequately meet the requirements for wetland monitoring given their short latency (of less than a day) and a revisit period that can range from about a month to 10 days. Numerous studies have been performed to monitor seasonal changes in water storage by wetlands. These studies have proven the power of using satellite nadir altimeters in assessment of hydrological water balance and anthropogenic impact on wetland water storage.

Success Stories for Monitoring of Wetland Dynamics

Ahmad et al. (2020) recently quantified variations in volumetric storage for the seasonal wetlands of northeastern Bangladesh, locally known as "haors." Haors support waterfowl populations, as well as subsistence and commercial fisheries, rice-growing activities, and rich grazing lands for domestic livestock, among other benefits. The study team used three radar altimeters, Jason-3, Sentinel-3A, and Sentinel-3B, to monitor haors heights, and developed an operational monitoring system for the Bangladesh Water Development Board (see http://depts.washington.edu/saswe/haors). The altimetry data complemented the water height data gathered by a NASA-supported citizen science-based project, the Lake Observations by Citizen Scientists & Satellites (LOCSS) to understand water storage variations in ungauged water bodies (see www.locss.org).

---

[1] © American Meteorological Society. Used with permission.

Reservoir Monitoring in Ungauged Regions

Problem

Understanding reservoir operation has become increasingly important with growing population, increasing water demands, and climate variability. However, monitoring reservoir operation in ungauged regions is hindered by limited, outdated, or nonexistent hydrologic data. The issue is exacerbated for international river basins, owing to the lack of data-sharing mechanisms between nations (Plengsaeng et al., 2014).

Solution and Proof of Concept

Various techniques have been developed and validated to provide accurate estimation of reservoir storage levels using altimeter measurements. Most reservoir operations are typically carried out at weekly or biweekly timescales. Therefore, most altimeters with overpasses are able to meet the latency and spatial resolution requirement for reservoirs with a surface area larger than 10 km^2. These altimeter-based water levels are currently made available globally through various operational satellite altimetry databases and archives, such as Hydroweb, and the Physical Oceanography Distributed Active Archive Center (PO.DAAC), maintained publicly by international space agencies. The use of such satellite-driven methods in modeling reservoir operations and river systems has been successful in operational environments for transboundary basins such as the Mekong basin, the Ganges–Brahmaputra–Meghna basin in Southeast Asia, the Indus River basin, and the Nile River basin. These studies concluded that, in most basins, satellite-based estimation of storage levels is in a high level of agreement with in-situ-data.

Success Stories for Reservoir Monitoring in Ungauged Regions

The Hydroweb database, developed at the LEGOS laboratory (Laboratoire d'Etudes en Géophysique et Océanographie Spatiales) in Toulouse, France, contains time series of water levels over large rivers, lakes, and wetlands at the global scale (https://hydroweb.next.theia-land.fr/). The water levels are based on six radar altimetry missions including the US Navy Geosat Follow On (GFO), the US and French TOPEX/Poseidon (T/P), Jason-1, Jason-2, and Jason-3, and Sentinel-3 missions, and the European Space Agency (ESA) ENVISAT (ENVIronmental SATellite). It should be mentioned that this service has now evolved to an operational time series with more than a decade of records that is constantly growing.

The Global Reservoirs and Lakes Monitor (G-REALM) database, developed by the US Department of Agriculture's Foreign Agricultural Service (USDA-FAS), in cooperation with NASA and the University of Maryland, routinely monitors lake and reservoir height variations every 10 days based on Jason-series nadir altimeter data. Surface elevation time series for large inland water bodies around the world are produced operationally using a semi-automated process. The products are provided to the USDA and available for public viewing at https://ipad.fas.usda.gov/cropexplorer/global_reservoir/.

6.4   Volumetric Estimation for Water Bodies Using Satellite Data

Now that we can estimate the surface area as well as the elevation of the water surface, we can estimate the volume change in a lake or a reservoir based on two successive overpasses of satellites. And if we are lucky, we may even be able to track total storage or volume using data from successive satellite overpasses, if the first observation occurred when the reservoir was completely dry or empty. In this section, we shall see how it is possible to estimate volume changes using satellite data, with some assumptions, of course.

The first scenario is if we only have satellite data that tracks water extent but not elevation (altimeter), or elevation but not water extent (e.g. Landsat, Sentinel-1, Sentinel-2). Can we still track the volume change of a water body? All this time, we have been saying that we need to have both water extent and elevation to do so. And yes, SWOT is the first mission to provide that. Well, it turns out that we can still estimate volume changes if we know the bathymetry of the water body beforehand. In other words, if there was a digital elevation model (DEM) or topographic maps, all we need is just one (either extent mapping or elevation estimates). It turns out that there is a global DEM from satellite interferometric observations by the Shuttle Radar Topographic Mission (SRTM; https://science.jpl.nasa.gov/projects/srtm/). Using the shuttle, the whole Earth's topography was scanned in the early 2000s to produce a DEM for the world at 30-m resolution (each pixel is 30 × 30 m). Since SRTM, there have been many other satellite-based DEMs. In this book, we will concern ourselves only with SRTM DEM, as it is not only global and publicly available, it has already been used extensively in water management scenarios to establish confidence among practitioners for use.

Using SRTM DEM data, we can construct what we call an "area–elevation curve" (AEC), which is also known as the hypsometric relationship of a lake or reservoir. Figure 6.15 shows an idealized example of an AEC that could be derived from SRTM DEM data and then, using either satellite-based area or altimeter-based height, we can estimate the other. This way, and using only one type of observation, we can have both surface area and elevation at the same time and then estimate volume change (ΔS) between successive satellite overpasses (Equation 6.1 and Figure 6.16).

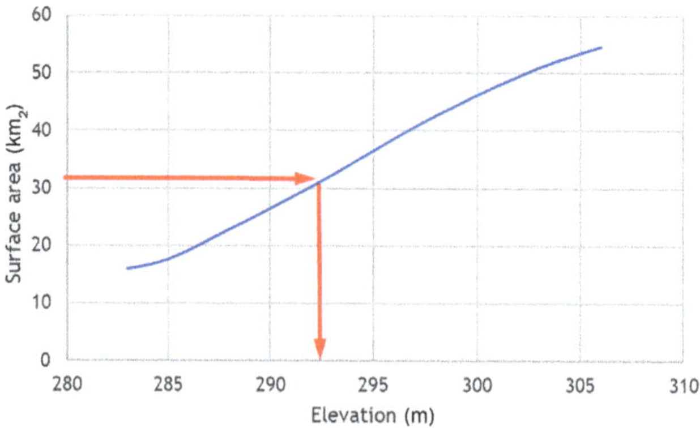

Figure 6.15 An idealized area–elevation curve (AEC) for a reservoir to capture the bathymetry.

$$\Delta S = A_{\mathrm{avg}} \times \Delta h = \frac{\left(A_2 + A_1\right)}{2} \times \left(h_2 - h_1\right) \tag{6.3}$$

Here ΔS in Equation 6.3 is the storage change between two successive satellite overpasses (or readings).

When using this a-priori DEM approach from SRTM data, we have to remember some potential limitations. If the water body existed before the DEM was created, then the SRTM data will have bathymetry mapped from above the water level that existed during February of 2000 when SRTM surveyed Earth's terrain. Of course, if the water body appeared after February 2000, such as a reservoir built in 2010, SRTM data can reveal the complete bathymetry and

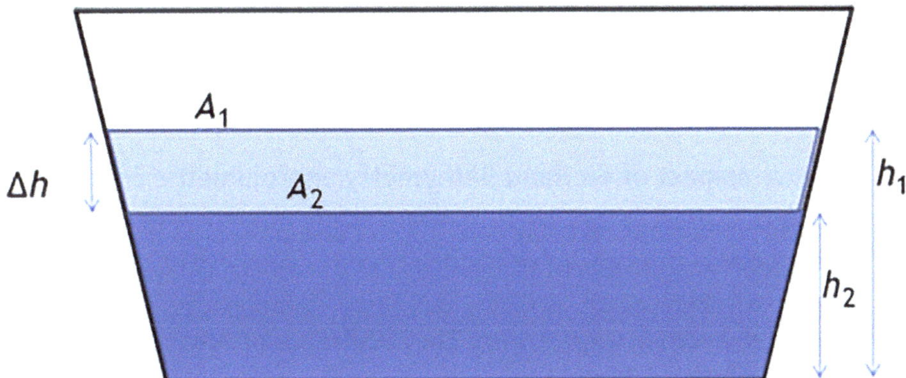

Figure 6.16 Example of volume change computation assuming a trapezoidal bathymetry between two successive satellite overpasses measuring height or area.

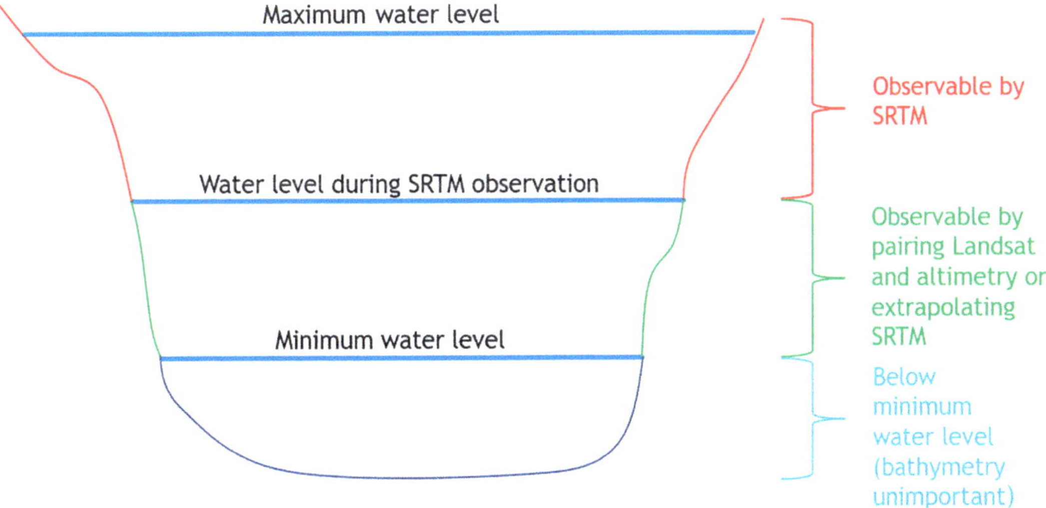

Figure 6.17 Schematic showing what SRTM typically measures as bathymetry for reservoirs that existed before February 2000.

AEC from the very bottom. There is also the issue of terrain changing due to development, erosion, land cover change, etc. The terrain that existed in 2000 may not exist exactly that way in 2010 or 2020 or during present times, as land is constantly deforming. Water managers need to be mindful of all these issues (Figure 6.17).

If we have satellite data on water extent/area and elevation, then we could use two pairs of such information to estimate volume change without the need for an AEC relationship. There are some caveats here, though. What cross-section should we assume – trapezoidal, inverse pyramid, or just plain rectangular? It all depends on the water manager's perception and understanding of the likely bathymetry of the water body. The assumed cross-section can make a big difference. The other issue is that the area and height estimates should be around the same time, which is very rare with the current constellation of satellites. Fortunately, SWOT mitigates a lot of these challenges by providing simultaneous water area and height information for a water body.

Case Study 6.3: Impact of Wetland Bathymetry on Volumetric Estimates

We derived volumetric estimates using two different assumptions of bathymetry for large wetlands called "haors" in northeastern Bangladesh: (A) trapezoidal (which is more suitable for flat terrain like northeast Bangladesh, as shown in the area–elevation relationship) (Figure 6.18), and (B) rectangular cross-section (Figure 6.19). The calculations of volumetric estimates for the two bathymetries are illustrated below.

(A) **Trapezoidal bathymetry**

Maximum volume stored,

$$V_{stored} = \left(h_{max} - h_{min}\right) \times \left(A_{max} + A_{min}\right)/2 \tag{6.4}$$

Volume change between two consecutive acquisitions, ΔV_t, is given as,

$$\Delta V_t = \left(h_t - h_{t-1}\right) \times \left(A_t + A_{t-1}\right)/2 \tag{6.5}$$

Maximum volume change, $V_{change} = \left(\Delta V_t\right)$

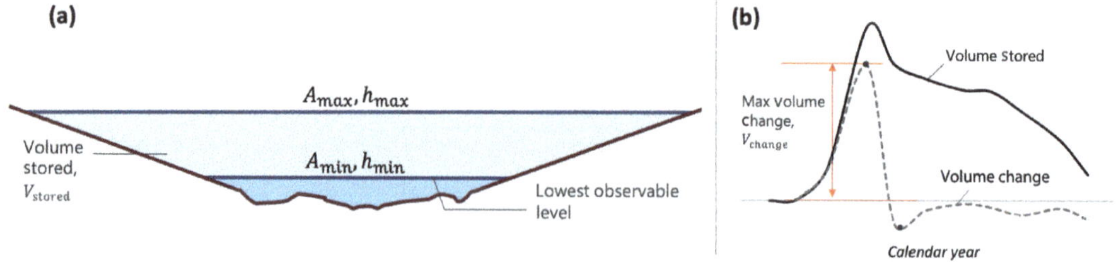

Figure 6.18 Illustration of (a) total volume stored relative to the lowest observable levels (light blue shaded area), and (b) maximum volume change over the year for a trapezoidal bathymetry.

(B) Rectangular bathymetry

Maximum volume stored,

$$V_{\text{stored}} = \left(h_{\max} - h_{\min}\right) \times A \qquad \left(A = A_{\max}\right) \tag{6.6}$$

Volume change between two consecutive acquisitions,

$$\Delta V_t = \left(h_t - h_{t-1}\right) \times A \tag{6.7}$$

Maximum volume change,

$$V_{\text{change}} = \left(\Delta V_t\right) \tag{6.8}$$

Thus, the volume storage and change estimate along with the associated uncertainty for the two bathymetries are shown in Table 6.2. These results show that uncertainty due to the assumption of haor (water body) bathymetry is critical. Estimated volume storage increased by up to 47% on assuming a rectangular bathymetry as compared with the trapezoidal cross-section.

Table 6.2 **Volumetric storage and change (along with the associated intervals at 95% confidence level) with the assumptions of trapezoidal and rectangular bathymetries for haors**

Satellite acquisition dates	Volume stored (km³)		Percent difference from rectangular	Volume change (km³)	
	Trapezoidal	Rectangular		Trapezoidal	Rectangular
May 22, 2019	6.5 ± 0.4	12.2 ± 0.8	46.7	5.1 ± 0.5	9.5 ± 1.1
June 1, 2019	7.8 ± 0.5	14.7 ± 0.9	46.9	6.2 ± 0.7	11.0 ± 1.4
June 25, 2019	19.3 ± 1.3	36.4 ± 2.3	47.0	15.2 ± 1.6	28.3 ± 3.3
July 16, 2019	30.9 ± 2.0	58.2 ± 3.6	46.9	24.3 ± 2.6	45.3 ± 5.4

Difference is shown as percent difference of volume stored in a trapezoidal bathymetry compared with rectangular bathymetry. Note, here the satellite data provides area, and elevation data is provided from in-situ gauges from a citizen science project (see Chapter 11 on citizen science for water storage tracking.

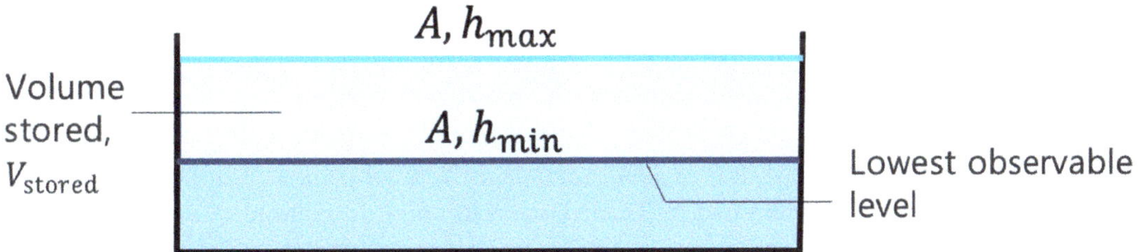

Figure 6.19 Same as Figure 6.18 but for volume storage calculation for a rectangular bathymetry.

Case Study 6.3 (cont.)

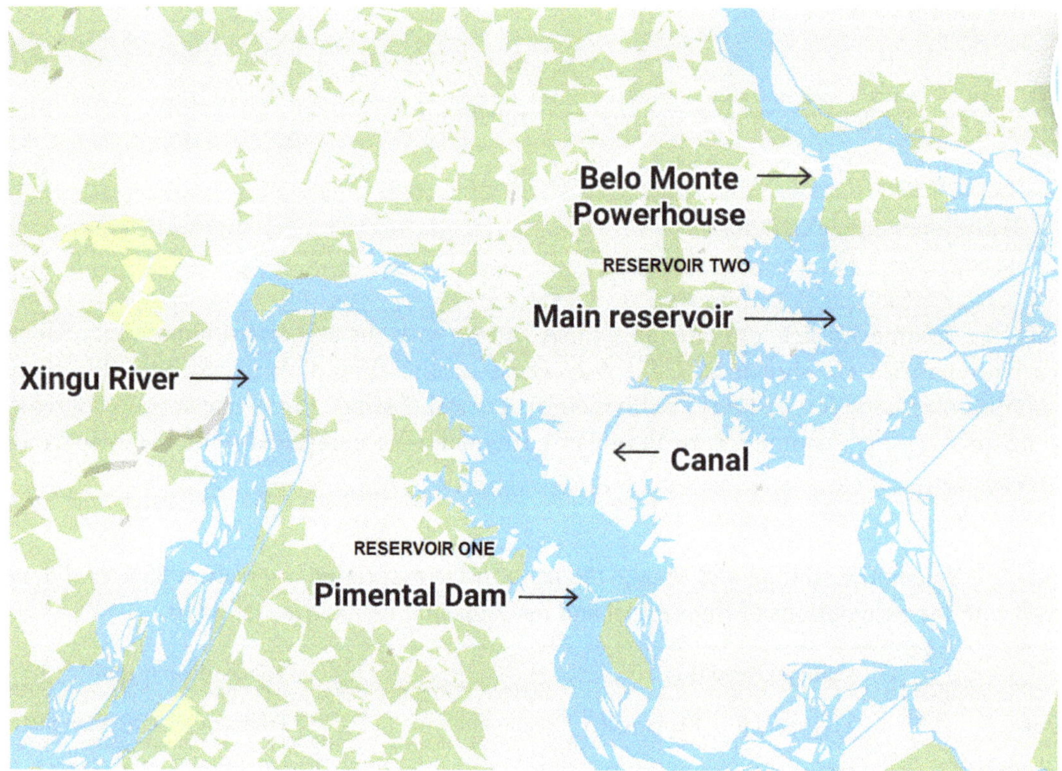

Figure 6.20 The Belo Monte Dam project showing the two reservoirs created by the Pimental and Belo Monte dams. [Credit: Openstreetmaps]

The Area–Elevation Curve (AEC) for Belo Monte Dam Project

Recall the earlier discussion about the Belo Monte Project, which was built and filled around 2017 and comprises two sets of reservoirs – Pimental Dam and Belo Monte Dam (Figure 6.20). SRTM DEM data can capture its total bathymetry (ignoring the canopy and any other terrain modification issues after 2000).

Figure 6.21 shows the AECs (right panel) for each of the reservoirs (1 – Pimental; 2 – Belo Monte). The left panel shows the histogram of SRTM elevation pixels. Since reservoir 1 (upper left panel) already had water, we see a peak in the histogram frequency at the elevation of the water with greatest number of pixels in that region of interest. For reservoir 2 (lower left panel), which did not exist when SRTM flew in 2000, there is no distinct peak at a single elevation. The lower right panel shows the combined AEC for both reservoirs. In Chapter 8 on reservoir management, we will learn how to leverage this AEC curve with satellite data on water extent, elevation, and SWOT to operationally track a reservoir state, infer its rule curve, explore its historical pattern and downstream impact, and even forecast release.

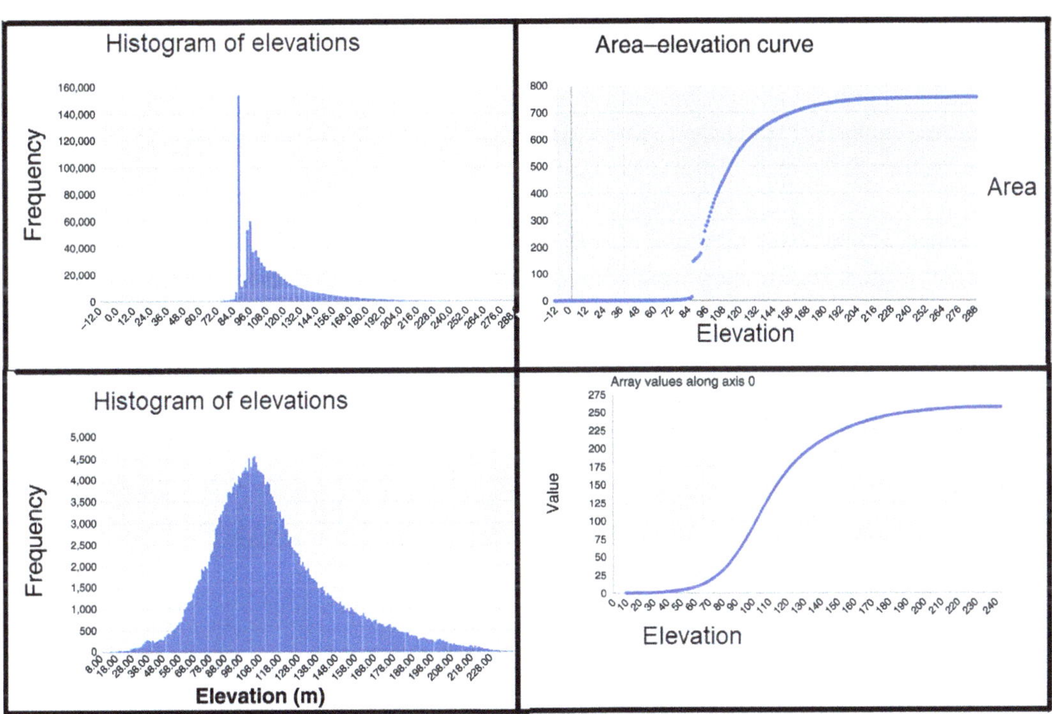

Figure 6.21 Upper panel: left – distribution or histogram of elevations for Reservoir 1, due to Pimental Dam, which existed when SRTM flew in February 2000. A distinct peak in frequency of elevations is seen, as this reservoir had water when SRTM flew. Lower panel: left – distribution or histogram of elevation for Reservoir 2, due to Belo Monte Dam, which was filled from the bottom in 2017. Hence there is a wide range of elevations around the peak frequency. Right panels show the area–elevation curve (AECs).

6.5 The Surface Water and Ocean Topography (SWOT) Mission

It is now appropriate to describe the SWOT mission, and the data we have available for water management. This section will be a precursor to future chapters on discharge estimation and reservoir management.

The SWOT mission (Biancamaria et al., 2016) is jointly developed by NASA and CNES with contributions from the Canadian and UK space agencies. It is designed to provide, for the first time, spatially extensive, high-resolution measurements of water elevation for the hydrology and oceanography communities. Even though the sampling frequency of SWOT is latitude-dependent (lowest at the equator, highest over the poles), a large fraction of the Earth's ungauged surface will benefit from multiple SWOT observations in a 21-day repeat cycle. SWOT satellite data on surface water is expected to be useful for applications such as disaster management, reservoir operations, flood management, ecosystem services planning, wetlands monitoring, hydropower, navigation, and fisheries (freshwater and marine). More details on the SWOT mission are provided at http://swot.jpl.nasa.gov or at www.aviso.altimetry.fr/swot.

For water management, SWOT can be expected to play a major role in building the world's space-based observational network for surface water, similar to the Global Precipitation Measurement (GPM) mission with its constellation of precipitation sensors. While the SWOT mission has all the ingredients to become an enduring mission for scientific discovery and societal applications owing to its unique capability to observe water surface topography, a key challenge can be the structure and format of its data products, which are probably unfamiliar to many water practitioners and decision-makers. Using SAR interferometry, SWOT provides data on water surface elevations, in the form of point clouds, that will then be packaged into other more widely used geospatial formats such as feature polygons and raster data (see https://podaac.jpl.nasa.gov/SWOT?tab=datasets). There remains a sizeable community of water scientists and practitioners who are more accustomed to point-based data from in-situ gauges, which have a longer heritage and record of availability compared with remotely sensed data. For them, such spatially distributed formats may present some initial challenges for uptake. This is particularly relevant for water management applications, while the challenge for ocean applications may be more on the use of the image information from SWOT. With this in mind,

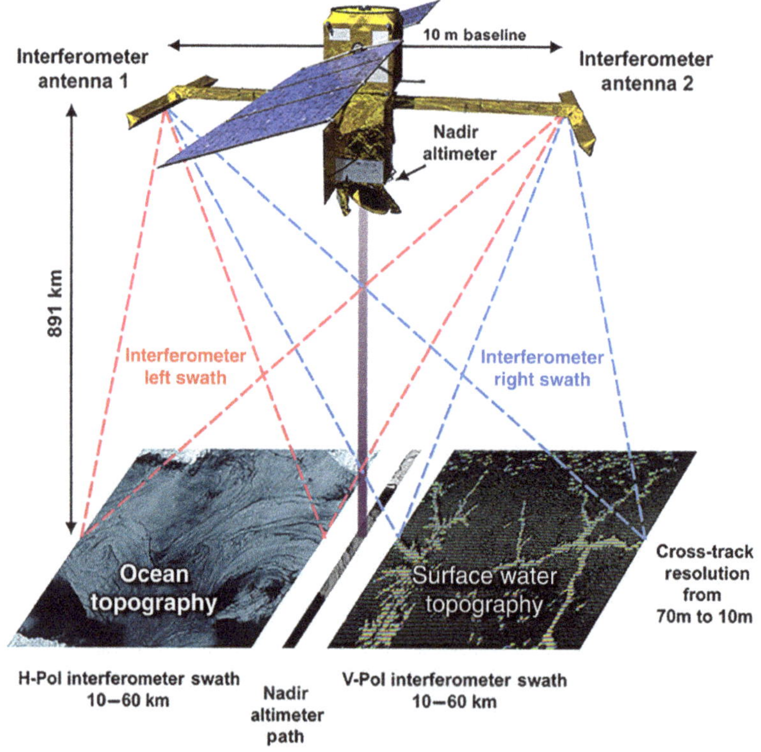

Figure 6.22 The Surface Water and Ocean Topography (SWOT) mission showing the key instruments of nadir altimeter and KaRIN instruments with altitude and swath. [Credit: NASA-JPL and CNES]

we review the data products of SWOT that are expected to be of highest value in water management. To help us understand the data format and the data types, it is useful to quickly review the physical architecture of the SWOT Mission, shown in Figure 6.22.

6.5.1 SWOT Data Products for Water Management

Four main products for SWOT have been defined. The pixel cloud and raster products are in NetCDF format, while the river and the lake are in shapefile format, compatible with GISs. River- and lake-specific products are linked to a-priori databases, which will be updated during the mission.

"River" Products

Each river over 30 m wide is defined in an a-priori database, as a line along the center of the river divided into reaches of 10 km long, and as nodes every 200 m along this line.

River products are provided according to these two types of breakdown, as defined in the a-priori database: for the "node" file with mean values around each node for parameters such as width, height, flooded area, etc., and for the "reach" file the mean height, mean slope, mean flow, etc. on each of these river reaches. These products are provided for each SWOT overpass (Figures 6.23 and 6.24).

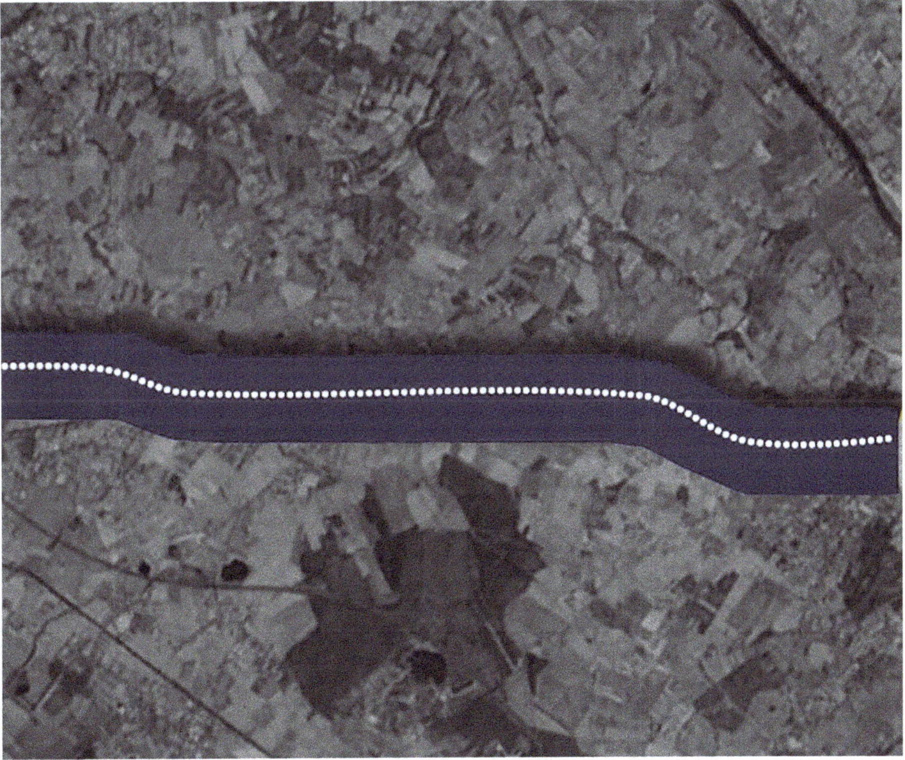

Figure 6.23 Graphical definition of river nodes. [Credit: JPL NASA, http://swot.jpl.nasa.gov]

Figure 6.24 Graphical definition of a river reach. [Credit: JPL NASA, http://swot.jpl.nasa.gov]

The "reach" file product is also provided by cycle, in which case the values of the variables are provided on each of the different passes that measured all or part of the section of the river in question during an entire satellite cycle (of 21 days). Here a cycle represents the complete sampling of Earth in 21 days involving multiple orbits. For example, cycle number 4 for SWOT represents the fourth such cycle since the launch of SWOT (from day 64 to day 84, if launch day is considered day 1).

"Lake" Products

Water surfaces that are not known rivers, and that are more than one hectare in size, are treated as "lakes" by SWOT data products. The product for each individual satellite pass gives information (height, surface area, contour, etc.) for each observation of these surfaces. For each object that can be linked to an element of the a-priori lakes database, the identification information is added.

The product for each cycle includes a median contour and the average height of the lakes for passes when they were seen in their entirety (observations of only part of the lake are not taken into account). Other more complex scenarios (views of different parts of the lake at each passage) will be taken into account.

"Raster" Products

A "raster" product is systematically generated from the "pixel cloud" product of SWOT. This is in NetCDF 2D format, and cover four point-cloud tiles (the two swaths and 120 km along

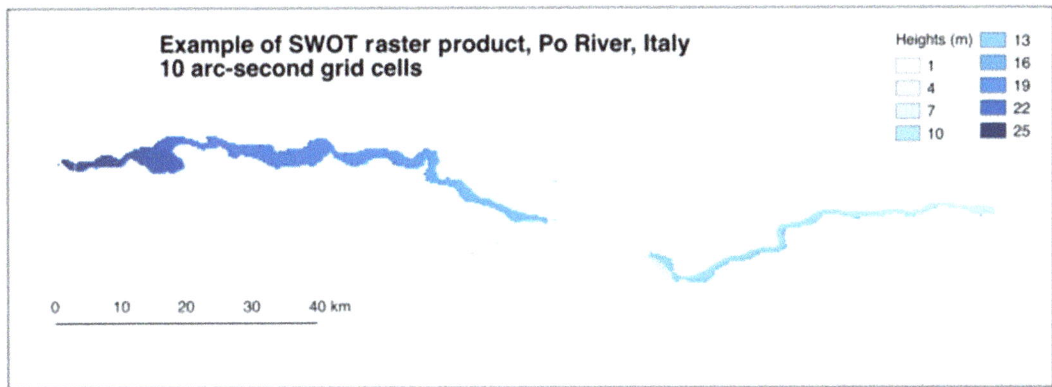

Figure 6.25 SWOT data simulated for a river: example of the Po (Italy). [Credit: JPL NASA, http://swot .jpl.nasa.gov]

the ground track) and will come in two resolutions: 100 m and 250 m. The raster product will be given on a regular grid according to a Universal Transverse Mercator (UTM) projection. This will be a one-per-pass product (Figure 6.25).

It can also be generated on demand (via the data dissemination portal):

- Format: NetCDF, GeoTIFF, etc.
- Subset of variables
- Resolution (>100 m) and user-defined area, with a limit on the volume of the data to be downloaded.

"Pixel Cloud" Product

This is the rawest product for land surfaces. A file, in NetCDF format, covers a tile, which corresponds to a swath on one of the two sides (left or right) over 60 km along the satellite's ground track.

The product provides longitude, latitude, height, pixel size, corrections, and uncertainties for each point classified as water, for points on a buffer zone around these water zones as well as on systematically included areas (defined according to an a-priori mask). This product is generated for each pass of the satellite. There is also a companion product called the pixel cloud vector product which includes height-constrained geolocation and assignment to river and lake database.

6.6   Conclusion

In this chapter, we explored how remote sensing from space can be used to detect water on land. We learned that both passive visible and infrared are quite suited for water detection given their ability to discriminate water as a target from land. We also learned that active microwave in the form of angle-looking synthetic aperture radar (SAR) can also detect water on land during cloudy conditions. Radar altimeters were also introduced in this chapter to estimate the elevation of a water surface. Using satellite-based digital elevation model, we learned how the bathymetry of reservoirs can be quantified in the form of area–elevation relationships. In this chapter, we also learned how consecutive satellite overpasses estimating area or elevations can be used to estimate storage change. This is going to be important in Chapter 8 when we cover the topic of reservoir management. We ended the chapter with an introduction to the newly launched Surface Water and Ocean Topography (SWOT) mission that is dedicated for tracking water movement on land. SWOT data products cover both lakes and rivers. For rivers, the SWOT satellite will measure river width and slope, which can be used to estimate discharge. In the next chapter, we will therefore cover the topic of discharge and how it can be estimated from satellites.

TUTORIAL ON SATELLITE SURFACE WATER DATA[2]

In this tutorial, we will learn how we can apply satellite data on surface water to understand the inundation impact of the recently constructed Belo Monte Dam in the Amazon. As in the previous chapter, we will learn to do this in two ways – (a) using cloud computing (Google Earth Engine) platform and (b) using local computing (in Python). The goal here is to experience the potential of satellite data in tracking surface water changes due to a large water infrastructure such as the Belo Monte Dam.

About Belo Monte Dam: The Belo Monte Dam, one of the world's largest hydroelectric projects, was completed in 2016 on Brazil's Xingu River. This dam has had a profound effect on Indigenous communities. Redirecting river water to its reservoir, it has disrupted their lives, culture, and environment. This impact underscores the delicate balance between development and preserving Indigenous heritage (see Figure 6.20).

Part 1: Google Earth Engine (GEE)

Normalized Difference Water Index (NDWI) and Modified Normalized Difference Water Index (MNDWI) Techniques

Please note that all the GEE scripts for this tutorial are available at: https://code.earthengine.google.com/?scriptPath=users%2Fskhan7%2FTextBook%3ATutorial%2FChapter6_Tutorial1_NDWI_MNDWI

> **Step 1: Delineating region of interest (ROI).** We already know what an ROI is (we learned it in the Chapter 4 tutorial to map the inundation during the 2018 floods in Kerala, India). For the Belo Monte Dam, we need to delineate the ROI. We can just click on any required shape; for simplicity, the tutorial has used the rectangle box (Figures 6.26–6.28).
>
> **Step 2: Defining the thresholds for the surface water classification techniques.** Early in this chapter, we briefly covered the normalized difference ways of classifying a satellite pixel as either water or non-water. The two difference techniques were NDWI

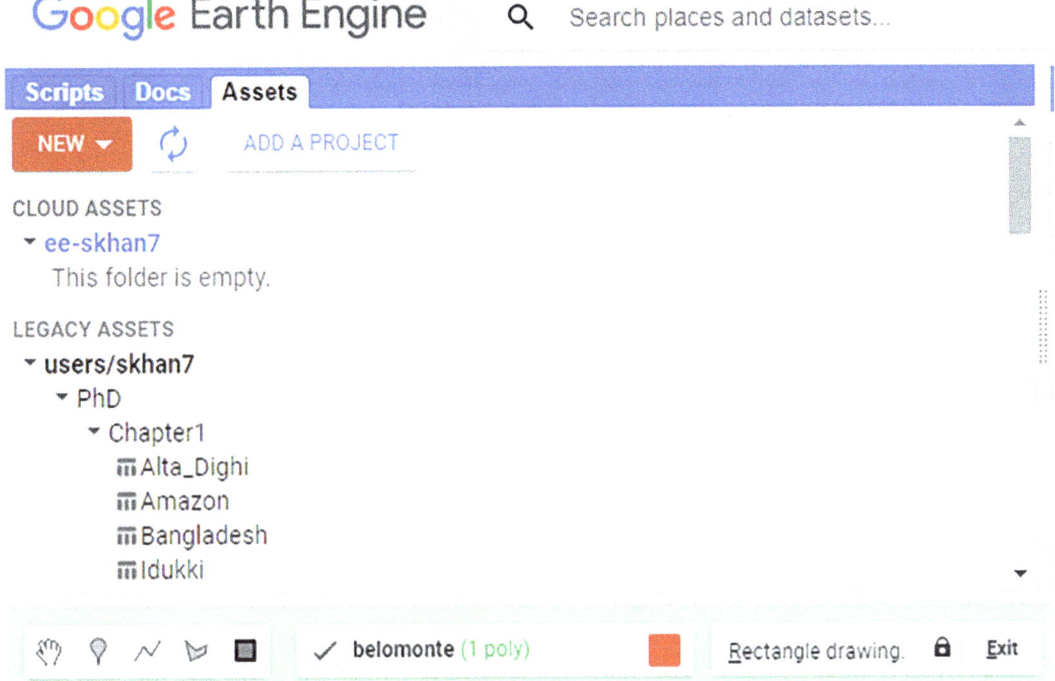

Figure 6.26 Delineating ROI in GEE for Belo Monte Dam. [Credit: Google Earth Engine]

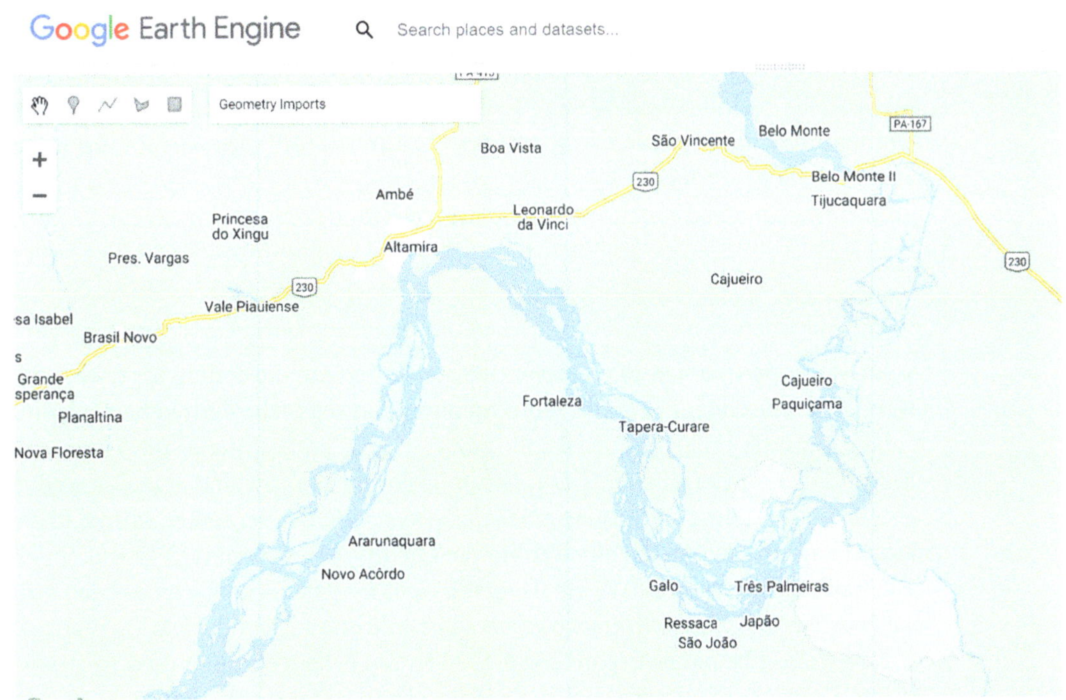

Figure 6.27 Before delineating ROI. [Credit: Google Earth Engine]

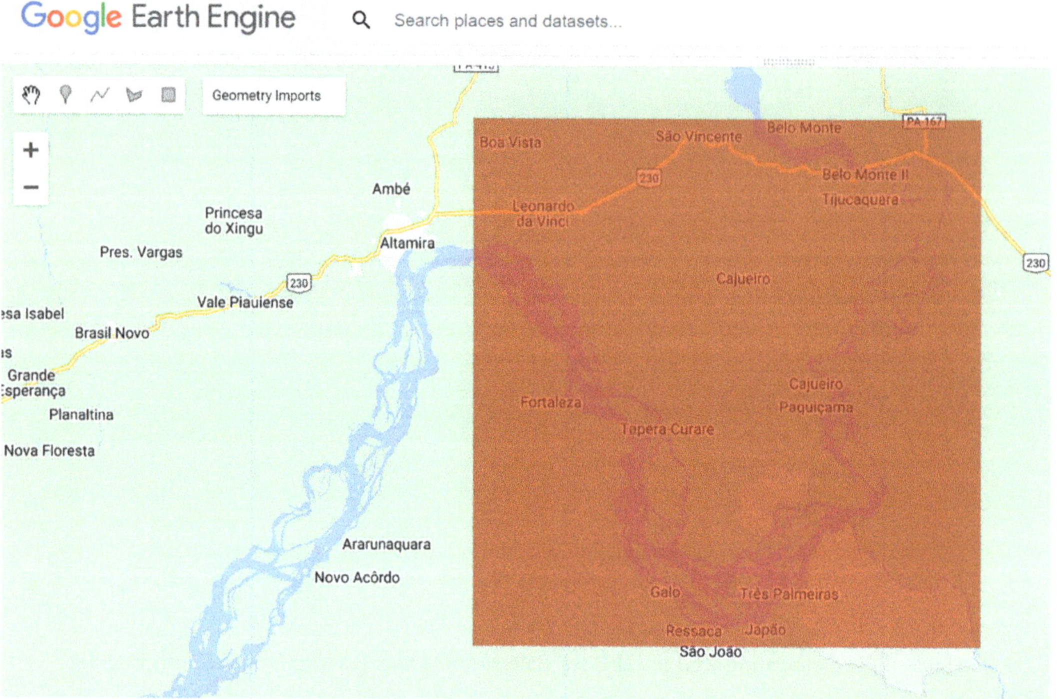

Figure 6.28 After delineating ROI. [Credit: Google Earth Engine]

(Normalized Difference Water Index) and MNDWI (Modified NDWI). The thresholds of two techniques can vary from place to place. By default, the tutorial here demonstrates an example using zero as the threshold values, although we can easily change them as we see fit.

```
// Defining Thresholds for water classification techniques
var threshold_ndwi = 0;
var threshold_mndwi = 0;
```

Step 3: Getting the overview of the script. Before we dig into the coding part, we should get an overview of the code. The script has a function that select the desired bands from Landsat 5 and Landsat 8 Imageries, masks the cloud (if any) and estimates the water surface area using NDWI and MNDWI. The function needs an input which is the desired date. The designed script automatically searches the lowest cloud-cover image within 30 days of the desired date and masks the cloud using the Quality Assessment (QA) band.

Step 4: Selecting the dataset. We are using the Landsat 5 dataset which has been atmospherically corrected surface reflectance values. The collection ID in GEE is "LANDSAT/LT05/C02/T2_L2". The dataset spans from 1984 to 2011, hence can be used to study the Belo Monte Dam. More information on the dataset can be found using the collection ID in the search bar in GEE (Figure 6.29).

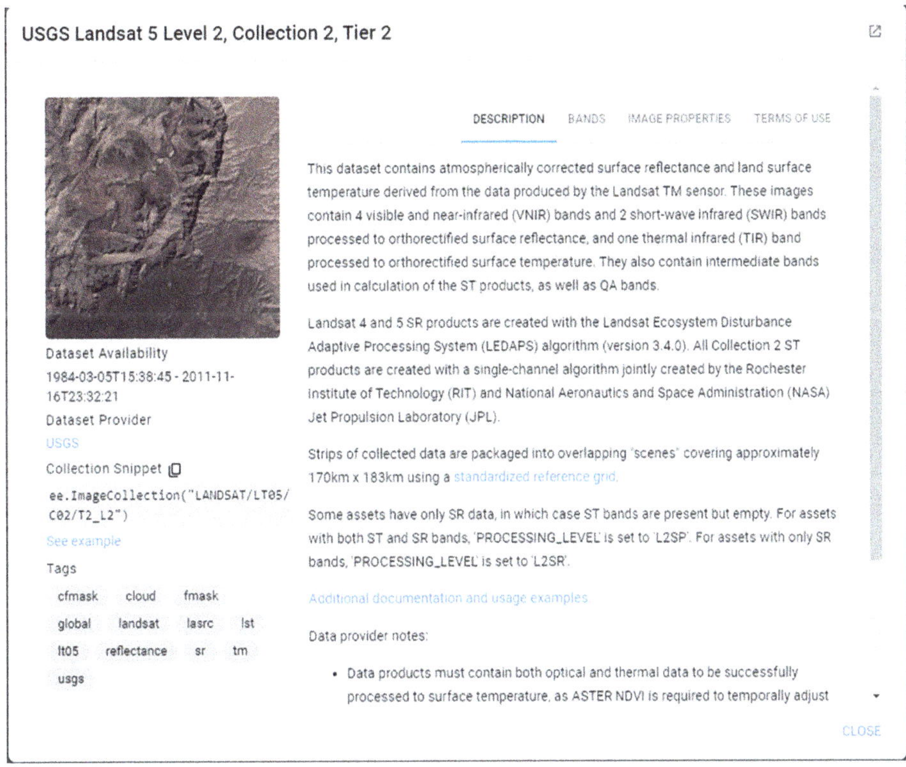

Figure 6.29 Information on data collection from Landsat 5. [Credit: Google Earth Engine]

Figure 6.30 Information on bands of interest (i.e. wavelengths used by the sensor) for water classification. [Credit: Google Earth Engine]

We are interested in the following bands – SWIR1, NIR, Green, Red, Blue, Pixel_QA. The required information on bands can be accessed by clicking on the "BANDS", given in the screenshot in Figures 6.30; 6.31.

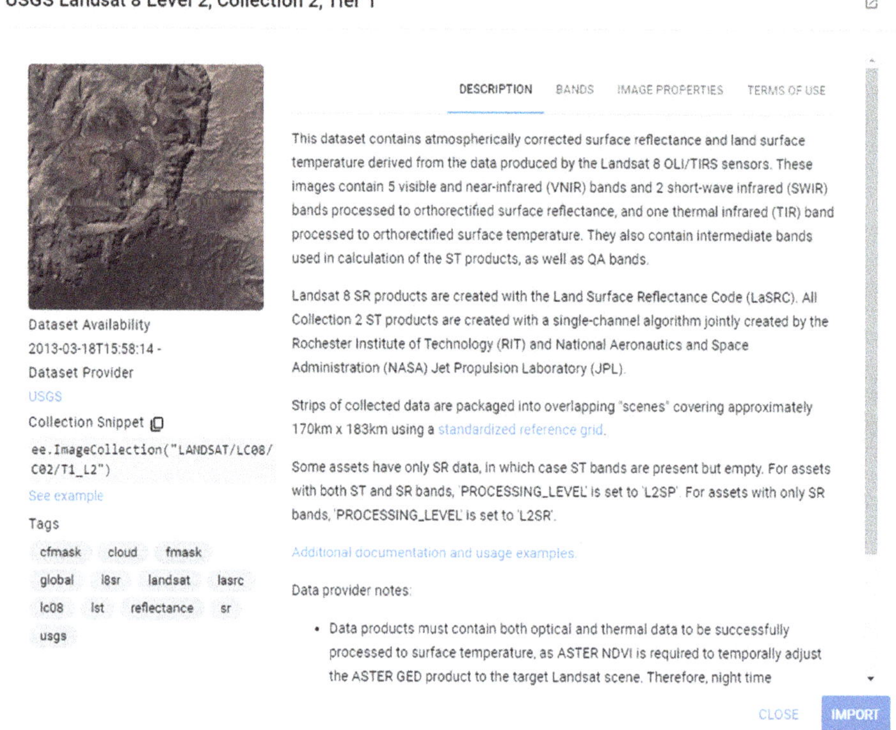

Figure 6.31 Same as Figure 6.30 but for Landsat 8. [Credit: Google Earth Engine]

We also intend to use Landsat 8 data which spans from 2013 to present (2023-09-15 at the time of writing). Landsat 8 data will be used to see the effect of the dam after it was constructed. More information on the Landsat 8 dataset can be found using the collection ID "LANDSAT/LC08/C02/T1_L2" on the GEE search bar.

Step 5: Merging the dataset. After calling the two datasets and selecting the required bands, we use the merge function as below:

```
l5sr = ee.ImageCollection(l5sr).filterBounds(ROI).select(L5_BANDS, L5_NAMES)
var l8sr = ee.ImageCollection("LANDSAT/LC08/C02/T1_L2").filterBounds(ROI).select(L8_
  BANDS, L8_NAMES)
var image = ee.ImageCollection(l5sr.merge(l8sr))
                  .sort('system:time_start')
                  .filterDate(date_range)
                  .sort('CLOUD_COVER').first()
```

The details on the merge function can be found using the "Docs" tab on the left pane of GEE, as shown in Figure 6.32. The merged image will have all the images from Landsat 5 and Landsat 8 sensors within the selected date range. The merged image will be sorted as per their cloud-cover percentage, and only one image, with the lowest cloud cover within 30 days, will be selected.

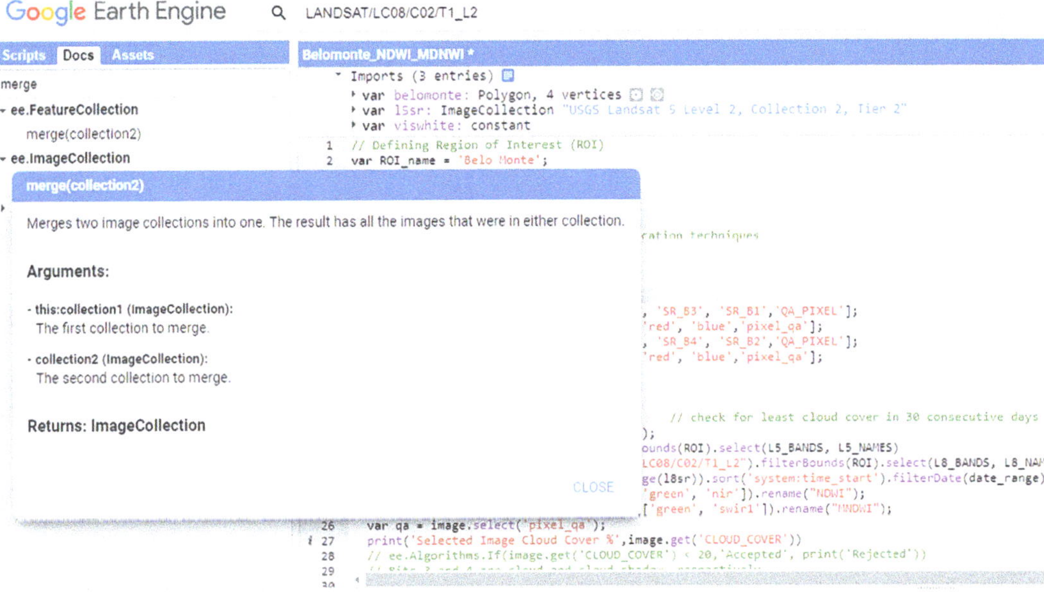

Figure 6.32 Details on the merge function in GEE. [Credits: Google Earth Engine]

Step 6: Defining the NDWI and MNDWI indices.

NDWI: As briefly discussed earlier in this chapter, NDWI uses the NIR and Green spectral bands to detect the water content of the water bodies. We have the threshold of 0, which means pixel values greater than 0 are considered as water surface and pixel values less than 0 are non-water values.

NDWI can be calculated using the following formula:

$$NDWI = \frac{Green - NIR}{Green + NIR}$$

Bands	Landsat 8	Landsat 5
Green	SR_B3	SR_B2
NIR	SR_B5	SR_B4

MNDWI: MNDWI uses the Green and SWIR spectral bands for enhancement of open water features. It also diminishes the built-up area features that are often correlated with the open water in other indices.

MNDWI can be calculated using the following formula:

$$MNDWI = \frac{Green - SWIR}{Green + SWIR}$$

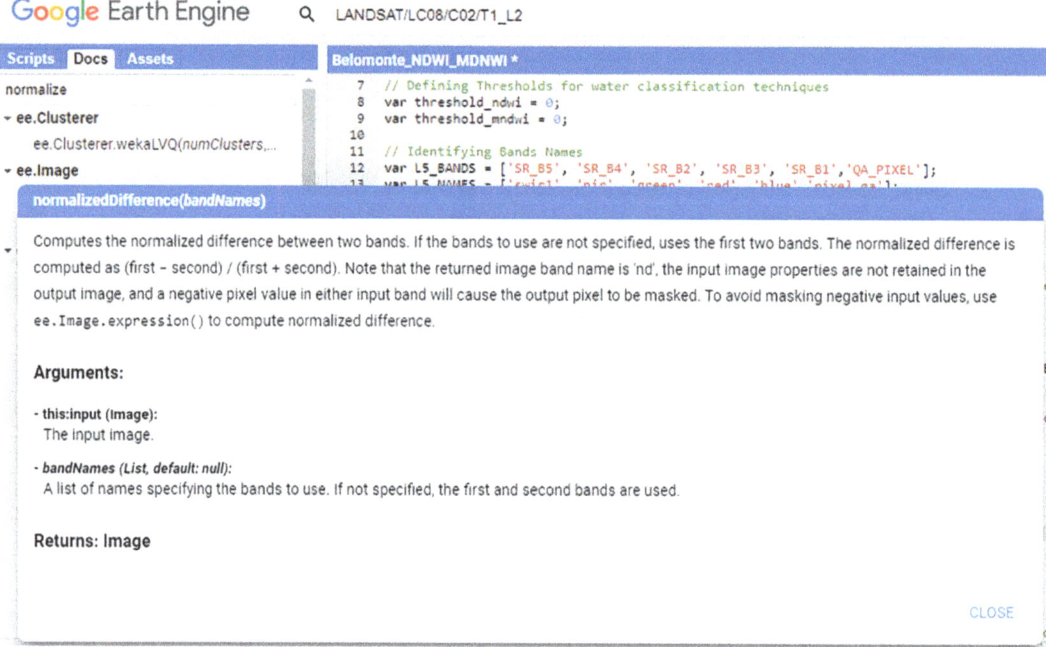

Figure 6.33 Information on GEE's normalized difference function. [Credit: Google Earth Engine]

Bands	Landsat 8	Landsat 5
Green	SR_B3	SR_B2
SWIR1	SR_B6	SR_B5

We intend to use GEE's inbuilt "normalizeDifference" feature. Figure 6.33 shows a screenshot of the "docs" page about the normalizeDifference feature.

The screenshot captured below from the script shows the use of the normalizeDifference feature.

```
// Defining the required indices
var ndwi = image.normalizedDifference(['green', 'nir']).rename("NDWI");
var mndwi = image.normalizedDifference(['green', 'swirl']).rename("MNDWI");
```

Step 7: Masking the clouds and estimating the cloud cover.

The Amazon can be a cloudy region depending on the time of the year and specific region. Visible bands from the Landsat mission will not work during high cloud cover. So it is important to know how much cloud cover a specific satellite scene had at the time and region of interest. To mask the clouds, we need to use the pixel QA band of the merged image. We are also interested in knowing the cloud cover of the obtained merged image. Since the sensor is Landsat 5 or Landsat 8, the bands do not have a high wavelength to

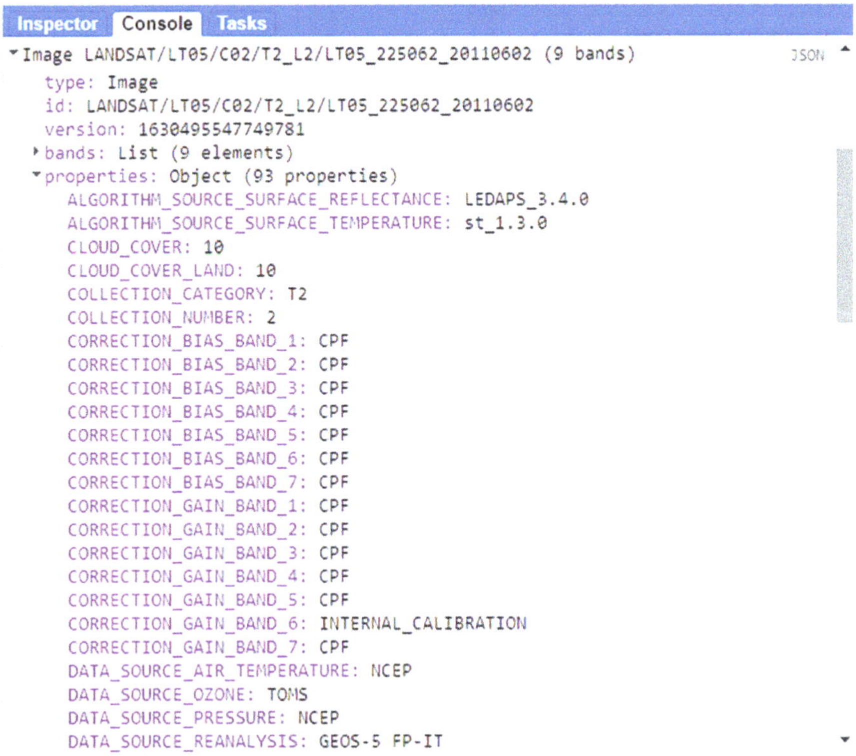

Inspector | Console | Tasks
▾Image LANDSAT/LT05/C02/T2_L2/LT05_225062_20110602 (9 bands)   JSON
    type: Image
    id: LANDSAT/LT05/C02/T2_L2/LT05_225062_20110602
    version: 1630495547749781
  ▸bands: List (9 elements)
  ▾properties: Object (93 properties)
      ALGORITHM_SOURCE_SURFACE_REFLECTANCE: LEDAPS_3.4.0
      ALGORITHM_SOURCE_SURFACE_TEMPERATURE: st_1.3.0
      CLOUD_COVER: 10
      CLOUD_COVER_LAND: 10
      COLLECTION_CATEGORY: T2
      COLLECTION_NUMBER: 2
      CORRECTION_BIAS_BAND_1: CPF
      CORRECTION_BIAS_BAND_2: CPF
      CORRECTION_BIAS_BAND_3: CPF
      CORRECTION_BIAS_BAND_4: CPF
      CORRECTION_BIAS_BAND_5: CPF
      CORRECTION_BIAS_BAND_6: CPF
      CORRECTION_BIAS_BAND_7: CPF
      CORRECTION_GAIN_BAND_1: CPF
      CORRECTION_GAIN_BAND_2: CPF
      CORRECTION_GAIN_BAND_3: CPF
      CORRECTION_GAIN_BAND_4: CPF
      CORRECTION_GAIN_BAND_5: CPF
      CORRECTION_GAIN_BAND_6: INTERNAL_CALIBRATION
      CORRECTION_GAIN_BAND_7: CPF
      DATA_SOURCE_AIR_TEMPERATURE: NCEP
      DATA_SOURCE_OZONE: TOMS
      DATA_SOURCE_PRESSURE: NCEP
      DATA_SOURCE_REANALYSIS: GEOS-5 FP-IT

Figure 6.34 Estimating cloud cover in GEE. [Credit: Google Earth Engine]

penetrate the clouds. If the cloud-cover percentage is high, we need to select a different date. We can print the cloud cover of the image using the "get" feature. This feature obtains the values set in the properties of the image, which can be obtained by printing the merged image and looking into properties as shown (Figure 6.34).

To mask the clouds, we need the information of the bits. To obtain the required information, we need to call back to the collection ID of our Landsat 5 or Landsat 8 data and go to bands. A screenshot of the required information is shown in Figure 6.35.

Consider a four-bit pattern. In that pattern, the command "1<<8" pushes "1" eight spaces to the left. Our test image QA is 00 00 00 00 00 00 00 00. Hence the mediumConfidenceCloud will make the following arrangements: a medium-confidence cloud image should have QA 00 00 00 01 00 00 00 00, whereas a high-confidence cloud image will have QA 00 00 00 11 00 00 00 00.

```
var qa = image.select('pixel_qa');
print('Selected Image Cloud Cover %',image.get('CLOUD_COVER'))
var mediumConfidenceCloud = qa.bitwiseAnd(1 << 8).neq(0) // Bit 8 is set
.and(qa.bitwiseAnd(1 << 9).eq(0)); // Bit 9 is not set

image = image.updateMask(mediumConfidenceCloud.not()); // Invert the mask to keep
non-cloudy pixels
```

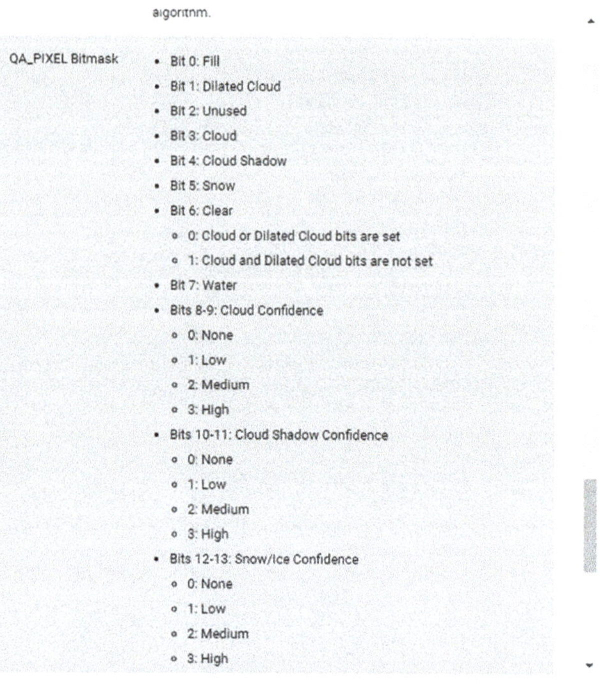

Figure 6.35 Estimating cloud cover in GEE. [Credits: Google Earth Engine]

```
// var cloudShadowBitMask = 1 << 4;
var highConfidenceCloud = qa.bitwiseAnd(1 << 8).neq(0) // Bit 8 is set
                        .and(qa.bitwiseAnd(1 << 9).neq(0)); // Bit 9 is set

var image_masked = image.updateMask(highConfidenceCloud.not());
```

Hence any image whose QA bits match with a medium-confidence cloud image or with a high-confidence cloud image shall be masked.

Step 8: Adding NDWI and MNDWI bands to masked image

Here we add the estimated NDWI and MNDWI values of each pixel as a band. Further, we also select only those pixels whose indices values are greater than or equal to our threshold limit. The screenshot of the script performing this is given below.

```
image_masked = image_masked.addBands(ndwi);
image_masked = image_masked.addBands(mndwi);
var image_masked_ndwi = image_masked.updateMask(ndwi.gte(threshold_ndwi));
var image_masked_mndwi = image_masked.updateMask(ndwi.gte(threshold_mndwi));
image_masked_ndwi = image_masked_ndwi.addBands(ndwi)
                    .set("system:time_start",ee.Date(image.get('system:time_
                    start'))
                    .millis()).clip(ROI)
image_masked_mndwi = image_masked_mndwi.addBands(mndwi)
```

```
                 .set("system:time_start", ee.Date(image.get('system:time_start'))
                 .millis()).clip(ROI)
```

Step 9: Estimating water surface area

To estimate the water surface area, reducers are used. In the following snapshot of the script ee.Reducer.sum() is used. The function of the sum reducer is to sum all the pixels that have indices greater than the threshold limit. Later, the sum of all the pixels is multiplied by the pixel area and divided by 100,000 to estimate the water surface area in square kilometers.

```
var ndwiarea = image_masked_ndwi.select("NDWI")
                 .gt(threshold_ndwi).multiply(ee.Image.pixelArea()).divide(1e6).
                 reduceRegion({
                 'reducer': ee.Reducer.sum(),
                 'geometry': ROI,
                 'scale': 30,
                 'maxPixels': 1e9
                 }).get('NDWI');

var mndwiarea = image_masked_mndwi.select("MNDWI")
                 .gt(threshold_mndwi).multiply(ee.Image.pixelArea())
                 .divide(1e6).reduceRegion({
                 'reducer': ee.Reducer.sum(),
                 'geometry': ROI,
                 'scale': 30,
                 'maxPixels': 1e9
                 }).get('MNDWI');
```

Step 10: Applying the function

The input of the function is a date (given as year-month-day), so a variable 'Date_Start' has been defined and is used as an input to the function.

```
// // User Inputs Date
var Date_Start = ee.Date('2011-06-01');

// If the Cloud Cover % is more than 20, please change the date to a cloud free
date.
// The script by defaults pick the date having the minimum cloud cover within 30
days of the selected data.

var water_classified = detectWater_ndwi(Date_Start);
```

Results

Before 2013, What Did the Region of Belo Monte Dam Look Like?

See Figures 6.36–6.38.

After 2013, What Does the Region Look Like, with the Belo Monte Dam Project?

See Figures 6.39–6.41.

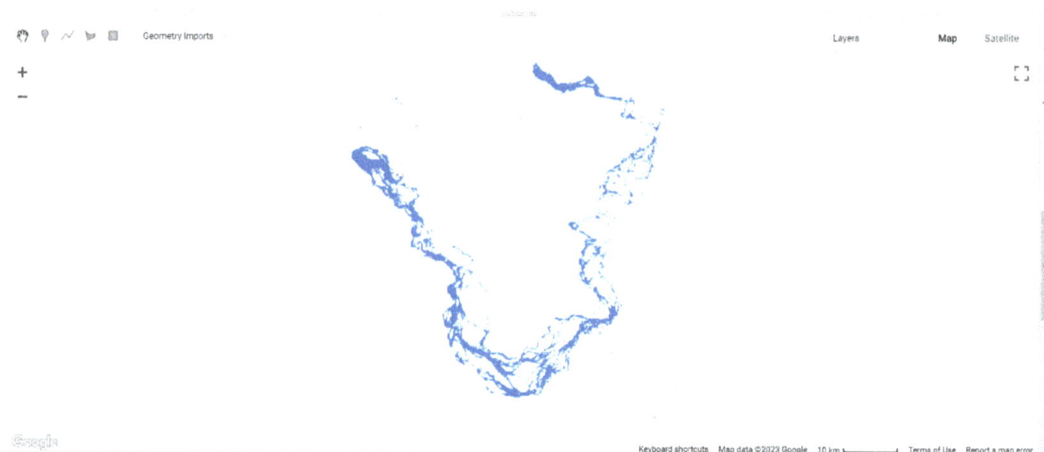

Figure 6.36 NDWI water surface area estimated on 2011-06-02. [Credit: Google Earth Engine]

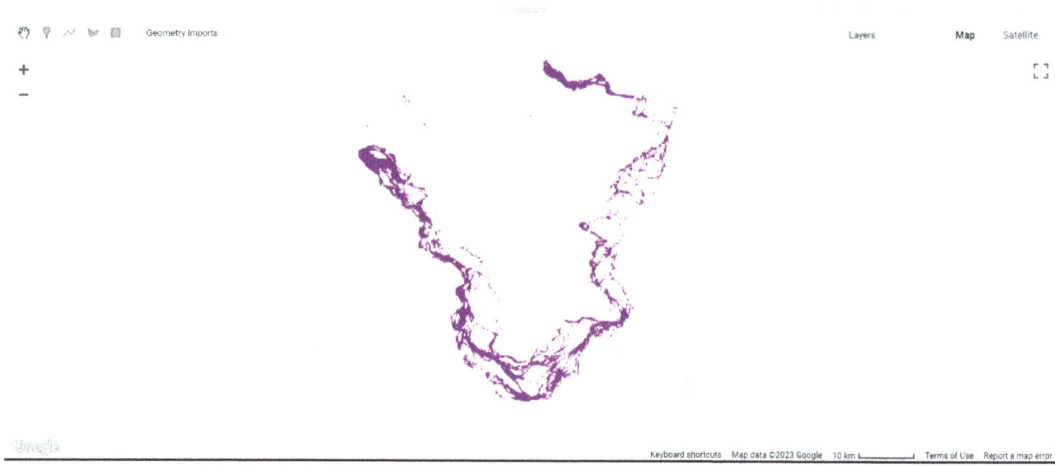

Figure 6.37 MNDWI water surface area estimated on 2011-06-02. [Credit: Google Earth Engine]

Figure 6.38 Final answers on pre-dam water area in GEE. [Credit: Google Earth Engine]

Inspector Console Tasks
Use print(...) to write to this console.

Selected Image Cloud Cover %   JSON
10

NDWI Area (km2) for Date 2011-06-02   JSON
253.7763917201502

MNDWI Area (km2) for Date 2011-06-02   JSON
254.05911462096827

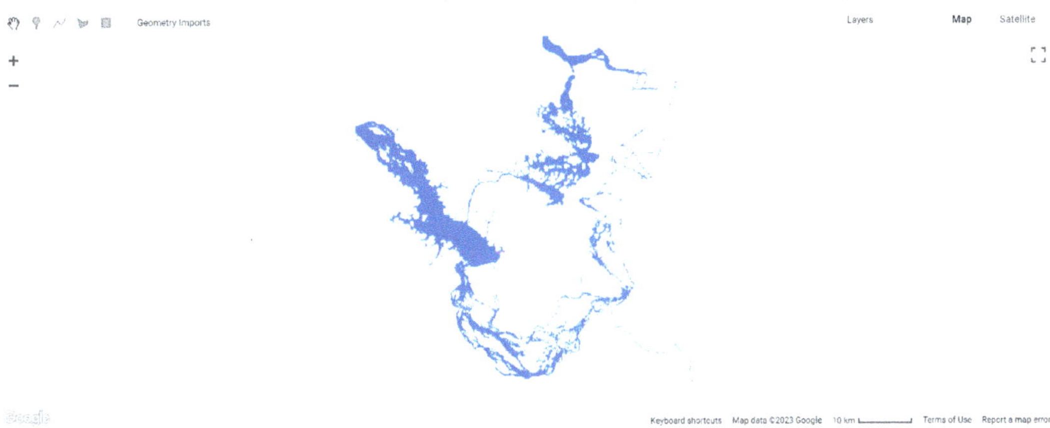

Figure 6.39 NDWI water surface area estimated on 2020-07-28. [Credits: Google Earth Engine]

Figure 6.40 MNDWI water surface area estimated on 2020-07-28. [Credits: Google Earth Engine]

Figure 6.41 Final results on post-dam water area in GEE. [Credits: Google Earth Engine]

Dynamic Surface Water Extent

Similar to NDWI and MNDWI, there is another method called Dynamic Surface Water Extent (DSWE) that we can use to classify or detect surface water extent. In this part of the tutorial, we will learn how to repeat the above exercise using the DSWE algorithm for surface water classification.

The link to GEE scripts can be found here for the DSWE tutorial example https://code .earthengine.google.com/?scriptPath=users%2Fskhan7%2FTextBook%3ATutorial%2F Chapter6_Tutorial1_DSWE

The DSWE algorithm performs analysis at a sub-pixel level and uses the five decision rule-based tests to classify a pixel into water with low, moderate, high confidence, potential wetland, and "not water" pixels. For further information, readers can refer to the paper on DSWE by Jones (2019) and www.usgs.gov/landsat-missions/landsat-dynamic-surface-water-extent-science-products.

Step 1: Delineating the ROI. This step was mentioned earlier while using NDWI and MNDWI.

Step 2: Defining water pixel thresholds. Only the class of pixels that fall under water with moderate confidence, water with high confidence, and potential wetland will be considered. Below is the snapshot captured from the Algorithm Description Document (ADD) showing the bands developed from the DSWE algorithm.

```
The interpreted class DSWE band has the following possible values:

0 -> Not Water
1 -> Water - High Confidence
2 -> Water - Moderate Confidence
3 -> Potential Wetland
4 -> Low Confidence Water or Wetland
255 -> Fill (no data)
```

Step 3: Overview of the script. The script first preprocesses the input elevation data to produce a terrain shadow mask. The five independent tests are applied to each pixel to detect water. Test results are recoded into categories that indicate the likelihood of water. The likelihood of water is filtered using the land cover/terrain-shadow-based masking. Cloud and cloud shadows are labeled based on the QA bands. Finally, masked bands are shown in DSWE as a result, and only pixels falling under our desired water threshold are considered for estimating water surface area.

Step 4: Preprocessing DEM data. Slope and Hill shade are required to produce an image structure representing the extent of terrain-produced shadow for each pixel.

```
// Apply slope algorithm to DEM.
var slope = ee.Terrain.slope(srtm);
// Map.addLayer(slope)
// Apply hillshade algorithm to DEM.
var hillshade = ee.Terrain.hillshade(srtm);
// Map.addLayer(hillshade)
```

Step 5: Defining the threshold of the five different tests.

```
// Define thresholds
var wigt = 0.0124
var awgt = 0.0
var pswt_1_mndwi = -0.44
var pswt_1_swir1 = 900
var pswt_1_nir = 1500
var pswt_1_ndvi = 0.7
var pswt_2_mndwi = -0.5
var pswt_2_blue = 1000
var pswt_2_nir = 2500
var pswt_2_swir1 = 3000
var pswt_2_swir2 = 1000

// thresholds for slope, hillshade
var percent_slope_high = 30
var percent_slope_moderate = 30
var percent_slope_wetland = 20
var percent_slope_low - 10
var hillshade_threshold = 110
```

Step 6: Creating dummy images for recoding and identifying bands. Required bands can be found using the Jones or the ADD paper mentioned earlier.

```
var diagn_img = ee.Image(0)
var out_img = ee.Image(255)
var final_dswe = ee.Image(5)

var L5_BANDS = ['SR_B7', 'SR_B5', 'SR_B4', 'SR_B2', 'SR_B3', 'SR_B1','QA_PIXEL'];
var L5_NAMES = ['swir2' ,'swir1', 'nir', 'green', 'red', 'blue','pixel_qa'];
var L8_BANDS = ['SR_B7', 'SR_B6', 'SR_B5', 'SR_B3', 'SR_B4', 'SR_B2','QA_PIXEL'];
var L8_NAMES = ['swir2' ,'swir1', 'nir', 'green', 'red', 'blue','pixel_qa'];
```

Step 7: Calculating different indices. DSWE algorithm uses multiple indices such as MNDWI (discussed earlier), Multi-Band Spectral Relationship Visible (MBSRV), Multi-Band Spectral Relationship Near-Infrared (MBSRN), Automated Water Extent Shadow (AWESH), and Normalized Difference Vegetation Index (NDVI). The aforementioned indices can be calculated using the following formulas:

MNDWI:

$$MNDWI = \frac{Green - SWIR1}{Green + SWIR1}$$

MBSRV:

$$MBSRV = Green + Red$$

Bands	Landsat 8	Landsat 5
Green	SR_B3	SR_B2
Red	SR_B4	SR_B3

MBSRN:

$$MBSRN = NIR + SWIR1$$

Bands	Landsat 8	Landsat 5
NIR	SR_B5	SR_B4
SWIR1	SR_B6	SR_B5

AWESH:

$$AWESH = Blue + (2.5 \times Green) - (1.5 \times MBSRN) - (0.25 \times SWIR2)$$

Bands	Landsat 8	Landsat 5
Blue	SR_B2	SR_B1
SWIR2	SR_B7	SR_B7

NDWI:

$$NDWI = \frac{NIR - Red}{NIR + Red}$$

Bands	Landsat 8	Landsat 5
NIR	SR_B5	SR_B4
Red	SR_B4	SR_B3

```
var mndwi = image.normalizedDifference(['green', 'swir1'])
var mbsrv = image.select('green').add(image.select('red'))
var mbsrn = image.select('nir').add(image.select('swir1'))
var awesh = image.select('blue')
            .add(image.select('green').multiply(ee.Number(2.5)))
            .subtract(mbsrn.multiply(ee.Number(1.5)))
            .subtract(image.select('swir2').multiply(ee.Number(0.25)))
            var ndvi = image.normalizedDifference(['nir', 'red']);
```

Step 8: Applying the diagnostic tests. Following is the snapshot from the ADD, giving information about the diagnostic tests (Figure 6.42).

```
var d1 = diagn_img.where(mndwi.gt(wigt), ee.Image(1));
var d2 = diagn_img.where(mbsrv.gt(mbsrn), ee.Image(1));
var d3 = diagn_img.where(awesh.gt(awgt), ee.Image(1));
var d4 = diagn_img.where(mndwi.gt(pswt_1_mndwi)
                .and(image.select('swir1').lt(pswt_1_swir1))
                .and(image.select('nir').lt(pswt_1_nir))
                .and(ndvi.lt(pswt_1_ndvi)), ee.Image(1));
var d5 = diagn_img.where(mndwi.gt(pswt_2_mndwi)
                .and(image.select('blue').lt(pswt_2_blue))
                .and(image.select('swir1').lt(pswt_2_swir1))
                .and(image.select('swir2').lt(pswt_2_swir2))
                .and(image.select('nir').lt(pswt_2_nir)), ee.Image(1));
```

Test 1: Compare MNDWI to the wigt Wetness Index threshold, where the threshold ranges from 0.0 to 2.0 and is defaulted to a value of 0.124.

```
if (mndwi > wigt) set the ones digit  (Example 00001)
```

Test 2: Compare the MBSRV and MBSRN values to each other.

```
if (mbsrv > mbsrn) set the tens digit  (Example 00010)
```

Test 3: Compare AWESH to the awgt Automated Water Extent Shadow threshold, where the threshold ranges from -2.0 to 2.0 and is defaulted to a value of 0.0.

```
if (awesh > awgt) set the hundreds digit  (Example 00100)
```

Test 4: Compare the MNDWI and NDVI along with the NIR and SWIR1 bands to the following thresholds.

- Partial Surface Water Test-1 MNDWI (pswt_1_mndwi) threshold, where the threshold ranges from -2.0 to 2.0 and is defaulted to a value of -0.44.
- Partial Surface Water Test-1 SWIR1 (pswt_1_swir1) threshold, where the threshold ranges from 0 to data maximum and is defaulted to a value of 0.0900.
- Partial Surface Water Test-1 NIR (pswt_1_nir) threshold, where the threshold ranges from 0 to data maximum and is defaulted to a value of 0.15.
- Partial Surface Water Test-1 NDVI (pswt_1_ndvi) threshold, where the threshold ranges from 0 to 2.0 and is defaulted to a value of 0.7.

```
if (mndwi > pswt_1_mndwi
    && SWIR1 < pswt1_1_swir1
    && NIR < pswt_1_nir
    && NDVI < pswt_1_ndvi)
    set the thousands digit  (Example 01000)
```

Test 5: Compare the MNDWI along with the Blue, NIR, SWIR1, and SWIR2 bands to the following thresholds.

- Partial Surface Water Test-2 MNDWI (pswt_2_mndwi) threshold, where the threshold ranges from -2.0 to 2.0 and is defaulted to a value of -0.5.
- Partial Surface Water Test-2 Blue (pswt_2_blue) threshold, where the threshold ranges from 0 to data maximum and is defaulted to a value of 0.1000.
- Partial Surface Water Test-2 NIR (pswt_2_nir) threshold, where the threshold ranges from 0 to data maximum and is defaulted to a value of 0.2500.
- Partial Surface Water Test-2 SWIR1 (pswt_2_swir1) threshold, where the threshold ranges from 0 to data maximum and is defaulted to a value of 0.3000.
- Partial Surface Water Test-2 SWIR2 (pswt_2_swir2) threshold, where the threshold ranges from 0 to data maximum and is defaulted to a value of 0.1000.

Figure 6.42 Diagnostic test using DSWE in GEE. [Credits: Google Earth Engine]

Step 9: Recoding the diagnostic test results. Readers are advised to refer to section 3.6 (Recode to Interpreted DSWE) in ADD.

Step 10: Estimating the water surface area. Water surface area is estimated using the previously mentioned reducer function.

Step 11: Using the function. To use the function, we need to give a start date. The function will look for the lowest cloud-cover image within 30 days and estimate the water surface area.

```
var Date_Start = ee.Date('2011-06-01');
var water_classified = detectWater(Date_Start);

water_classified = calcWaterPix(water_classified)
// User Inputs end

print('DSWE Area (km2) for Date '+
ee.Algorithms.Date(water_classified.get('system:time_start')).format("YYYY-MM-
dd").getInfo(),water_classified.get('water_area'))
```

Results

Before 2013, What Did the Region of Belo Monte Dam Look Like?

See Figures 6.43, 6.44.

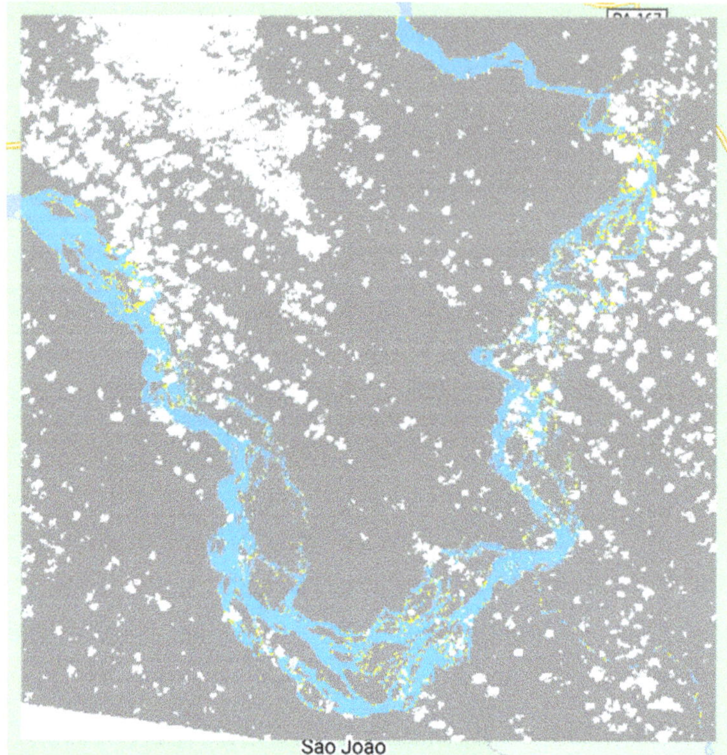

Figure 6.43 DSWE water surface area estimated on 2011-06-02. [Credits Google Earth Engine]

Figure 6.44 Final results in GEE using DSWE. Notice how the results differ from Figure 6.38 using NDWI or MNDWI. [Credit: Google Earth Engine]

Inspector Console Tasks

Use print(...) to write to this console.

Selected Image Cloud Cover %   JSON
10

DSWE Area (km2) for Date 2011-06-02   JSON
322.3245988235295

After 2013, What Does the Region Look Like, with the Belo Monte Dam Project?

See Figures 6.45, 6.46.

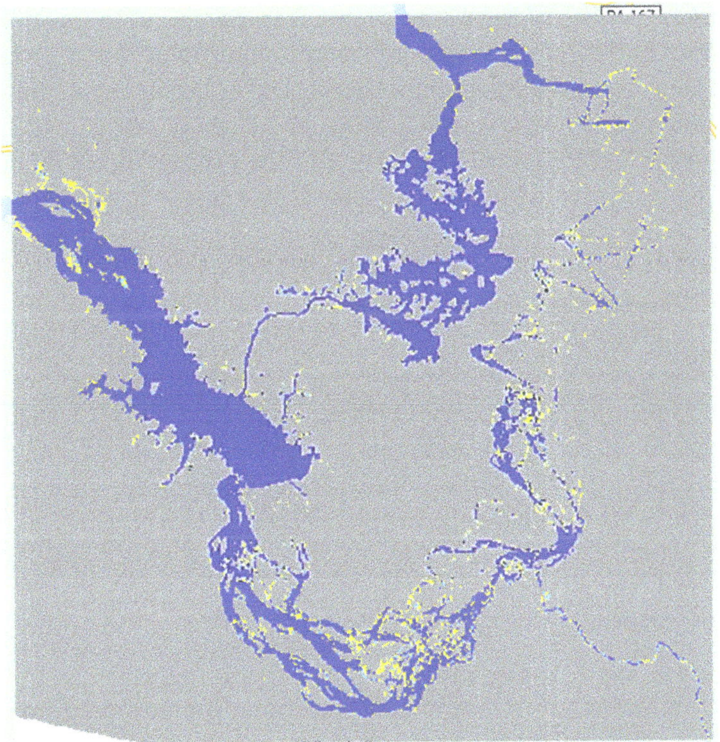

Figure 6.45 DSWE water surface area estimated on 2020-07-01. [Credit: Google Earth Engine]

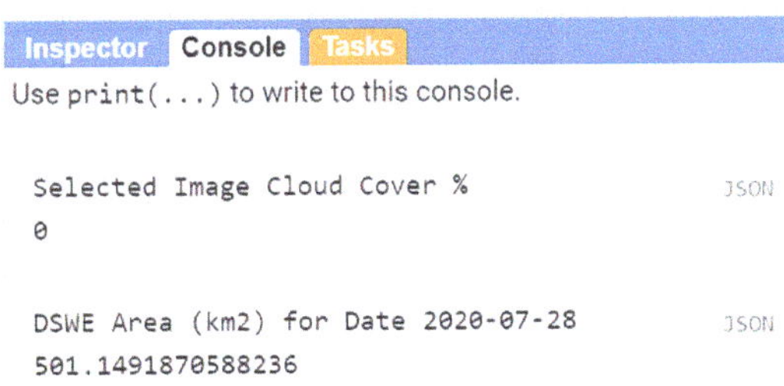

Figure 6.46 Final results in GEE using DSWE. Compare the results with those using NDWI and MNDWI in Figure 6.41. [Credit: Google Earth Engine]

Part 2: Python (Local Computing)

The Python code for carrying out a local computing version of the above tutorial is available in this link: www.cambridge.org/hossain (copy and paste on your browser if the link does not work directly)

Tutorial1_NDWI_MNDWI.ipynb and Tutorial1_DSWE.ipynb notebooks are provided in this link. In the Python version of the tutorial, we will carry out the computing/processing locally. However, we will access the data "as is" in the cloud archived by GEE. This is somewhat different from what we did in Chapter 5 where we downloaded all the data on precipitation to run our Python code locally. For the Python approach here, users need to install Earth Engine Python API using the following command in the Anaconda command prompt:

conda install -c conda-forge earthengine-api

Similar to the Python version of the tutorial in Chapter 5, this chapter's Python .code will be applied a Jupyter Notebook environment.

NDWI and MNDWI Techniques

Step 1: Importing the required libraries. To run the NDWI and MNDWI specific Python script, the Earth Engine library is required. An additional library import warning is also being imported. The work of the warnings.filterwarnings('ignore') is to ignore any warning that does not hinder the running of script.

```python
# Importing required libraries
import ee
import warnings
warnings.filterwarnings("ignore")
```

Step 2: Authenticating/initializing the GEE. If you are running the Earth Engine library for the first time on your local machine, you will need to authenticate it. To authenticate the GEE, uncomment line 2 (ee.Authenticate()) and run the cell.

```python
# Initialize the GEE
#ee.Authenticate() # If you are running earth engine for the first time in your
machine, you need to authenticate it.

ee.Initialize()

# Defining ROI
belomonte = ee.Geometry.Polygon(
        [[[-52.13500521161592, -3.0693661390892055],
          [-52.13500521161592, -3.6396739782713268],
          [-51.570582604194044, -3.6396739782713268],
          [-51.570582604194044, -3.0693661390892055]]])
ROI = belomonte
ROI_Name = 'Belo Monte'
```

This will lead you to choose a Google account through which you want to access the GEE. An example is shown in Figure 6.47.

Click on "**Generate Token**" and copy and paste the authentication code generated to your notebook (Figure 6.48).

Step 3: Defining the thresholds for the techniques. The thresholds of two techniques can be dynamic. By default, the tutorial is demonstrated using zero as the threshold values.

Figure 6.47 Google account selection to access GEE via Python API. [Credits: Google Earth Engine]

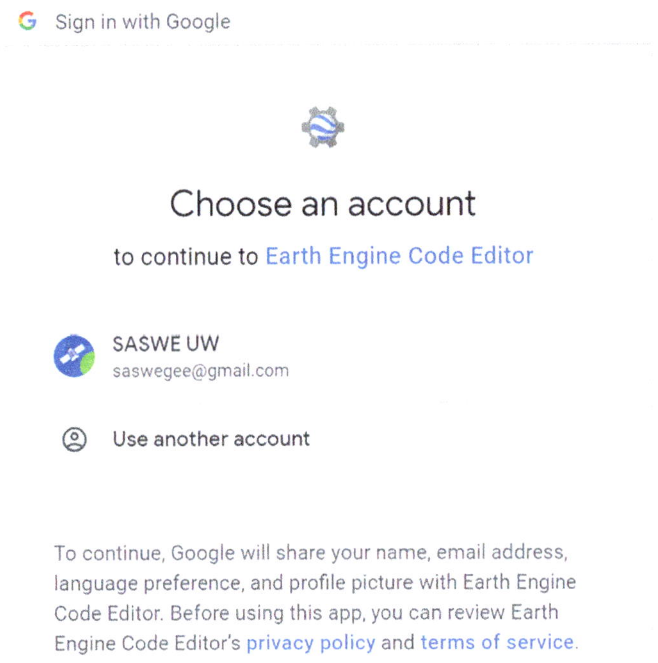

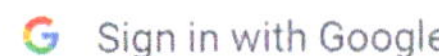

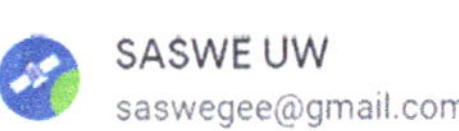

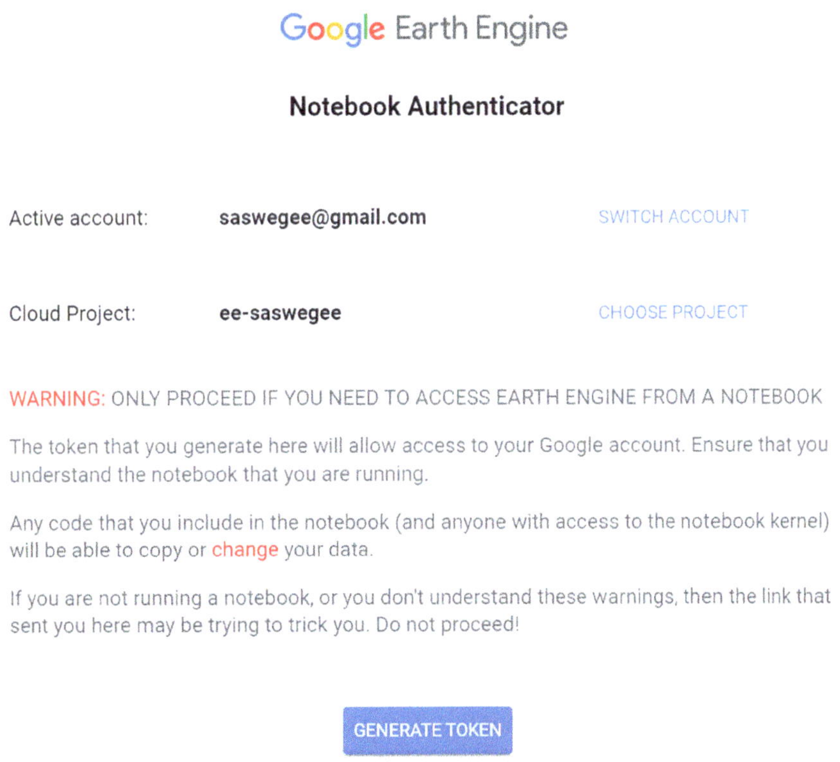

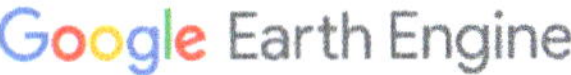

Figure 6.48 Token generation for using Python API for GEE. [Credit: Google Earth Engine]

```
# Defining thresholds
threshold_ndwi = 0
threshold_mndwi = 0
```

Step 4: Getting the script overview. Again, before we dig into the coding part, we should get an overview of the code. The script has a function that selects the desired bands from Landsat 5 and Landsat 8 Imageries, masks the cloud (if any), and estimates the water surface area using NDWI and MNDWI. The function needs an input, which is the desired date. The designed script automatically searches the lowest cloud-cover image within 30 days of the desired date and masks the cloud using the QA band.

Step 5: Selecting the dataset. We want to estimate the water surface area before 2013 and after 2016. Hence, we will use Landsat 5 and Landsat 8 data. The collection ID in GEE is "LANDSAT/LT05/C02/T2_L2" for Landsat 5 and "LANDSAT/LC08/C02/T1_L2" for Landsat 8. More information on the dataset can be found using the collection ID in the search bar in GEE.

```
l5sr = ee.ImageCollection("LANDSAT/LT05/C02/T2_L2").filterBounds(ROI).select(L5_
    BANDS, L5_NAMES)
l8sr = ee.ImageCollection("LANDSAT/LC08/C02/T1_L2").filterBounds(ROI).select(L8_
    BANDS, L8_NAMES)
```

Step 6: Merging the dataset. To merge the two Landsat datasets, we use the command Image.merge(Image).

```
image = ee.ImageCollection(l5sr.merge(l8sr))//
        .sort('system:time_start').filterDate(date_range).sort('CLOUD_
            COVER').first()
```

Step 7: Defining the NDWI and MNDWI indices.

Indices	Formula
NDWI	$\dfrac{NIR - Red}{NIR + Red}$
MNDWI	$\dfrac{Green - SWIR1}{Green + SWIR1}$

Step 8: Masking the clouds. To mask the clouds, we need to use the pixel QA band of the merged image and we need the information of the bits. To obtain the required information, we need to call back to the collection ID of our Landsat 5 or Landsat 8 data and go to bands. A screenshot of the required information is shown in Figure 6.49.

Consider a four-bit pattern. In that pattern, the command "1<<8" pushes "1" eight spaces to the left. Our test image QA is 00 00 00 00 00 00 00 00. Hence the mediumConfidenceCloud will make the following arrangements: Medium-confidence cloud image should have QA 00 00 00 01 00 00 00 00 whereas High-confidence cloud image will have QA 00 00 00 11 00 00 00 00.

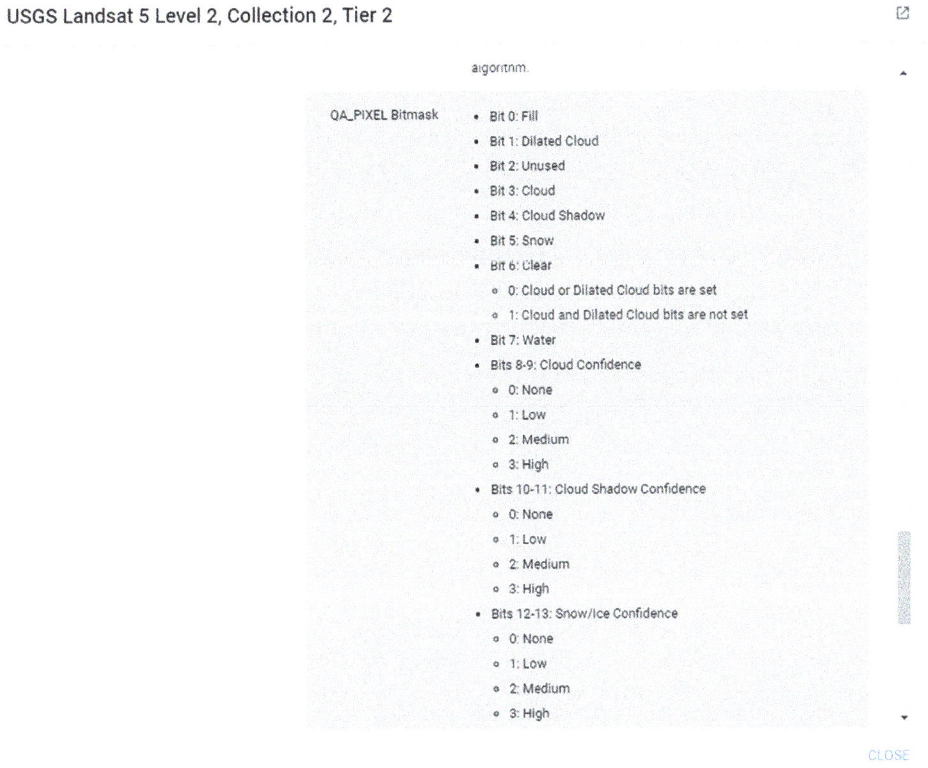

USGS Landsat 5 Level 2, Collection 2, Tier 2

Figure 6.49 Masking for clouds. [Credit: Google Earth Engine]

```
qa = image.select('pixel_qa')
print('Selected Image Cloud Cover % =',image.get('CLOUD_COVER').getInfo())
mediumConfidenceCloud = qa.bitwiseAnd(1 << 8).neq(0).And(qa.bitwiseAnd(1 <<
9).eq(0));

image = image.updateMask(mediumConfidenceCloud.Not()); # Invert the mask to keep
non-cloudy pixels
highConfidenceCloud = qa.bitwiseAnd(1 << 8).neq(0).And(qa.bitwiseAnd(1 <<
9).neq(0));

image_masked = image.updateMask(highConfidenceCloud.Not())
```

Step 9: Adding NDWI and MNDWI bands to masked image. Here we add the estimated
NDWI and MNDWI values of each pixel as a band. Further, we also select only those
pixels whose indices are greater than or equal to our threshold limit. The screenshot of the
script performing the same is mentioned below.

```
image_masked = image_masked.addBands(ndwi)
image_masked = image_masked.addBands(mndwi)
image_masked_ndwi = image_masked.updateMask(ndwi.gte(threshold_ndwi))
image_masked_mndwi = image_masked.updateMask(ndwi.gte(threshold_mndwi))
```

```
image_masked_ndwi = image_masked_ndwi.addBands(ndwi).set("system:time_start",
ee.Date(image.get('system:time_start')).millis()).clip(ROI)
image_masked_mndwi = image_masked_mndwi.addBands(mndwi).set("system:time_start",
ee.Date(image.get('system:time_start')).millis()).clip(ROI)
```

Step 10: Estimating water surface area. To estimate the water surface area, reducers are used. In the following snapshot of the script ee.Reducer.sum() is used. The function of the sum reducer is to sum all the pixels that have indices values greater than the threshold limit. Later, the sum of all the pixels is multiplied by the pixel area and divided by 100,000 to estimate the water surface area in square kilometers.

```
ndwiarea = image_masked_ndwi.select("NDWI").gt(threshold_ndwi)
        .multiply(ee.Image.pixelArea()).divide(1e6).reduceRegion(
        reducer = ee.Reducer.sum(),
        geometry = ROI,
        scale = 30,
        maxPixels = 1e9
        ).get('NDWI')

mndwiarea = image_masked_mndwi.select("MNDWI").gt(threshold_mndwi)
        .multiply(ee.Image.pixelArea()).divide(1e6).reduceRegion(
        reducer = ee.Reducer.sum(),
        geometry = ROI,
        scale = 30,
        maxPixels = 1e9
        ).get('MNDWI')
```

Step 11: Applying the function. The input of the function is a date, so a variable 'Date_Start' has been defined and is used as an input to the function. Users need to change the Date_ Start variable to estimate the water surface area using the two techniques.

```
# ###### User Inputs the date#############
## Running the function

# Both NDWI and MNDWI shows pretty good picture of river flow on '2011-06-01'before
the Dam was built.
# Changing the river direction is clearly visible on '2020-07-01', a clear sky day
Date_Start = ee.Date('2011-06-01')

# If the Cloud Cover % is more than 20, please change the date to a cloud free
date.
# The script by defaults pick the date having the minimum cloud cover within 30
days of the selected data.

water_classified = detectWater(Date_Start)
```

Step 12: Visualizing the water surface extent. The function returns the links of the water surface extent estimated using the NDWI and MNDWI techniques. The visual graphs can be used to compare the change in the dynamics due to the building of the dam (Figures 6.50 and 6.51).

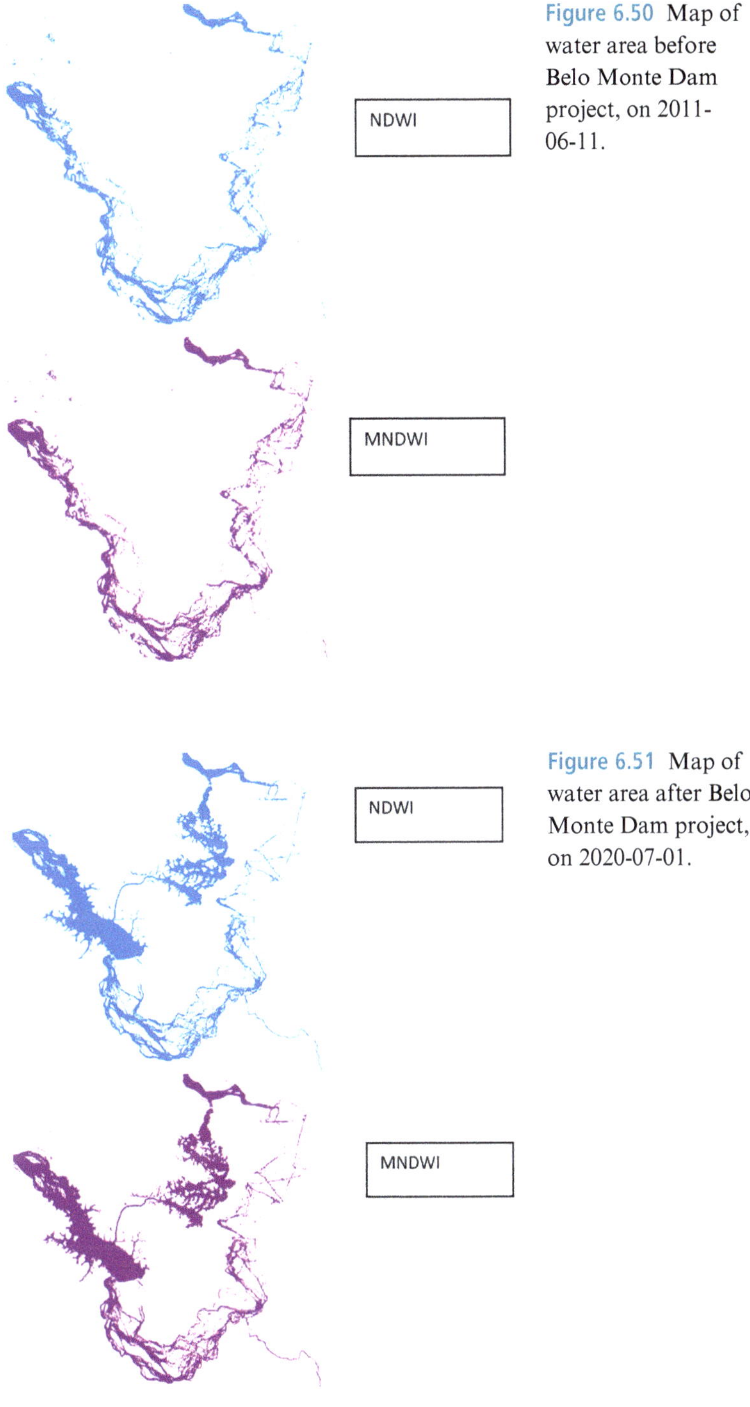

Figure 6.50 Map of water area before Belo Monte Dam project, on 2011-06-11.

Figure 6.51 Map of water area after Belo Monte Dam project, on 2020-07-01.

```
# printing the links of the TIF files. First link is of NDWI and seconds is of
MDNWI
water_classified
```

Result

Using the Python code approach, the map of surface water before the Belo Monte Dam project should look like that in Figure 6.50.

DSWE Technique

Step 1: Importing the required libraries. To run the DSWE specific python script, the Earth Engine library is required. An additional library import warning is also being imported. The work of the warnings.filterwarnings('ignore') is to ignore any warning which does not hinder the running of script.

```
# Importing required libraries
import ee
import time
import warnings
warnings.filterwarnings("ignore")
```

Step 2: Authenticating/initializing the GEE. If you are running the Earth Engine library for the first time on your local machine, you will need to authenticate it. To authenticate the GEE, uncomment the line 2 (ee.Authenticate()) and run the cell.

```
# Initialize the GEE
#ee.Authenticate() # If you are running earth engine for the first time in you
machine, you need to authenticate it.

ee.Initialize()

# Defining ROI
belomonte = ee.Geometry.Polygon(
        [[[-52.13500521161592, -3.0693661390892055],
          [-52.13500521161592, -3.6396739782713268],
          [-51.570582604194044, -3.6396739782713268],
          [-51.570582604194044, -3.0693661390892055]]])
ROI = belomonte
ROI_Name = 'Belo Monte'
```

This will lead you to choose a Google account through which you want to access the GEE and is very similar to the NDWI and MNDWI example of Figures 6.47 and 6.48.

Click on **"Generate Token"** and copy and paste the authentication code generated to your notebook.

Step 3: Defining water pixel thresholds. We define the two variables water_min and water_max. These two variables will be used to consider a particular band from the resulting DSWE technique.

```
# Defining water pixel threshold
wat_min = 1
wat_max = 3
```

A snapshot from the Algorithm Description Document (ADD) of the DSWE method by Jones shows the pixels from the resulting DSWE technique.

```
The interpreted class DSWE band has the following possible values:

0   -> Not Water
1   -> Water - High Confidence
2   -> Water - Moderate Confidence
3   -> Potential Wetland
4   -> Low Confidence Water or Wetland
255 -> Fill (no data)
```

Step 4: Overview of the script. The script first preprocesses the input elevation data to produce a terrain shadow mask. The five independent tests are applied to each pixel to detect water. Test results are recoded into categories that indicate the likelihood of water. The likelihood of water is filtered using the land cover/terrain-shadow-based masking. Cloud and cloud shadows are labeled based on the QA bands. Finally, masked bands are shown in DSWE as a result, and only pixels falling below our desired water threshold are considered for estimating water surface area. The final DSWE generated image is exported to Google Drive.

Step 5: Preprocessing DEM data. Slope and Hill shade are required to produce an image structure representing the extent of terrain-produced shadow for each pixel. Variables slope and hillshade create a query to GEE via the local machine.

```
# Precprocessing DEM data
# Apply slope algorithm to DEM.
slope = ee.Terrain.slope(srtm)
# Map.addLayer(slope)

# Apply hillshade algorithm to DEM.
hillshade = ee.Terrain.hillshade(srtm)
```

Step 6: Defining the threshold of the five different tests. Thresholds for different-different indices are set from line 26 to line 43.

```
# Define thresholds
wigt = 0.0124
awgt = 0.0
pswt_1_mndwi = -0.44
pswt_1_swir1 = 900
pswt_1_nir = 1500
pswt_1_ndvi = 0.7
pswt_2_mndwi = -0.5
pswt_2_blue = 1000
pswt_2_nir = 2500
pswt_2_swir1 = 3000
pswt_2_swir2 = 1000

# thresholds for slope, hillshade
percent_slope_high = 30
```

```
percent_slope_moderate = 30
percent_slope_wetland = 20
percent_slope_low = 10
hillshade_threshold = 110
```

Step 7: Defining the bands. Bands that are required to estimate the various indices can be found in the Jones paper or from the ADD.

```
# Identifying bands
L5_BANDS = ['SR_B7', 'SR_B5', 'SR_B4', 'SR_B2', 'SR_B3', 'SR_B1','QA_PIXEL']
L5_NAMES = ['swir2' ,'swir1', 'nir', 'green', 'red', 'blue','pixel_qa']
L8_BANDS = ['SR_B7', 'SR_B6', 'SR_B5', 'SR_B3', 'SR_B4', 'SR_B2','QA_PIXEL']
L8_NAMES = ['swir2' ,'swir1', 'nir', 'green', 'red', 'blue','pixel_qa']
```

The table for mapping different bands of Landsat 8 and Landsat 5 is shown below.

Bands	Landsat 8	Landsat 5
Red	SR_B4	SR_B3
Blue	SR_B2	SR_B1
Green	SR_B3	SR_B2
NIR	SR_B5	SR_B4
SWIR1	SR_B6	SR_B5
SWIR2	SR_B7	SR_B7

Step 8: Calculating different indices. The DSWE algorithm uses multiple indices such as MNDWI (discussed earlier), Multi-Band Spectral Relationship Visible (MBSRV), Multi-Band Spectral Relationship Near-Infrared (MBSRN), Automated Water Extent Shadow (AWESH), and Normalized Difference Vegetation Index (NDVI). The aforementioned indices can be calculated using the following formulas:

```
mndwi = image.normalizedDifference(['green', 'swir1'])
mbsrv = image.select('green').add(image.select('red'))
mbsrn = image.select('nir').add(image.select('swir1'))
awesh = image.select('blue').add(image.select('green')
          .multiply(ee.Number(2.5))).subtract(mbsrn.multiply(ee.Number(1.5)))
          .subtract(image.select('swir2').multiply(ee.Number(0.25)))
ndvi = image.normalizedDifference(['nir', 'red'])
```

Indices	Formula
MNDWI	$\dfrac{\text{Green} - \text{SWIR1}}{\text{Green} + \text{SWIR1}}$
MBSRV	$\text{Green} + \text{Red}$
MBSRN	$\text{NIR} + \text{SWIR1}$
AWESH	$\text{Blue} + (2.5 \times \text{Green}) - (1.5 \times \text{MBSRN}) - (0.25 \times \text{SWIR2})$
NDWI	$\dfrac{\text{NIR} - \text{Red}}{\text{NIR} + \text{Red}}$

Step 9: Applying diagnostic tests and creating dummy images. Information on diagnostic tests can be found in the ADD document or by referring to Step 8 of estimating DSWE water surface area using cloud computing. In this step readers should pay attention to the creation of dummy image i.e. out_img = ee.Image(255). Other dummy image variables are in lines 71, 80, and 138 of the code. These dummy images are required when we recode the diagnostic test results.

```
# comparisons with threshold
diagn_img = ee.Image(0)
d1 = diagn_img.where(mndwi.gt(wigt), ee.Image(1))
d2 = diagn_img.where(mbsrv.gt(mbsrn), ee.Image(1))
d3 = diagn_img.where(awesh.gt(awgt), ee.Image(1))
d4 = diagn_img.where(mndwi.gt(pswt_1_mndwi).And(image.select('swir1').lt(pswt_1_
swir1))// .And(image.select('nir').lt(pswt_1_nir))
        .And(ndvi.lt(pswt_1_ndvi)), ee.Image(1))
d5 = diagn_img.where(mndwi.gt(pswt_2_mndwi).And(image.select('blue')//
        .lt(pswt_2_blue)).And(image.select('swir1').lt(pswt_2_swir1))//
        .And(image.select('swir2').lt(pswt_2_swir2)).And(image.select('nir')
        .lt(pswt_2_nir)), ee.Image(1))

# Recode
# not water
out_img = ee.Image(255)
```

Step 10: Recoding the diagnostic test results. Readers are advised to refer to section 3.6 in ADD (Recode to Interpreted DSWE).

```
# Recode
# not water
out_img = ee.Image(255)
out_img=out_img.where(d1.eq(0).And(d2.eq(0)).And(d3.eq(0)).And(d4.eq(0)).And(d5.
eq(0)), 0)
out_img=out_img.where(d1.eq(0).And(d2.eq(0)).And(d3.eq(0)).And(d4.eq(0)).And(d5.
eq(1)), 0)
out_img=out_img.where(d1.eq(0).And(d2.eq(0)).And(d3.eq(0)).And(d4.eq(1)).And(d5.
eq(0)), 0)
out_img=out_img.where(d1.eq(0).And(d2.eq(0)).And(d3.eq(1)).And(d4.eq(0)).And(d5.
eq(0)), 0)
out_img=out_img.where(d1.eq(0).And(d2.eq(1)).And(d3.eq(0)).And(d4.eq(0)).And(d5.
eq(0)), 0)

# water high confidence
out_img=out_img.where(d1.eq(0).And(d2.eq(1)).And(d3.eq(1)).And(d4.eq(1)).And(d5.
eq(1)), 1)
out_img=out_img.where(d1.eq(1).And(d2.eq(0)).And(d3.eq(1)).And(d4.eq(1)).And(d5.
eq(1)), 1)
out_img=out_img.where(d1.eq(1).And(d2.eq(1)).And(d3.eq(0)).And(d4.eq(1)).And(d5.
eq(1)), 1)
out_img=out_img.where(d1.eq(1).And(d2.eq(1)).And(d3.eq(1)).And(d4.eq(0)).And(d5.
eq(1)), 1)
out_img=out_img.where(d1.eq(1).And(d2.eq(1)).And(d3.eq(1)).And(d4.eq(1)).And(d5.
eq(0)), 1)
out_img=out_img.where(d1.eq(1).And(d2.eq(1)).And(d3.eq(1)).And(d4.eq(1)).And(d5.
eq(1)), 1)
```

```
# water moderate confidence
out_img=out_img.where(d1.eq(0).And(d2.eq(0)).And(d3.eq(1)).And(d4.eq(1)).And(d5.
eq(1)), 2)
out_img=out_img.where(d1.eq(0).And(d2.eq(1)).And(d3.eq(0)).And(d4.eq(1)).And(d5.
eq(1)), 2)
out_img=out_img.where(d1.eq(0).And(d2.eq(1)).And(d3.eq(1)).And(d4.eq(0)).And(d5.
eq(1)), 2)
out_img=out_img.where(d1.eq(0).And(d2.eq(1)).And(d3.eq(1)).And(d4.eq(1)).And(d5.
eq(0)), 2)
out_img=out_img.where(d1.eq(1).And(d2.eq(0)).And(d3.eq(0)).And(d4.eq(1)).And(d5.
eq(1)), 2)
```

Step 11: Estimating water surface area. Water surface area is estimated using the reducer function. The syntax of the reducer function in GEE Python API is different than the usual GEE platform. The input of the calcWaterPix is an image. The function filters the pixels based on our earlier water threshold values, sums it up and then multiplies by the area, to estimate water surface area.

```
def calcWaterPix(img):
  # calculate area
  mt1 = img.gte(wat_min).And(img.lte(wat_max))
  mt2 = mt1.multiply(ee.Image(1))
  mArea = mt2.reduceRegion(
    reducer = ee.Reducer.sum(),
      geometry = ROI,
      scale = 10,
      maxPixels = 1e9
)
wAreaDSWE = ee.Number(mArea.get('constant')).multiply(10*10).divide(1e6)
return img.set("water_area", wAreaDSWE)
```

Step 12: Using the function. The input of the function is a date. We need to give a date of interest. The function will look for the lowest cloud-cover image within 30 days and estimate the water surface area.

```
# A low cloud cover image is on '2011-06-01', before the Dam was built.
# Changing the river direction is clearly visible on '2020-07-01', a clear sky
day
Date_Start = ee.Date('2020-07-01')
water_classified = detectWater(Date_Start)
water_classified = calcWaterPix(water_classified)
```

Step 13: Exporting the TIF image. GEE gives the freedom to download the generated image via Google drive, and links. Below is the use of export image to Google Drive feature.

```
# Exporting TIF image to google drive
dswe_image = ee.Image(water_classified).clip(ROI).select('constant')
# Map.addLayer(dswe_image, dsweVis,'DSWE')
task = ee.batch.Export.image.toDrive(image=dswe_image,
                                 description='DSWE '+
```

```
ee.Algorithms.Date(water_classified.get('system:time_start')).format("YYYY-MM-
dd").getInfo(),
                                    scale=30,
                                    region=ROI,
                                    fileNamePrefix='DSWE '+
ee.Algorithms.Date(water_classified.get('system:time_start')).format("YYYY-MM-
dd").getInfo(),
                                    crs='EPSG:4326',
                                    fileFormat='GeoTIFF')

task.start()
```

Step 14: Checking the status of export. The status of export can be found using the task. status() command. However, the task.status() gives us the status of the export only at that instant. Hence a for loop is used to know the status frequently. The range of the for loop is selected to be 9 and runs every 20 seconds, as per the experience that an image takes around 3 minutes to export completely.

```
for _ in range(9):
    print('Checking status of the export')
    print(task.status())
    time.sleep(20) # Sleep for 20 seconds

print("If the state is still running wait for some time and then check the
'Recent' in Google Drive. Download the TIF and view it in QGIS or ArcGIS")
```

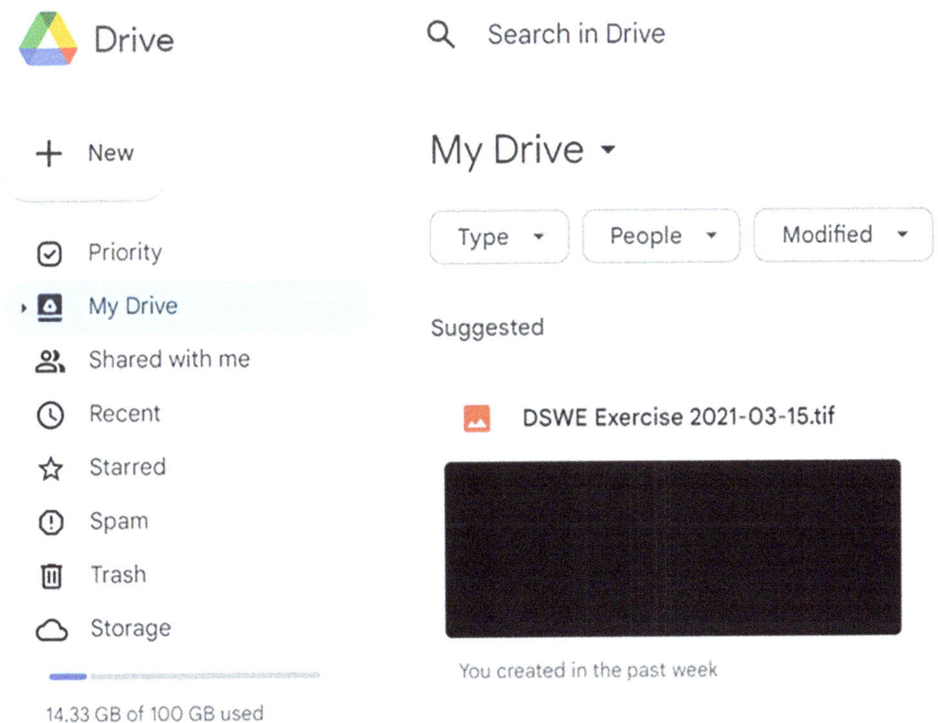

Figure 6.52 Downloading the TIFF file locally for visualizing results.

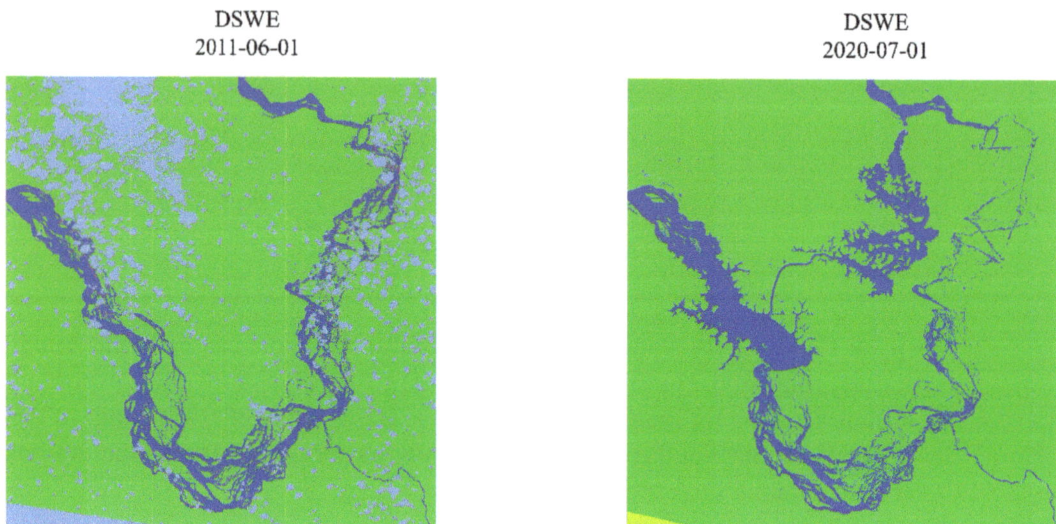

Figure 6.53 DSWE maps before and after the Belo Monte Dam project using Python code (local computing) with API.

Step 15: Downloading the TIF file. The TIFF file can be downloaded from the "Recent" tab on Google Drive (Figure 6.52). Right click on the image and click on the download button. Users can visualize the TIF using Qgis/ArcGIS or Python as per their own choice (Figure 6.53).

Results

Water surface area (km^2):

```
Selected Image Cloud Cover % = 10
DSWE Area (km2) for Date 2011-06-02 = 322.3
```

```
Selected Image Cloud Cover % = 0
DSWE Area (km2) for Date 2020-07-28 = 476.2
```

EXERCISES: CHAPTER 6

Q6.1   Name all the E-M wavelengths that can be used for detecting water on the Earth's surface. List the limitations of each band, the nature of remote sensing (active or passive), and the spatial resolution from orbiting platforms.

Q6.2   Describe how visible wavelength can be used for detecting water and the specific wavelengths that could be used in combination. Name some methods using visible wavelengths.

Q6.3   How can temperature remote sensing be used to detect surface water using thermal infrared (IR) or microwave (MW)? What are the pros and cons of each approach?

Q6.4   What is a radar altimeter and how does it work?

Q6.5   Name the typical bands and sensors an altimeter carries.

Q6.6   What is the datum used for altimeters? What is the difference between a geoid and ellipsoid datum?

Q6.7   How does an angle-looking MW synthetic aperture radar (SAR) work in detecting water on surface? What are its key limitations?

Q6.8   How can an area–elevation curve or bathymetry of a water body be derived using satellite data on terrain elevation? What would be the limitations of using a digital elevation model (DEM) based on the Shuttle Radar Topography Mission (SRTM) for mapping the bathymetry of reservoirs?

Q6.9   Explain the steps for estimating storage change in a lake or a reservoir using satellite data. State your assumptions clearly.

Q6.10   **Data-based exercise (contributed by Shahzaib Khan, University of Washington)**
Using your understanding from the tutorial provided in this chapter, estimate the change in water surface area using NDWI, MNDWI, and DSWE techniques for the Grand Ethiopian Renaissance Dam (GERD) using the option of your choice (GEE, Python, or any other coding language). Use the following coordinates to make the region of interest (ROI):

ee.Geometry.Polygon([[[35.02604478386646, 11.286767556898942],
    [35.02604478386646, 10.900335400985146],
    [35.29692643669849, 10.900335400985146],
    [35.29692643669849, 11.286767556898942]]])

Estimate the water surface area on the following dates: 2021-03-01 and 2023-03-01.
Tip: While following the tutorial, remember to adjust the ROI and the dates as needed for your specific requirements.

REFERENCES

Ahmad, S. K., F. Hossain, T. Pavelsky, et al. (2020). Understanding volumetric water storage in monsoonal wetlands of northeastern Bangladesh. *Water Resources Research*, vol. 56, e2020WR027989. https://doi.org/10.1029/2020WR027989

Biancamaria, S., D. P. Lettenmaier, and T. M. Pavelsky (2016). The SWOT mission and its capabilities for land hydrology. *Surveys in Geophysics*, vol. 37, 307–337.

Biswas, N. K., F. Hossain, M. Bonnema, M. A. Okeowo, and H. Lee (2019). An altimeter height extraction technique for dynamically changing rivers of South and South-East Asia. *Remote Sensing of Environment*, vol. 221, 24–37. https://doi.org/10.1016/j.rse.2018.10.033

Costanza, R., R. dArge, R. DeGroot, et al. (1997). The value of the world's ecosystem services and natural capital. *Nature*, vol. 387, 253–260. https://doi.org/10.1038/387253a0

Crétaux, J. F., S. Calmant, R. A. Del Rio, et al. (2011). Lake studies from satellite altimetry. In *Coastal Altimetry*, 509–533. Springer.

Eldardiry, H., F. Hossain, M. Srinivasan, and V. Tsontos (2022). Success Stories of Satellite Radar Altimeter Applications. *Bulletin of the American Meteorological Society*. https://doi.org/10.1175/BAMS-D-21-0065.1

Khan, S., K. F. Hossain, T. Pavelsky, et al. (2023). Understanding volume estimation uncertainty of lakes and wetlands using satellites and citizen science. *IEEE Journal of Selected Topics in Applied Earth Observations and Remote Sensing*, https://doi.org/10.1109/JSTARS.2023.3250354

Jones, J. W. (2019). Improved automated detection of subpixel-scale inundation – revised Dynamic Surface Water Extent (DSWE) partial surface water tests. *Remote Sensing*, vol. 11, Art. no. 4. https://doi.org/10.3390/rs11040374

McFeeters, S. K. (1996). The use of the Normalized Difference Water Index (NDWI) in the delineation of open water features. *International Journal of Remote Sensing*, vol. 17, 1425–1432. https://doi.org/10.1080/01431169608948714

Mitsch, W. J., B. Bernal, and M. E. Hernandez (2015). Ecosystem services of wetlands. *International Journal of Biodiversity Science, Ecosystem Services & Management*, vol. 11. https://doi.org/10.1080/21513732.2015.1006250

Mobley, C. R. (1994). *Light and Water. Radiative Transfer in Natural Waters*. Academic.

Plengsaeng, B., W. Wehn, and P. van der Zaag (2014). Data-sharing bottlenecks in transboundary integrated water resources management: a case study of the Mekong River Commission's procedures for data sharing in the Thai context, *Water International*, vol. 39, https://doi.org/10.1080/02508060.2015.981783

Ramsar Convention Secretariat (2010). *International Cooperation: Guidelines and other support for International Cooperation under the Ramsar Convention on Wetlands. Ramsar Handbooks for the Wise Use of Wetlands*, 4th edn, vol. 20. Ramsar Convention Secretariat.

Smith, R. C. and K. S. Baker (1981). Optical properties of the clearest natural waters (200–800 nm). *Applied Optics*, vol. 20, 177–184.

Vermote, E., J. C. Roger, B. Franch, and S. Skakun (2018). LaSRC (Land Surface Reflectance Code): overview, application and validation using MODIS, VIIRS, Landsat and Sentinel 2 Data's. In *IGARSS 2018–2018 IEEE International Geoscience and Remote Sensing Symposium*, 8173–8176. https://doi.org/10.1109/IGARSS.2018.8517622

Xu, H. (2006). Modification of Normalized Difference Water Index (NDWI) to enhance open water features in remotely sensed imagery. *International Journal of Remote Sensing*, vol. 27, 3025–3033. https://doi.org/10.1080/01431160600589179

SUGGESTED READING

Alsdorf, D. E., E. Rodríguez, and D. P. Lettenmaier (2007). Measuring surface water from space. *Reviews of Geophysics*, vol. 45, RG2002. https://doi.org/10.1029/2006RG000197

Biancamaria, S., D. P. Lettenmaier, and T. M. Pavelsky (2016). The SWOT mission and its capabilities for land hydrology. In Cazenave, A., N. Champollion, J. Benveniste, and J. Chen (eds.) *Remote Sensing and Water Resources*. Space Sciences Series of ISSI, vol. 55. Springer.

Cooley, S. W., J. C. Ryan, and L. C. Smith (2021). Human alteration of global water storage variability. *Nature*, vol. 591, 78–81.

Pekel, J.-F., A. Cottam, N. Gorelick, and A. S. Belward. (2016). High-resolution mapping of global surface water and its long-term changes. *Nature*, vol. 540, 418–422.

Satellite Remote Sensing of River Discharge

7.1 Chapter Overview

In the previous chapter, we covered how satellite remote sensing can be used to detect water on the surface in terms of its spatial extent, elevation, and change in storage. In this chapter, we will cover how discharge can be estimated using satellite-based observables, such as those from the newly launched Surface Water and Ocean Topography (SWOT) mission mentioned in Chapter 6.

7.2 Introduction

Discharge in a river is the volume of water passing through it over a unit of time and unit cross-sectional area. It is a key parameter for water management. The traditional and in-situ methods of estimating discharge are many, and they will not be covered here. The gist of these methods is as follows. Once discharge is accurately measured for various water levels in a river using in-situ means, a stage discharge relationship is usually established. This relationship can then be used against water level gauging, which is easier to measure round the clock, to provide a more continuous estimate of river discharge. One method of in-situ measurement of discharge is using an acoustic Doppler current profiler (ADCP) where a 3D or 2D field of velocity at a river cross-section can be derived to estimate the discharge. This is discussed later, in Section 7.3. Another, much older, method is to use current meters, where the river cross-section is divided into vertical sections and the average velocity estimated for each, then the discharge for each segment estimated and finally summed over (Figure 7.1).

We can also estimate discharge from space. Researchers have worked on this field for over two decades to develop methodologies for discharge estimation that can work purely using satellite data or in combination with ground data. One of the first and most intuitive approaches is to use existing hydrologic models with a river routing scheme to estimate discharge from surface runoff where precipitation events and continuous evapo-transpiration and groundwater seepage are also modeled. A hydrologic model is typically forced (fed) with meteorological forcing data that includes precipitation. If this precipitation data is from satellites using a combination of microwave (MW) and infrared (IR) sensor data (which we have covered in Chapter 5) and if all other data to force the model is also from satellites, then we can also estimate streamflow discharge almost entirely from space. This would be an indirect and complex way of deriving discharge entirely from satellite data, which is not the primary focus of this chapter.

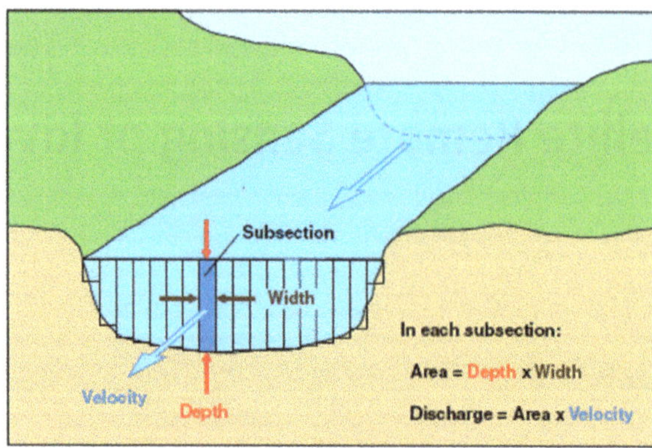

Figure 7.1 Measuring river discharge at a cross-section. Using a current meter, the average velocity is measured for each vertical segment and then summed over to derive the total discharge. The use of an acoustic Doppler current profiler (ADCP) is similar in concept except that the ADCP instrument can generate a much higher-resolution profile of velocity variation at a cross-section to allow for more accurate discharge estimation. [Credit: US Geological Survey]

However, we do need to be aware of this quite common approach and understand the limitations. The reliability and accuracy of such discharge derived from a hydrologic model would depend on the quality of the precipitation data, and the representativeness and calibration of the hydrologic model, among other factors. Here we will provide a quick overview of issues the reader should be aware of in estimating discharge via a hydrologic model.

7.2.1 Using Satellite-Based Precipitation and Hydrologic Model to Estimate Discharge

For starters, there are many satellite precipitation products out there that are produced operationally, as seen from Chapter 5. Readers can also refer to the International Precipitation Working Group at http://ipwg.isac.cnr.it/. Each product has different levels of error characteristics and uncertainty. Figure 7.2 shows how three different satellite precipitation products (generated during the era before the Global Precipitation Measurement (GPM) mission was launched) can estimate three different estimates of rainfall totals over a large region such as the USA. Hopefully, this difference narrows as the region of interest becomes smaller, but its impact on streamflow estimation via a hydrologic model can be quite stark (Figure 7.2).

When these same satellite precipitation products are plugged into the same hydrologic model calibrated against in-situ gauge data, the hydrographs simulated can be quite diverse. The variability, or uncertainty, observed in discharge estimation can often be attributable to the satellite precipitation error components of total bias, hit bias, false and missed precipitation (see Chapter 5). Figure 7.3 provides one such example for the Mississippi river using the widely used variable infiltration capacity (VIC) model (Hamman et al., 2018). Of course, we

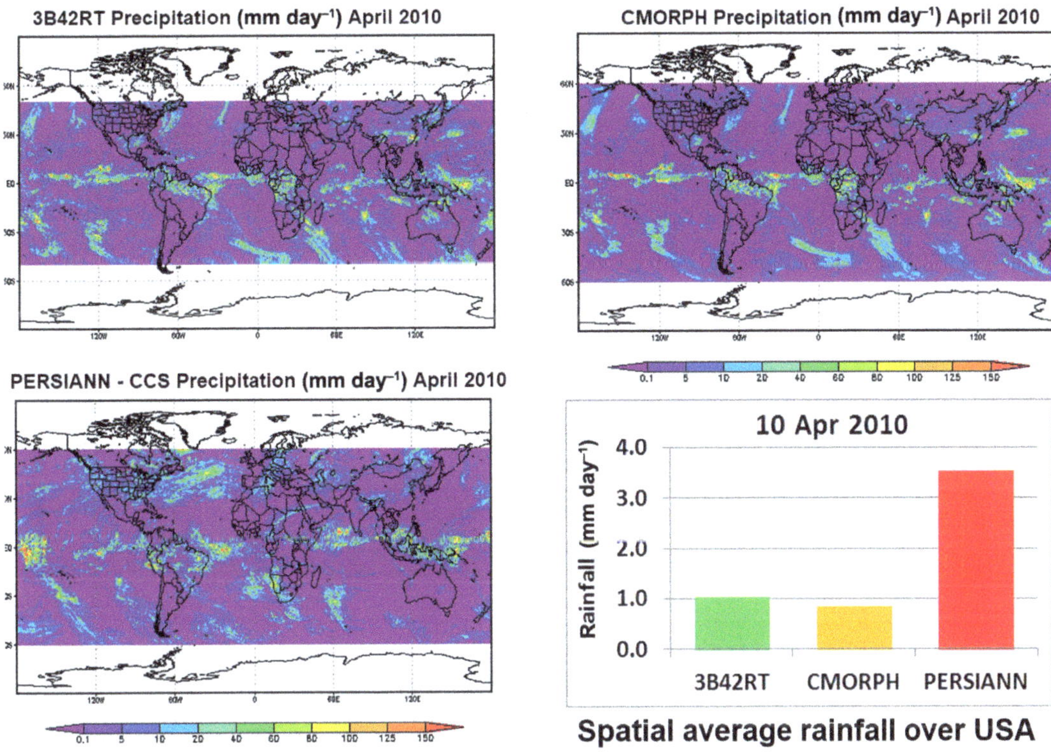

Figure 7.2 Multisensor precipitation products available globally, and their variability across products. The lower right panel shows this variation for a particular day when averaged over the contiguous United States (CONUS). [Credit: image courtesy of Dr. Abebe Gebregiorgis]

continue to make progress in this field of discharge estimation using hydrologic models through development of more accurate precipitation products, improved model calibration techniques, improved hydrologic modeling, and improved space-time resolution of hydrologic data. There is a vast body of literature on this that readers can refer to, such as Durand et al. (2021) and Lettenmaier et al. (2015).

7.3   Estimating Discharge from Space[1]

Remote sensing measurements can be used to estimate river flow in multiple ways. As mentioned in Section 7.2.1, one common approach uses satellite precipitation measurements to drive hydrologic models and predict discharge. Other approaches measure rivers directly, using remotely sensed river measurements to estimate discharge; these "direct" approaches are the subject of this section.

Temporal variations in river flow at a river cross-section drive accompanying changes in river elevation, slope, and river width, and these changes are large enough in many cases to

---

[1] Contributed by Dr. Mike Durand, The Ohio State University.

Impact of satellite rainfall uncertainty in stream flow simulation

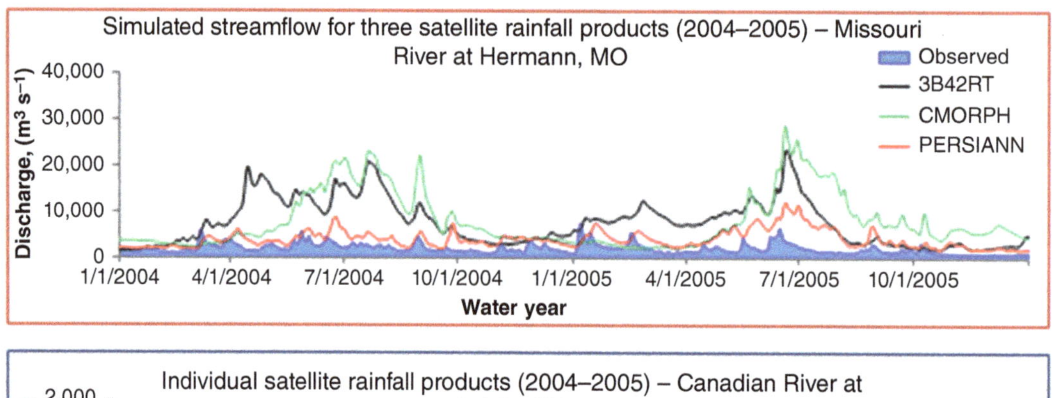

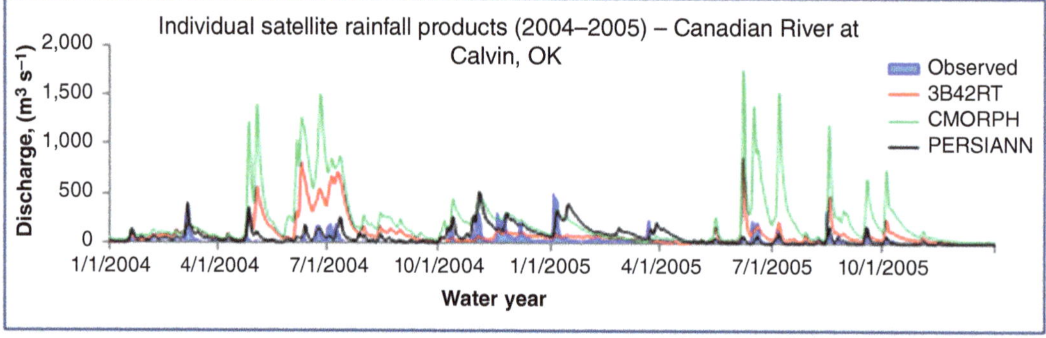

Figure 7.3 Simulation of river discharge (in $m^3\ s^{-1}$) using a hydrologic model forced with different types of satellite precipitation products. The uncertainty in discharge can be significant across different satellite precipitation products.

be visible from remote platforms. The paucity of in-situ discharge data globally (Hannah et al., 2011) has motivated research into remote sensing of discharge for decades, leading to literally hundreds of papers on remote sensing of river discharge. In this section, we thus only cite examples. For a more comprehensive review, readers can refer to Gleason and Durand (2020).

The process to compute river discharge from direct remote sensing measurements is similar in principle to calibrating a rating curve (Durand et al., 2023). First, a relationship between the remote sensing measurements must be established. Rating curves tend to refer to stage-discharge relationships. More generally, we can use the term "flow laws" to refer to any mathematical relationships between remote sensing measurements, parameters, and river discharge. Rating curves or flow laws in general require calibration, that is, estimation of empirical or physical quantities that appear in the flow law but cannot be measured remotely. Estimation of flow law parameters or calibration of rating curves typically proceeds by identifying a set of measurements of river discharge and remotely sensed measurements that occur at the same place and time. Parameters are then estimated using standard curve-fitting techniques. Second, once parameters are estimated, the flow law can be evaluated to compute discharge on each satellite overpass. We refer to these two steps as "flow law parameter estimation" and "discharge monitoring".

An exemplary use case is one where a well established set of in-situ river discharge measurements is available and overlaps with the remotely sensed measurements such that the two can be directly related to estimate flow law parameters, but there is no continuous in-situ measurement. For example, suppose a gauge is discontinued along a river, but that a set of discharge measurements are available that overlap in time with a spaceborne radar altimeter (for more on altimeters, see Chapter 6). Then the rating curve can be calibrated using these measurements, and the radar altimeter can be used to compute discharge forward in time, even though the gauge has been retired. This case study was explored, for example, by Birkinshaw et al. (2010) but can only be deployed in the situation where historical gauge data is available.

7.3.1 Theoretical Basis for Flow Laws

Estimates of river discharge and surface water storage changes generally require ancillary information beyond surface elevations alone. In-situ measurement of river discharge is time- and resource-intensive, and generally requires measurement of the full velocity profile of a river cross-section (see Section 7.2). Relating river discharge and water elevation has many decades of theoretical and empirical basis, and these are explored in turn here. Propagation of discharge downstream in an open channel can be approximated by the 1D Saint-Venant or shallow water equation (Saint-Venant, 1871):

$$S_f = -\frac{\partial Z}{\partial x} - \frac{\partial Y}{\partial x} - \frac{v}{g}\frac{\partial v}{\partial x} - \frac{1}{g}\frac{\partial v}{\partial t} \tag{7.1}$$

where S_f is the so-called friction slope, or slope of the energy grade line, Z is the river bathymetric elevation, Y is the flow depth, x is the distance downstream, g is the acceleration due to gravity, v is the flow velocity, and t is time. In practice, for large rivers, the two rightmost terms are generally negligible (Moussa and Bocquillon, 1996), leading to the so-called diffusion wave equation. Note that, by definition, the elevation of the water surface, H, is equal to the sum of Z and Y. Thus the diffusion wave approximation holds that $S_f = -\mathrm{d}H/\mathrm{d}x$.

It has long been known that S_f can be related to discharge; this is the basis of the Manning–Strickler equation (Dingman, 1984). For rivers with large width-to-depth ratios, Tinkler (1982) showed that the so-called wide-rectangular assumption can be invoked, leading to a convenient form of Manning's equation:

$$Q = \frac{1}{n}\left(H - Z\right)^{5/3} W \frac{\mathrm{d}H}{\mathrm{d}x}^{1/2} \tag{7.2}$$

where n is the so-called roughness coefficient, W is the river width, and $\mathrm{d}H/\mathrm{d}x$ is the river slope. Manning's equation is usually recommended for use with so-called uniform flow, where $S_f = -\mathrm{d}Z/\mathrm{d}x$, in Equation 7.1. For uniform flow, $\mathrm{d}Z/\mathrm{d}x = \mathrm{d}H/\mathrm{d}x$, enabling use of measurements of the water surface elevation in Manning's equation. Indeed, for spatially varying river depth, the surface slope $\mathrm{d}H/\mathrm{d}x$ can be used in a Manning's approximation of the diffusion wave equation. Thus, dH/dx represents the sum of the friction slope due to the bed slope, and due to the spatial variations of pressure forces (Dingman, 1984), as shown in Equation 7.2. Manning's equation has some physical basis, as it is based on the proportionality of friction forces to the square of velocity, leading to the ½ power of slope. However, the 5/3 power of $H - Z$ is empirical; moreover, the roughness coefficient, although it can in theory be related to roughness of the bed surface and sediment grain size, is in practice often treated as an empirical parameter based on look-up tables (Dingman, 1984).

Operationally, discharge is generally monitored using rating curves. A rating curve relates periodic measurements of river discharge to river elevation or "stage." Water elevations can be measured continuously by automated sensors, which (combined with rating curves) are used to produce time series of river discharge. Rating curves often take a power law form, such as:

$$Q = a\left(H - b\right)^c \tag{7.3}$$

where H is water elevation, and a, b, and c are empirical constants, determined by regression of periodic, simultaneous measurements of H and Q in situ. By comparison, the form of Equation 7.3 has some basis in Manning's equation, Equation 7.2; thus, the power law equation for rating curves has some basis in hydraulic theory. The Q measurements themselves are typically obtained via the ADCP techniques mentioned in Section 7.2. For some locations, rating curve parameters are changed seasonally owing to dynamic hydraulic conditions, changes in vegetation impact upon the flow, and other factors.

River width increases monotonically with water elevation, owing to the physical geomorphological processes that shape riverbeds. Thus, many researchers have constructed width–discharge rating curves as well (Pavelsky, 2014).

At in-situ gauges, water levels are typically produced hourly or daily. For many management applications, at least daily measurements are needed. For hydro-climatological and large-scale water balance science questions, however, sparser temporal sampling offered by satellite remote sensing can make a huge impact, especially for areas that are not well covered by the gauge network. Papa et al. (2012) have stated that river discharge estimated from remote sensing to within 15–20% of monthly or annual values is often achievable, and that this discharge estimates with this level of accuracy are useful for large-scale studies. This is inferior to the accuracy that can be achieved by in situ stream gages, where accuracies of 5% are attainable for ideal cases. Nonetheless, accuracy in the 15–20% realm is a benchmark that is both realistically attainable and of high value to basin- and continental-scale water balance estimates and also for large-scale water management decisions.

7.3.2 Flow Law Parameter Estimation

Efforts have existed for many years to extend flow law parameter estimation beyond calibrating rating curves using in-situ data. Indeed, most rivers globally are ungauged (Pavelsky et al., 2014). Thus, the "extended gauging" approach of Birkinshaw et al. (2010) discussed above cannot be applied in most places. This has motivated efforts to calibrate flow laws using modeled river discharge. For example, Getirana et al. (2009) calibrated rating curves between radar altimeters and large-scale rainfall runoff model predictions of river discharge. This effort neatly avoids the problem of lack of gauges in most parts of the world. However, the remote sensing discharge predictions when computed in this way will have accuracy limited to the accuracy of the model itself, which can be a significant limitation.

New approaches are being developed based on flow law parameter estimation in ungauged basins. Motivation for these approaches is the limited spatial coverage of the global gage network on one hand and the limited accuracy of global modeling on the other. Gleason and Smith (2014) pioneered methods to estimate river discharge from Landsat measurements alone. Similar methods have been explored in great depth in context of the recently launched SWOT satellite mission. These approaches are all built on the basic idea of mass-conserved flow law inversion. Essentially, a system of equations is invoked equating river discharge in neighboring locations along a river network. River discharge is substituted for the flow laws, leaving a system of equations with remote sensing measurements and flow law parameters. Analysis by Garambois and Monnier (2015) found that such systems of equations can be inverted under some circumstances. SWOT simultaneous estimates of water surface elevation, width, and slope are expected to make inversion of flow law parameters more tractable.

Durand et al. (2023) described approaches for estimate discharge using SWOT, including a set of six inversion-based algorithms running to estimate flow law parameters, and discharge produced at low latency following the parameter estimation period.

7.3.3 Discharge Monitoring

Remote sensing of discharge studies have used observations from dozens of platforms, and the nature of these remote measurements shapes what can be done in a particular application. As with any remote sensing application (such as precipitation in Chapter 5 and surface water storage estimation in Chapter 6), capabilities of a remotely sensed discharge are determined by spatial and temporal coverage and resolution, and by measurement precision.

Spaceborne remote sensing of discharge has the longest history with approaches that respond to river width or more generally to the river inundated area within a remotely sensed pixel. A wide range of sensor types has been used, including active MW (radar), passive MW, and visible and near-IR remote sensing. An important consideration is the minimum river width that can be sensed, which in turn is linked to how the technique is implemented, and to the sensor resolution. Most of these approaches measure river width directly, following methods pioneered by Pavelsky and Smith (2008). For example, the work of Feng et al. (2021) leverages automated river width measurements derived from visible-band imagery. Similar approaches are often used on radar imagery. Indeed, some of the earliest work on remote sensing of discharge measured river width from radar images of braided rivers and related those to discharge (Smith et al., 1995).

Measurements of direct river width are constrained by remote sensing pixel size (Allen and Pavelsky, 2018). However, not all approaches actually measure river width directly. Some approaches are simply derived from the lower-level remote sensing data products such as the brightness (or reflectance) of a pixel relative to a pixel on the river bank. This approach is equally applicable to passive MW measurements, as shown by Brakenridge et al. (2012). Even though rivers occupy only a small fraction of typical spaceborne passive MW pixels (which are at minimum several kilometers in spatial scale), the measurement can be responsive to changes in fractional inundation extent. Monitoring of discharge from passive MW measurements is now done routinely. Approaches that monitor discharge using river width or inundation must take into account the temporal revisit time of observations. As an example of typical tradeoffs, consider two sensor examples: moderate-resolution sensors such as MODIS, which provides daily observations at 500-m scale, and higher-resolution sensors such as Landsat, which provides 30-m observations every 16 days. This tradeoff is largely determined by orbital dynamics and the need for satellite imagers to achieve a critical duration of observation referred to as "dwell time" to have adequate observation precision. However, remote sensing of rivers is in a time of rapid expansion of data availability, including commercial options such as Planet and Worldview. For water management, the choice of sensor must be based on the needs of particular decision making or planning, the desired temporal frequency, river size, and the length of observation record needed.

Many studies have used measurements of river water surface elevation (WSE), rather than inundated area, as a basis for discharge calculation. Most such studies have used radar altimeters originally designed for measuring ocean elevations. Use of radar altimeter observations to estimate river discharge has been exemplified in the work of Kouraev et al. (2004), who analyzed a decade of TOPEX/Poseidon altimeter water level measurements near the mouth of the Ob' River in Russia to estimate discharge by developing rating curves using nearby in-situ measured river discharge. The median error value across two "virtual stations" for altimetry-derived daily discharge values was 675 m^3 s^{-1}, or 8% of mean daily discharge. The Ob' is a nearly ideal riverine target, with a main-channel width of approximately 3 km at this location. Figure 7.4 shows a discharge time series from Kouraev et al. (2004), spanning 1991–2002. Note that the in-situ and remotely sensed discharge time series are averaged to monthly values, in this case. There are many other studies showing similar results in other rivers.

There are a number of practical considerations in any inland water altimetric study, the first of which is water body size. Small water bodies are often not measured by current generation altimeters since the smaller the river's spatial extent, the less the chance to be overpassed by a nadir-looking altimeter. This applies both to current radar and laser altimetry. Current-generation radar altimeters have a relatively wide beam width; the radar signal in principle responds to a quasi-circular area with a radius on the order of several kilometers. However, water is far brighter than land at frequencies utilized for radar altimetry (e.g. the Ku band) at nadir, and thus reflects significantly more of the radar signal back to the sensor. Thus it is possible to obtain accurate estimates of water level variations in water bodies significantly smaller than nominal footprint size, via reprocessing the altimetric waveform (see Chapter 6; Section 6.3.1), which is known as "retracking" (Berry et al., 2005). However, it is possible for radar altimetry measurements of smaller water bodies to be contaminated by land topography, even after retracking. Altimetric studies for smaller lakes, and for rivers, have been described in the literature and will be summarized below. Another critical consideration for radar altimeters is spatial coverage: while imagers such as MODIS and Landsat achieve global coverage, nadir-looking altimeters measure only a small fraction of the globe, missing most rivers (Alsdorf et al., 2007). Additional practical considerations with all inland water altimetry include the role of ice cover.

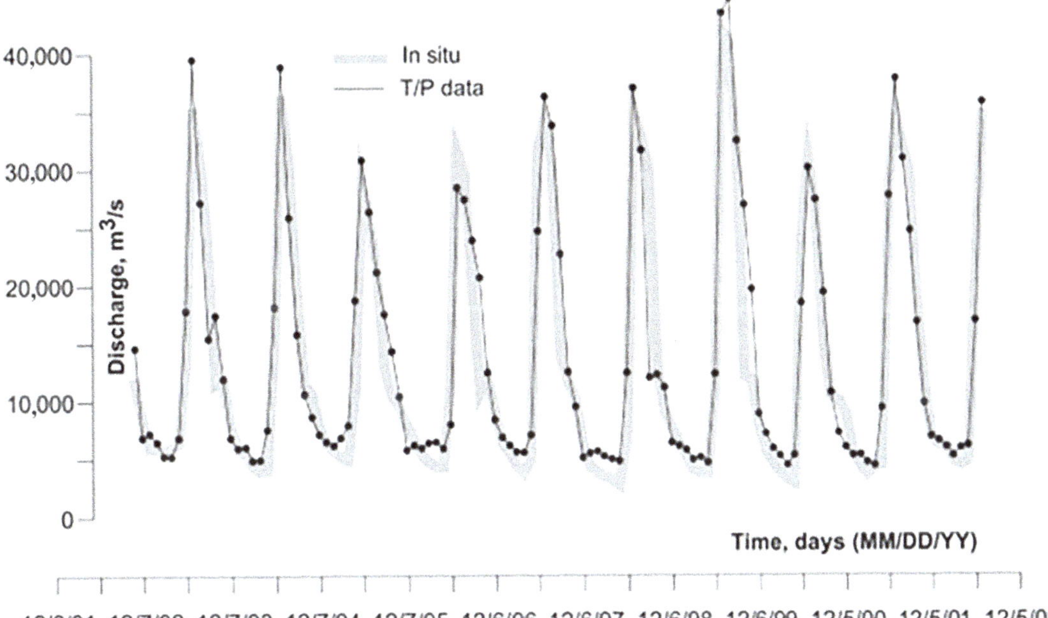

Figure 7.4 Mean monthly discharge on the Ob' River derived from TOPEX/Poseidon (T/P) data. [Taken from Figure 4 of Kouraev et al., 2004; reprinted from Elsevier Sciences with kind permission]

7.3.4 Summary and Perspective

This section has summarized efforts to relate measurable river phenomena (such as changes in river width and river water surface elevation) to river discharge. These efforts are firmly rooted in basic hydraulic theory and open the possibility to monitor river flow entirely from satellite remote sensing platforms. However, one of the most important drawbacks is that while the functional forms of the relationships are generally extendable from one location to another, each new location requires a new set of calibration parameters. In other words, there is no universal set of rating curve parameters for all the global rivers. The biggest limitation is thus the lack of in-situ discharge data to calibrate relationships between remote sensing measurements and discharge. However, once such relationships are calibrated, discharge can be monitored directly using remote sensors. Today, there are more and more sensors available that can be used to monitor discharge, including radar imagers, radar altimetry, and visible and NIR imagers, and from the major space agencies as well as commercial sources.

In Chapter 6, we briefly overviewed the SWOT mission that was launched in December 2022 and has started to provide simultaneous measurements of WSE, width, and slope globally. SWOT promises higher-precision WSE measurements than are currently available. These measurements have been hypothesized to provide the means to calibrate flow law parameters without the use of in-situ discharge measurements (Durand et al., 2023), an approach already demonstrated using measurements of width alone (Feng et al., 2021). These SWOT discharge measurements will be available for all rivers globally wider than 100 m, and perhaps as narrow as 50 m.

Case Study 7.1: Forecasting Downstream River Levels Using Upstream Altimeter Data at Transboundary Locations

Recall from Chapter 1 that in transboundary river basins, many countries sharing the same basin and located downstream are flood-prone and dependent on information on river levels in upstream nations. Yet often this information is unavailable in real-time, because of lack of either upstream measurements or information sharing mechanisms among riparian nations. So how can satellites help here? There are many ways to address this issue. One method could be to set up a hydrologic model for the entire river basin and force it with weather (precipitation) forecast information from global numerical weather models (see for example the GloFas system maintained by the European Union – www.globalfloods.eu/). Another could be a more data-based approach using data on river levels or discharge at the upstream transboundary location, as estimated by satellites, and trying to find the relationship with downstream river level or discharge at different lead times. This method will work for medium to large rivers where the travel time of a flood pulse is longer than the revisit time or frequency of the satellite-based upstream levels. And indeed, many regions and nations have adopted such a method in their real-world flood

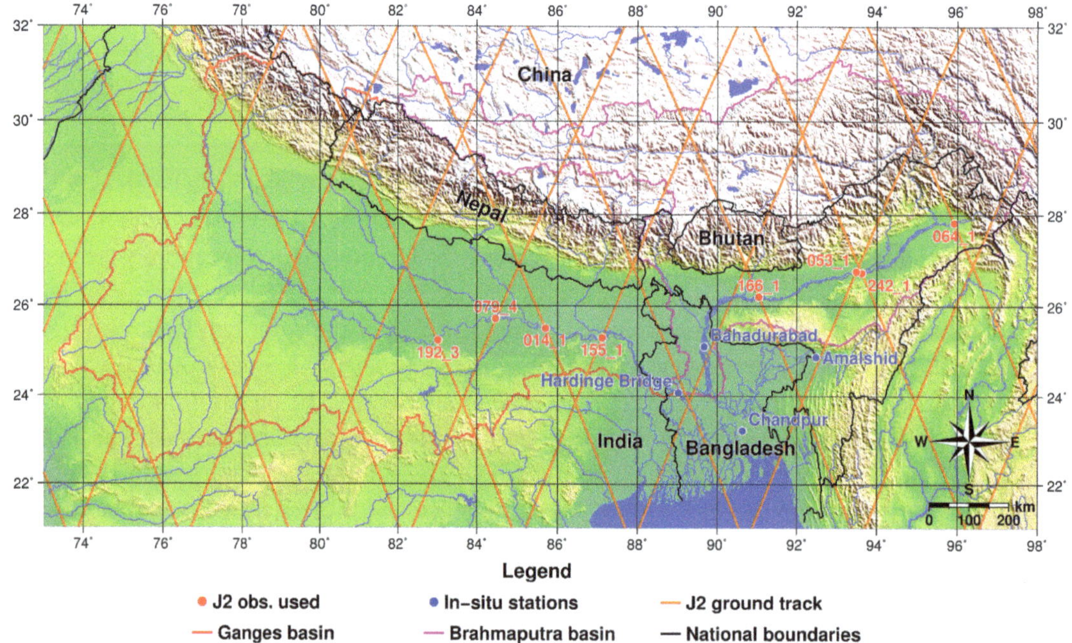

Figure 7.5 The Ganges–Brahmaputra transboundary river basin showcasing the upstream (India and Nepal) and downstream (Bangladesh) country dynamics for flood forecasting. For Bangladesh, the flood-prone nation downstream, to be effective in flood forecasting, information on river levels in upstream nations can be useful. In this case, the altimeter Jason-2 (J2) can be useful (see red circles). The number associated with each altimeter virtual station represents the pass number of the orbit. [© 2014 IEEE. Reprinted, with permission, from Hossain et al., 2014]

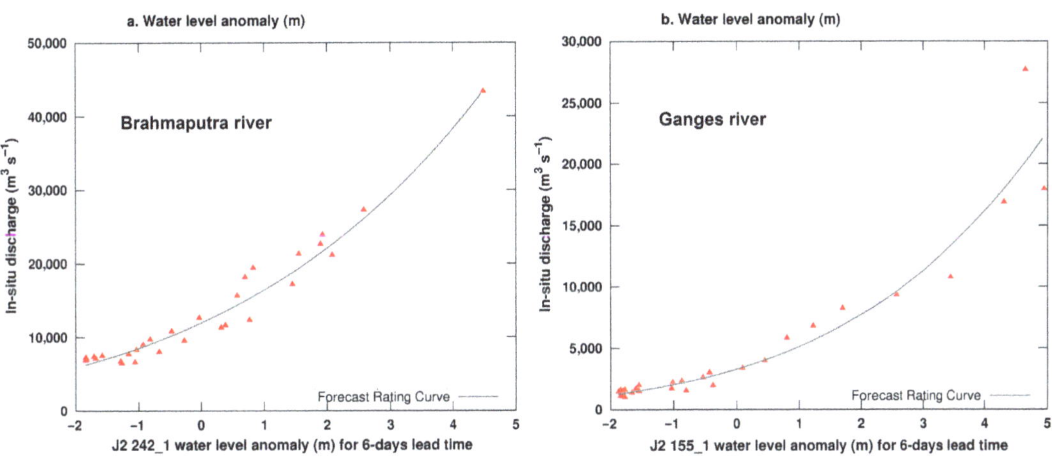

Figure 7.6 Example of the "forecasting" power of using upstream altimeter river levels (**a**, Brahmaputra; **b**, Ganges) from the Jason-2 (J2) altimeter to forecast discharge downstream at the border location of Bangladesh. The example here is for a 6-day lead time. The *x*-axis shows the location of the Jason-2 river locations (known as virtual gauge) in the upstream transboundary region of India. [© 2014 IEEE. Reprinted, with permission, from Hossain et al., 2014.]

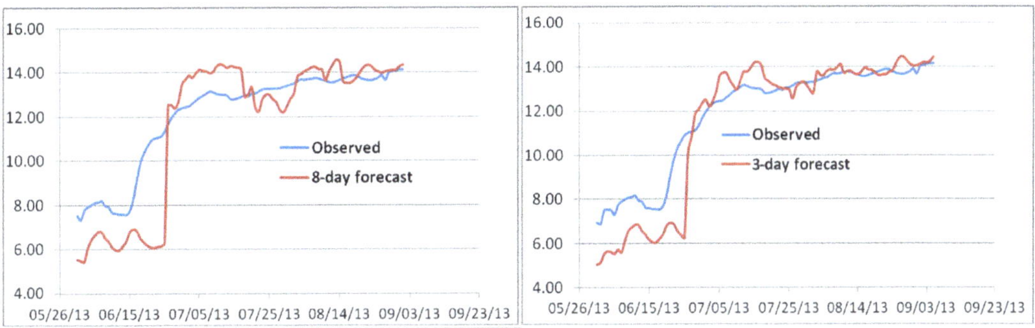

Figure 7.7 Satellite altimeter-based forecasting of downstream river level (*y*-axis in meters) inside Bangladesh (at a point on the Ganges River called Hardinge Bridge – see Figure 7.5, blue circle) using upstream river level measurements from Jason-2 in near real-time in Indian locations. Once the boundary condition is forecasted with a longer lead time, the Bangladesh water agency can then propagate these forecasts further downstream using a hydraulic or hydrologic model, and issue warnings for the public. [© 2014 IEEE. Reprinted, with kind permission, from Hossain et al., 2014]

forecasting systems, such as in Bangladesh (Hossain et al., 2014) and in the Mekong River basin (Chang et al., 2019).

The best way to understand the satellite data-based approach is to take the case of Bangladesh (the downstream and flood-prone nation) and India and Nepal (upstream and less flood-prone nations) of the Ganges–Brahmaputra River basins. For Bangladesh to improve its flood forecasting skill, information on upstream countries is critical as the country is small compared with

---

Case Study 7.1 (cont.)

the overall drainage area, where more than 90% of the flood water flowing into Bangladesh is transboundary. Figure 7.5 summarizes the situation clearly. A satellite altimeter mission called Jason-2 and its tracks where it crosses the main rivers of Ganges and Brahmaputra are also shown. The downstream boundary location for Bangladesh is shown in blue. If we could get our hands on Jason-2 river level data (or any current altimeter) in near real-time from upstream transboundary regions, then Bangladesh could potentially improve its forecasting lead time.

When we compare the Jason-2 river levels upstream in Indian locations with in-situ river level at the boundary location in Bangladesh lagged by a certain lead time, we clearly see the forecasting value of such altimeter river level data upstream. This is because it takes a few days for the flood pulse to show up at the borders of Bangladesh (Figure 7.6).

Now we can create such relationships between satellite altimeter-based river levels upstream and forecast river level/discharge downstream for various lead times until the relationship ceases to be tight. These relationships are typically called "forecast rating curves". Suppose we have such forecast rating curves for lead times ranging from 1 to 8 days, then we are in a position to forecast the river level or discharge at the boundary location inside Bangladesh up to that length of lead time (8 days). This is what Figure 7.7 shows during the monsoon season of May to September when the transboundary flooding happens.

7.4   Conclusion

In this chapter, we have learned the basics of discharge estimation using the traditional method. Next, we applied remote sensing variables related to a river's physical condition such as width and slope to understand how discharge can be estimated. The current methods for discharge estimation from space remain a work in progress as newer and better methods are developed. With the SWOT mission now flying, we can expect discharge estimation from space to experience a significant evolution. It is important for water management practitioners to be aware of satellite-based discharge products, as this may be the only source of information on how much water the river is carrying over time at many places around the world. In the next chapter, we will cover the topic of reservoir management by piecing together what we have learned in Chapters 4 (cloud computing), 5 (precipitation), 6 (surface water), and 7 (discharge).

EXERCISES: CHAPTER 7

Q7.1   What are the traditional methods of discharge measurement using in-situ monitoring?

Q7.2   Describe some methods by which satellite data can be used to estimate streamflow discharge.

Q7.3   Name and explain briefly some of the key methods of estimating discharge using satellite data only.

Q7.4   How would you forecast river levels downstream using upstream river levels?

Q7.5   What are the key data products of the SWOT mission for tracking water on land?

Q7.6   Describe briefly how SWOT will produce the river product.

Q7.7   What do you think are some of the key advantages of the SWOT mission compared with the current non-SWOT missions such as Landsat (visible) and Sentinel-1 (microwave synthetic aperture radar)?

REFERENCES

Allen, G. H. and T. M. Pavelsky (2018). Global extent of rivers and streams. *Science*, vol. 361, 585–587.

Alsdorf, D. E., E. Rodríguez, and D. P. Lettenmaier (2007). Measuring surface water from space. *Reviews of Geophysics*, vol. 45, RG2002. https://doi.org/10.1029/2006RG000197

Berry, P. A. M., J. D. Garlick, J. A. Freeman, and E. L. Mathers (2005). Global inland water monitoring from multi-mission altimetry. *Geophysical Research Letters*, vol. 32, https://doi.org/10.1029/2005GL022814

Birkinshaw, S. J., G. M. O'Donnell, P. Moore, et al. (2010). Using satellite altimetry data to augment flow estimation techniques on the Mekong River. *Hydrological Processes*, vol. 24, 3811–3825

Brakenridge, G. R., S. Cohen, A. J. Kettner, et al. (2012). Calibration of satellite measurements of river discharge using a global hydrology model. *Journal of Hydrology*, vol. 475, 123–136.

Chang, C.-H., H. Lee, F. Hossain, et al. (2019). A model-aided satellite-altimetry-based flood forecasting system for the Mekong River. *Environmental Modelling & Software*, vol. 112, 112–127. https://doi.org/10.1016/j.envsoft.2018.11.017

Dingman, S. L. (1984). *Fluvial Hydrology*. WH Freeman & Co.

Durand, M., A. Barros, J. Dozier, et al. (2021). Achieving breakthroughs in global hydrologic science by unlocking the power of multisensor, multidisciplinary Earth observations. *AGU Advances*, vol. 2, e2021AV000455. https://doi.org/10.1029/2021AV000455

Durand, M., C. J. Gleason, T. M. Pavelsky, et al. (2023). A framework for estimating global river discharge from the Surface Water and Ocean Topography satellite mission. *Water Resources Research*, vol. 59, https://doi.org/10.1029/2021WR031614

Feng, D. M., C. J. Gleason, P. R. Lin, et al. (2021). Recent changes to Arctic river discharge. *Nature Communications*, vol. 12, 6917.

Garambois, P.-A. and J. Monnier (2015). Inference of effective river properties from remotely sensed observations of water surface. *Advances in Water Resources*, vol. 79, https://doi.org/10.1016/j.advwatres.2015.02.007

Getirana, A. C. V. (2010). Integrating spatial altimetry data into the automatic calibration of hydrological models. *Journal of Hydrology*, vol. 387, 244–255. https://doi.org/10.1016/j.jhydrol.2010.04.013

Gleason, C. J. and M. T. Durand (2020). Remote sensing of river discharge: a review and a framing for the discipline. *Remote Sensing*, vol. 12, 1107.

Gleason, C. J. & L. C. Smith (2014). Toward global mapping of river discharge using satellite images and at-many-stations hydraulic geometry. *Proceedings of the National Academy of Sciences USA*, vol. 111, 4788–4791. https://doi.org/10.1073/pnas.1317606111

Hamman, J. J., B. Nijssen, T. J. Bohn, D. R. Gergel, and Y. Mao (2018). The Variable Infiltration Capacity model version 5 (VIC-5): infrastructure improvements for new applications and reproducibility. *Geoscientific Model Development*, vol. 11, 3481–3496. https://doi.org/10.5194/gmd-11-3481-2018

Hannah, D. M., S. Demuth, H. A. J. van Lanen, et al. (2011). Large-scale river flow archives: importance, current status and future needs. *Hydrological Processes*, vol. 25, 1191–1200.

Hossain, F., A. H. Siddique-E-Akbor, L. C. Mazumder, et al. (2014). Proof of concept of an altimeter-based river forecasting system for transboundary flow inside Bangladesh. *IEEE Journal of Selected Topics in Applied Earth Observations and Remote Sensing*, vol. 7, 587–601. https://doi.org/10.1109/JSTARS.2013.2283402

Kouraev, A. V., E. A. Zakharova, O. Samain, N. M. Mognard, and A. Cazenave (2004). Ob' river discharge from TOPEX/Poseidon satellite altimetry (1992–2002). *Remote Sensing of Environment*, 93, 238–245.

Lettenmaier D. P., D. Alsdorf, J. Dozier, et al. (2015). Inroads of remote sensing into hydrologic science during the WRR era. *Water Resources Research*, vol. 51, https://doi.org/10.1002/2015WR017616

Moussa, R. and Bocquillon, C. (1996). Criteria for the choice of flood-routing methods in natural channels. *Journal of Hydrology*, vol. 186, 1–30.

Papa, F., S. Biancamaria, C. Lion, and W. B. Rossow (2012). Uncertainties in mean river discharge estimates associated with satellite altimeter temporal sampling intervals: a case study for the annual peak flow in the context of the future SWOT hydrology mission. *IEEE Geoscience and Remote Sensing Letters*, vol. 9, 569–573.

Pavelsky, T. M. (2014). Using width-based rating curves from spatially discontinuous satellite imagery to monitor river discharge. *Hydrological Processes*, vol. 28, 3035–3040.

Pavelsky, T. M. and L. C. Smith (2008). RivWidth: a software tool for the calculation of river widths from remotely sensed imagery. *IEEE Geoscience and Remote Sensing Letters*, vol. 5, 70–73.

Saint-Venant, A. J. C. (1871). Théorie du mouvement non permanent des eaux, avec application aux crues des rivières et a l'introduction de marées dans leurs lits. *Comptes Rendus des Séances de Académie des Sciences*, 73, 148–154, 237–240.

Smith, L. C., B. L. Isacks, R. R. Forster, A. L. Bloom, and I. Preuss (1995). Estimation of discharge from braided glacial rivers using ERS-1 synthetic aperture radar: first results. *Water Resources Research*, vol. 31, 1325–1329.

Tinkler, K. J. (1982). Avoiding error when using the Manning equation. *Journal of Geology*, vol. 90, 326–328.

SUGGESTED READING

Gleason, C. J. and M. T. Durand (2020). Remote sensing of river discharge: a review and a framing for the discipline. *Remote Sensing* 12, 1107. https://doi.org/10.3390/rs12071107

8   Reservoir Management from Space

8.1   Chapter Overview

In this chapter, we will explore how reservoirs can be monitored from space for water management. Today it is possible to track the dynamic state of reservoirs at temporal and spatial scales of satellite remote sensing. This dynamic state comprises inflow, outflow, surface area, storage change, and evaporative losses of the reservoir. Most of these variables can be modeled using satellite data or directly estimated from satellite data. This chapter will introduce readers to the Reservoir Assessment Tool (RAT) that we have developed as an open-source and complete package for users to track reservoirs anywhere.

8.2   Introduction

First let us recall from Chapter 1 the ubiquity of reservoirs around the world and their growing numbers in regions that are developing. These regions have challenging circumstances for water measurement and data sharing across political boundaries. The challenges are many, but a few key ones that Chapter 1 mentioned are: (a) lack of in-situ measurement infrastructure; (b) lack of treaties to share data between different jurisdictions where water data and especially reservoir data is considered a national security risk; (c) difficult and remote terrain for measurement (e.g. high elevation, mountains); and lastly, (d) high cost of operation and maintenance of in-situ water measurement networks.

We will not be rehashing each of the above limitations that prove how satellite water data is a critical necessity to manage the multitude of the world's reservoirs. This chapter is not about reservoir management theory. The theory of reservoir management and applications is well established. It includes topics such as reservoir sizing, optimal allocation of water, capacity expansion plans, deriving rule curves, and standard operating procedures. There are many excellent books that readers can refer to for developing an understanding of the theory that forms the cornerstone of reservoir management in the twenty-first century. One such book that readers will find complementary to this chapter is *Water Resources Systems Planning and Management: An Introduction to Methods, Models and Applications* by Loucks et al. (2005).

This chapter assumes that readers are aware of the management theory for reservoirs or are more interested in understanding how satellite data can be applied to monitor and estimate a reservoir's dynamic state. Using satellite data, we will explore how we can explore and quantify the historical footprint of reservoir operations and predict downstream impact as well as

Figure 8.1 The key components that make up the reservoir's dynamic state: inflow (I), evaporation (E), precipitation over the reservoir (P), outflow (release and diversion) (O), and storage change (ΔS). Additionally, there is groundwater seepage, which is often ignored.

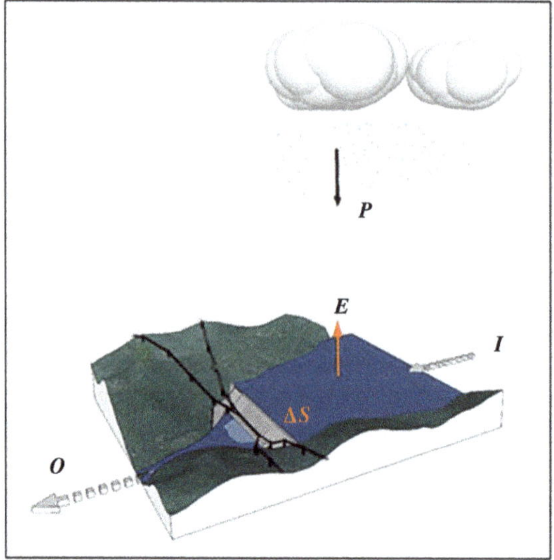

nowcast and forecast reservoir state. We will stress how all this can be done using publicly available satellite data to level the playing field for users and overcome the key challenges of water management outlined earlier in Chapter 1. Here, the reservoir state refers to the matrix of flux components with which a reservoir operation interacts. These are typically reservoir inflow, storage change, evaporation, and outflow (release and diversion) (Figure 8.1). In this chapter, we will ignore groundwater seepage and precipitation on the reservoir. However, those can be tracked if deemed important.

In the last two chapters (Chapters 6 and 7), we learned that satellite data can track inflow (discharge) either through the use of a calibrated hydrologic model forced with satellite precipitation data or by using satellite data directly. In addition, there is the potential of using discharge data created from the SWOT mission's surface water data. Similarly, storage change can be tracked using methods covered in Chapter 6 on surface water extent and/or elevation changes. Evaporation is mostly a weather-driven phenomenon dependent on surface area that can also be estimated using mostly satellite data. Finally, using simple mass balance and tracking as many of the input and output components as possible using satellite data, we can also track the amount of water over time that the reservoir has released (including lateral diversions) downstream (Equation 8.1).

8.2.1 Overview and Reservoir Mass Balance Approach

Satellite-based remote sensing data can be used to estimate reservoir outflow by employing a reservoir mass balance equation (see Equation 8.1), thereby completing the full monitoring of reservoir state (inflow, outflow, and storage change). For monitoring reservoir dynamics (fill, release, and storage change), this mass balance is the core component of most satellite-based reservoir frameworks. A schematic diagram of the mass balance concept is shown in Figure 8.2.

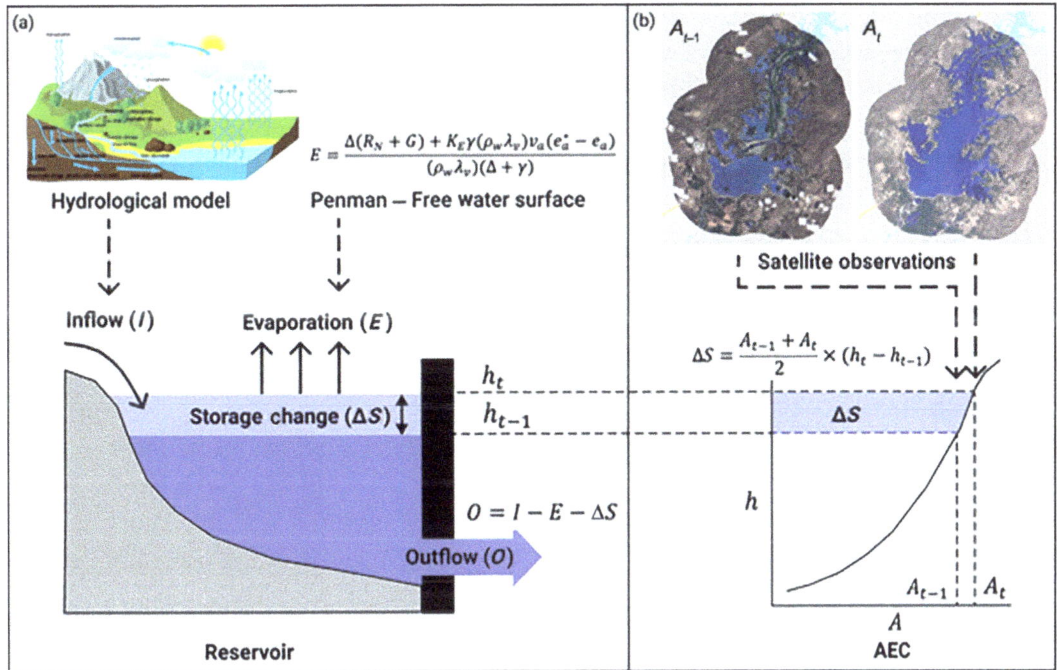

Figure 8.2 Concept of satellite data-based mass balance for reservoir monitoring. The reservoir parameters and corresponding satellite datasets used are as follows. **a**, Upper panel shows the hydrologic model that generates reservoir inflow and estimates reservoir evaporation, which is then used in mass balance to compute outflow (lower panel). **b**, Upper panel shows the satellite reservoir surface area estimation at two different times to compute storage change (lower panel). The storage change is then fed to the mass balance (a – lower panel). [After Das et al., 2022]

$$O = I - E - \Delta S \tag{8.1}$$

Here, the terms represent the following: O = outflow, I = inflow, E = evaporative loss, and ΔS = storage change (Figure 8.1). The term "outflow" is used as a proxy for "release," which can also include parallel diversions and other consumptive uses. Recall from Chapter 6 on surface water that the reservoir surface water extent areas A_{t-1} and A_t in Figure 8.2 can be extracted from visible/near-infrared (NIR) or synthetic aperture radar (SAR) imageries, while corresponding heights h_{t-1} and h_t can be extracted using an area–elevation curve (AEC). Alternatively, if surface elevations are available (such as from an altimeter), then surface areas can be estimated using the same AEC. ΔS can be calculated using Equation 8.2, which we have already covered in Chapter 6, while noting that this is also a data product from the SWOT mission.

$$\Delta S = \frac{A_{t-1} + A_t}{2} \times \left(h_t - h_{t-1} \right) \tag{8.2}$$

In order to carry out the above steps, a region of interest (ROI) for a reservoir first needs to be defined by following a reservoir size-dependent buffer distance. An example is shown in

Table 8.1. The ROI is then used to clip satellite observations on elevation (such as from SRTM) for preparing the area–elevation relationship and to extract the time series of surface water area. Previously, in Chapter 6, we showed examples of AEC for the Belo Monte Dam project on the Xingu River in the Amazon. Thus, in summary, the storage change ΔS of any reservoir can be computed using the reservoir water surface area or elevation time series and area–elevation relationship shown in Figure 8.2. Research has shown that this is a widely used technique that can yield acceptable skill (Biswas et al., 2021).

Next, meteorological observations, notably precipitation and land surface parameters such as land cover – all derived from satellite data – can be forced into a hydrologic model to derive reservoir inflow. An alternate approach is to directly estimate discharge at the inlet using methods outlined in Chapter 6. The inflow, evaporation, and storage change can then be used to infer the reservoir outflow using mass balance. Figure 8.2 outlines the overall procedure for tracking reservoir states using satellite data with the key equations that will be overviewed in more detail later.

Now that a big-picture conceptual overview of satellite-based reservoir state tracking has been provided, let us dive deeper into each component with the necessary details on datasets and specific techniques. We have built and made publicly available a comprehensive self-learning educational resource for readers who may be interested in learning how to operationally monitor reservoirs using satellite data at a region of their choice. This particular system of satellite-based tracking of reservoir state is called the "**Reservoir Assessment Tool**" or RAT for short. The resource can be accessed from the HELP menu on the following site: www.satellitedams.net. This menu (see Figure 8.3) provides readers with up-to-date access on three aspects of learning of satellite-based reservoir monitoring: (a) source code (if the reader wants to combine their knowledge of remote sensing and attempt to set up an operational system that tracks reservoirs); (b) documentation (hosted on http://ratdocs.io); and (c) tutorial

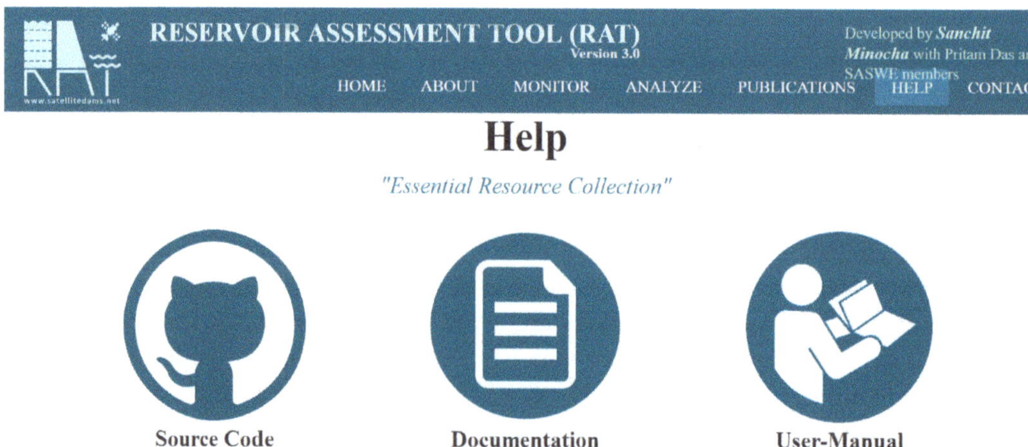

Figure 8.3 The online "HELP" resources to build literacy on satellite-based reservoir monitoring and setting up of the Reservoir Assessment Tool (RAT) system. For more details, see www.satellitedams.net or Minocha et al. (2023).

presentation that provides item-by-item and line-by-line explanation of code and setup of RAT at a region of the reader's choice. We strongly encourage readers to maintain access to these resources as they build literacy on this topic by reading the rest of this chapter.

8.3 Reservoir Tracking: The Details

8.3.1 Satellite Datasets

First, let's start with the reservoir's physical shape. This is often called "bathymetry" and can be summarized in the form of an area–elevation relationship as we have seen in Chapter 6. If local topographic maps are not available, then the land elevation dataset from the SRTM mission at 30-m resolution can be used. This is also something we have addressed, and provided some hands-on demonstrations using the Belo Monte Dam in the Amazon, in Chapter 6. Readers should go back to that section in Chapter 6 to refresh themselves about how digital elevation model (DEM) information is converted into a summary of surface water area for every elevation increment. There are also some hands-on educational resources in the HELP menu of www.satellitedams.net.

Now let's address the surface area of the reservoir, which is in a dynamic state of increasing or decreasing when a reservoir is storing or releasing water (or losing it through evaporation), respectively. Depending on the nature of bathymetry, reservoirs may show a modest to no change in surface area between two successive satellite observations when experiencing storage or release. Alternately, reservoirs may experience significant change that is detectable by the satellite sensors. For the former case, many reservoirs may be of a rectangular or deep and narrow gorge-type bathymetry where storage and release manifests in change in water surface elevation, which is detectable by radar altimeters or the SWOT mission. For other cases, reservoirs may experience changes in surface area but not to a level that is significantly larger than the satellite's pixel resolution. Whatever the reasons and issues are, readers should ponder a little on how exactly their reservoir of interest manifests storage or release, and what that means for the satellite's native resolution (spatial and temporal) of sampling (Figure 8.4).

The three common satellite sensors for estimating water extent–area time series are: (1) the Landsat series (currently 8 and 9 for operational activities); (2) Sentinel-1 SAR, which is a C-band synthetic aperture radar; and (3) Sentinel-2, which is a multi-spectral imager.

Although the topic of hydrologic modeling is beyond the scope of this book, let us mention a few more satellite data products one can use for setting up a hydrologic model to predict reservoir inflow. This inflow would be naturalized and would need to be adjusted for any regulation by upstream structures (Das et al., 2024). For a gridded hydrologic model such as Variable Infiltration Capacity (VIC) (Hamman et al., 2018), the key datasets needed are the FAO Harmonized World Soil Database, US Geological Survey Global Land Cover Characteristics (GLCC), and Land Cover Database. Satellite-based precipitation data to force the hydrologic model can be from Global Precipitation Measurement (GPM) products such as IMERG (Huffman et al., 2020). Other weather parameters such as maximum and minimum

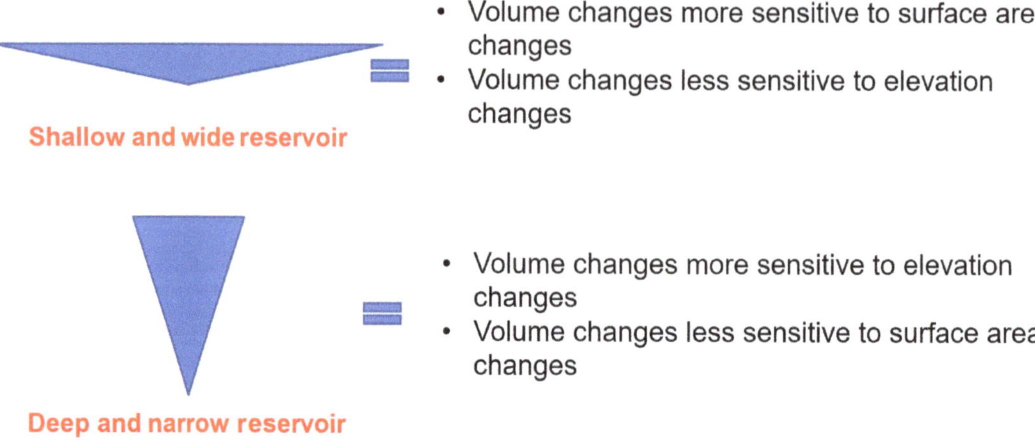

Figure 8.4 A generalized view of how reservoirs may experience storage or release/loss via surface area changes or elevation changes. A combination of both is likely for reservoirs with a bathymetry that is neither very shallow nor very deep and narrow.

temperature, and average wind speed, can be obtained from NOAA NCEP/Climate Prediction Center meteorological datasets.

8.3.2 Calculating Reservoir Storage Change

The method to follow for reservoir storage is already shown in Figure 8.2. Major steps in applying the mass balance equation (Equation 8.1) are (a) ROI generation, (b) AEC extraction, (c) time-series processing of water extent area, (d) storage change calculation, (e) simulation of reservoir inflow from the hydrological model (or from another source) and evaporation from the reservoir, and (f) reservoir outflow calculation. Aside from hydrologic modeling (items e and f), most components can now be executed in the cloud using Google Earth Engine (GEE) or some equivalent cloud system maintained by space agencies such as NASA. This can minimize internet bandwidth needed for downloading large datasets. Alternatively, if the datasets are available locally at the user's side, the same steps can be completed locally as well. In the following, we will overview the details assuming the reader has built some literacy of cloud computing using GEE from Chapter 4.

8.3.3 Region of Interest Delineation

As a starting point, the ROI of reservoirs can be defined according to the polygon defined in the GranD database (note that GranD is a database of dams with various information). The polygon area can be used to classify reservoirs into seven distinct classes to identify the appropriate buffering distance to create the ROI. Before deciding the buffering distance for each class, the frequency of occurrence map of the Global Surface Water Dataset (GSWD) prepared by Pekel et al. (2016) or any other "visual" source should be used for visual comparison with nearby reservoirs. By following the maximum water extent of the GSWD dataset, Table 8.1 provides a recommended buffer distance.

Table 8.1 Reservoir class, buffering distance, and computational scale in GEE

Reservoir surface area (km^2)	Buffering distance (m)
<2.0	500
2–10	750
10–50	1,000
50 200	1,250
200–500	1,500
500–1,000	1,750
>1,000	2,000

8.3.4 Area–Elevation Curve Extraction

Using the delineated ROI mentioned in Table 8.1 (and also demonstrated in Chapter 6), and SRTM DEM data, the AEC can be derived in two steps. First, the ROI of the selected reservoir should be used to clip the SRTM DEM. The SRTM DEM elevation is then used to generate the area–elevation relationship, which, since SRTM flew in February 2000, is valid for elevation above the water surface at the time of February 2000 (for reservoirs pre-dating 2000). The histogram of SRTM DEM is first populated to count the number of cells corresponding to each of the elevation data (see Chapter 6, Figure 6.21, for an example on the Belo Monte Dam reservoirs). The area of individual elevation is then calculated and incremented to get incremental area. The steepest slope of the AEC is used to identify elevations corresponding to reservoir surface areas. Areas that are less than the area within the reservoir with a water surface elevation are considered to be satellite noise and discarded from further analysis. Next, the relationship developed in the first step is extrapolated to the near-zero surface area in order

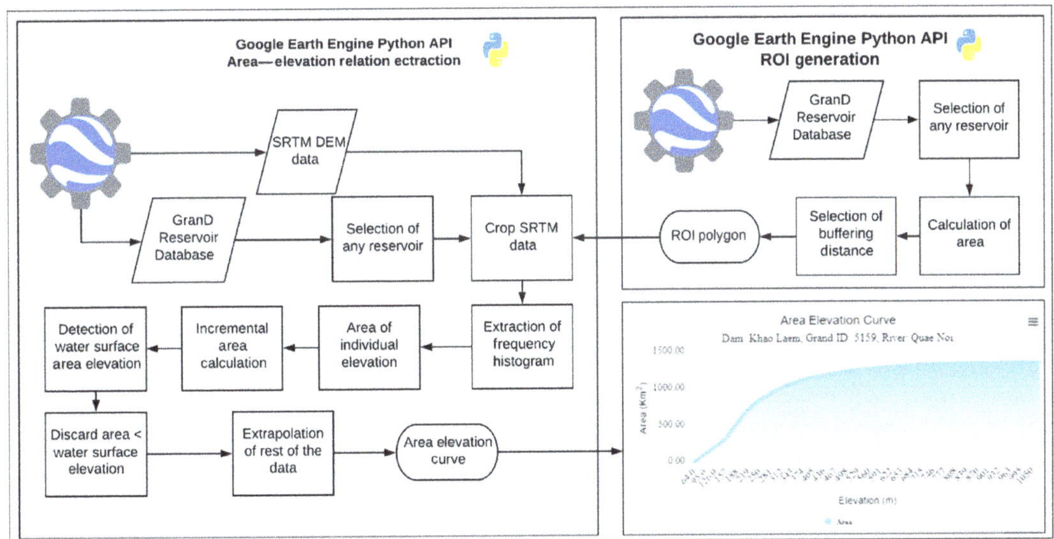

Figure 8.5 Reservoir ROI polygon delineation and AEC extraction procedure. [After Biswas et al., 2022]

to complete a virtual area–elevation relationship for elevations lower than the SRTM-observed water surface. Finally, these two area–elevation relationships (one data driven and the other extrapolated) are merged to create the complete area–elevation curve. The whole methodology of AEC development can be summarized in Figure 8.5 as a sequence of steps that a cloud computing platform like GEE would perform.

8.3.5 Surface Water Area Extraction

The surface area time series and the AEC are prerequisites to calculating the storage change of any reservoir. First, the ROI polygon and AEC need to be prepared by following the approach described earlier. All satellite imagery scenes need to be first filtered using a predefined date window and the ROI polygon, and then clipped using the ROI. In the case of Sentinel-2 and Landsat, which use visible and NIR wavelengths, cloudy pixels need to be removed from the ROI region of the scene. The area of scenes is typically calculated and areas that are less than 80% of the total (ROI) are filtered out (after the removal of cloudy and partially covered scene). In the case of Sentinel-1, scenes are filtered based on polarization, look angle, and date window, and pixels with less than –14 dB backscatter value can be treated as water. This threshold of –14 dB can be somewhat arbitrary, and users can do a sensitivity analyses to find an optimum threshold. Nevertheless, the threshold should be a very low number as the goal is to look for regions that are "dark" due to specular reflection. For Landsat and Sentinel-2, different index-based methods can be applied, such as Normalized Difference Water Index (NDWI),

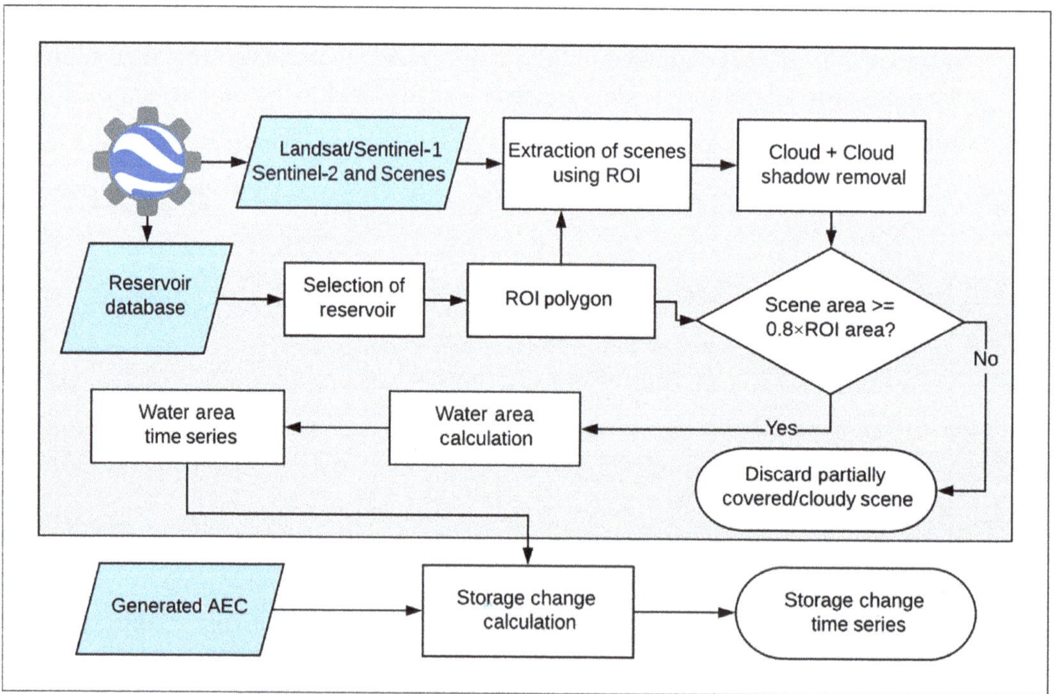

Figure 8.6 Workflow for surface water area time series and storage change. [After Biswas et al., 2021]

Modified Normalized Difference Water Index (MNDWI), or Dynamic Surface Water Extent (DSWE). The extracted time series are then used to calculate the storage change time series. Figure 8.6 shows the implementation in GEE cloud using satellite imageries that provide surface area of the reservoir at two successive times. Note – it is important for readers to remember that altimeters can also be used. In place of surface area, successive altimeter elevations of a reservoir, if available, could have yielded corresponding surface area and then the storage change (assuming a trapezoidal cross-section). Because altimeters can measure elevations with the precision of tens of centimeters (and SWOT will also be an improvement), we have found altimeter or elevation-based storage change to be more accurate than the same obtained via satellite-based surface area. The main reason for this advantage in accuracy is that no matter what the shape of a reservoir, it is always experiencing some elevation changes, even down to centimeter-scale variations that are often detectable by altimeters.

8.3.6 Calculation of Reservoir Evaporation

Evaporation can be modeled explicitly using the Penman equation with inputs also derived from satellite data. Also known as the combination equation, the Penman equation for free water surface (Penman, 1948) is defined as follows.

$$E = \frac{\Delta(R_N + G) + K_E \gamma (\rho_w \lambda_v) v_a (e_a^* - e_a)}{(\rho_w \lambda_v)(\Delta + \gamma)} \tag{8.3}$$

where E is the evaporation; Δ is the slope of the saturation vapor pressure–temperature relation; R_N is the net incoming radiation; G is the ground heat flux, which can be assumed to be 0 for daily calculations; K_E is the mass transfer coefficient, γ is the psychrometric constant; ρ_w is the water density; λ_v is the latent heat of vaporization; v_a is the wind speed; and e_a^* and e_a are the saturation and actual vapor pressure for the air temperature. The volumetric evaporative loss (dimensions $[L^3 T^{-1}]$) is estimated by multiplying the evaporation $[LT^{-1}]$ with the latest observed surface area of the reservoir. As an example, we show reservoir evaporation calculated as a function of inflow for many reservoirs located in the Mekong River basin (Figure 8.7). What we see is that, for the most part, reservoirs in humid tropical regions have negligible evaporative losses, unlike arid/semi-arid river basins, and thus the evaporation modeling is not that critical for the outflow calculation from mass balance. The same cannot be said for reservoirs in dry climates, such as Lake Mead in the United States or Lake Nasser in Egypt.

8.3.7 Outflow and Outflow Forecasting

The outflow in the form of release from the reservoir or diversion can be calculated using simple mass balance shown in Figure 8.2 using Equation 8.1. Here, we are assuming that the user will have access to reservoir inflow either from hydrologic modeling (driven by satellite datasets) or from in-situ measurements.

In addition to providing nowcast and historical (hindcast) estimates of outflow using the methods described above, we can also forecast reservoir outflow for a certain lead time

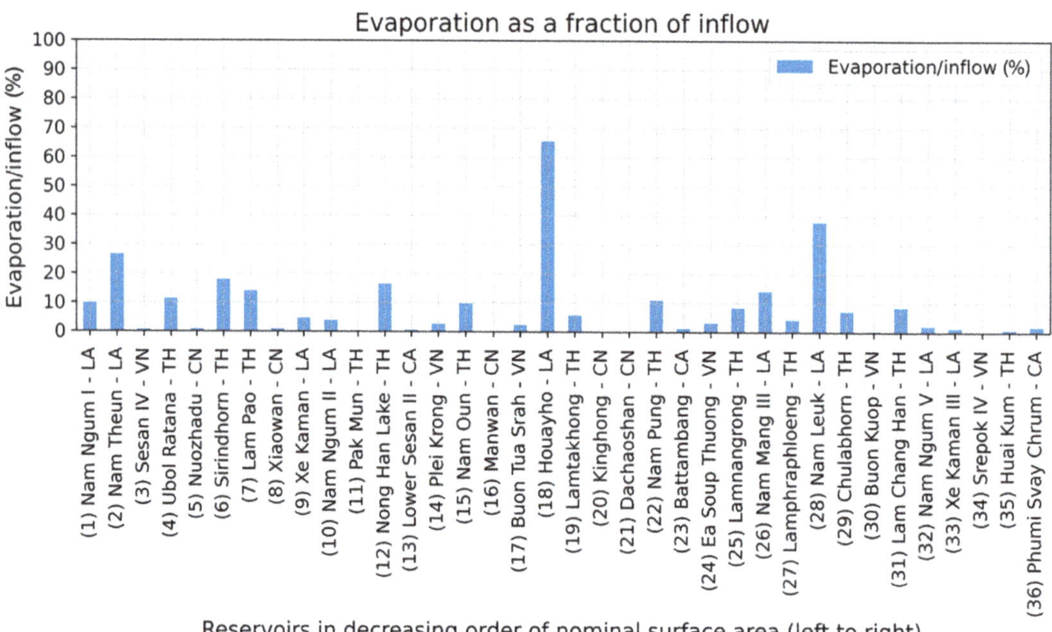

Figure 8.7 Average monthly evaporative losses compared to average monthly inflow at 36 selected reservoirs in the Mekong River basin. The suffixes denote the country where the reservoirs are located – CA: Cambodia; CN: China; LA: Lao PDR; TH: Thailand; and VN: Viet Nam. [After Das et al., 2022]

(at weather scales). The way to forecast reservoir outflow hinges on the forecasting of the inflow, and the likely response of the reservoir for that inflow and given the reservoir's initial storage. The likely response is obviously a human decision, and thus any forecasting of outflow will have to be conditioned on assumptions of the likely reservoir operating response. One such likely response can be summarized in the form of reservoir operating rules, which are basically a summary of how much a reservoir would store (and release) at a given time of the year, assuming expected inflow was climatologic and the objectives of the reservoir were maximized. Let us not get too much into the water management theory here, as the reader can refer to the book of Loucks et al. (2005) to learn more. Rather, for readers, let us just assume that they can explore various scenarios of storage and release to generate scenario-based outflow forecasts.

First, the forecasted meteorological conditions – minimum and maximum temperature, wind speed, and precipitation – that a hydrologic model like VIC would need can be weather forecast data from systems such as the Global Forecasting System (GFS). This way, the inflow can be forecasted using the hydrologic model. The forecasted evaporation can also be estimated using Equation 8.3 with forecasted meteorological conditions obtained from the GFS dataset.

When using the RAT software, we have two options for forecasting of outflow – (1) a satellite-derived rule curve by averaging satellite-estimated historical reservoir storage (Biswas et al., 2021); or (2) a user-defined rule curve or time-varying target storage. The rule curve can already be "inferred" by averaging the historical storage change patterns over a multiyear period for each month from remote sensing datasets, and then inferring the most likely storage target (S) that the reservoir is typically at based on past operation, as a fraction of the maximum storage (S_{max}). Reservoir outflow forecasting can be quite powerful for a water manager if it is skillful, as it can provide a range of likely scenarios to expect in the near future for which early decisions on storing/releasing and reallocation of water may be warranted. In this chapter, we have provided a hands-on tutorial opportunity to forecast outflow for a reservoir as part of learning to set up the RAT at a river basin.

8.4 Reservoir Rule Curves (or Operations Curves) from Satellite Data

As mentioned in the above section, the reservoir storage changes estimated from the entire record of a satellite data allow one to also compute reservoir total volume by identifying the lowest level below which the reservoir does not fall and using the AEC curve. If such volumes are grouped by month, and the average reservoir volume for each month is calculated using multiple years of data, then an approximation of the reservoir operations (rule) curve can be inferred. Figure 8.8 shows the average and range of estimated reservoir operations curves, normalized by their maximum storage so that comparisons can be made between reservoirs of different volumes. Although there is variation among the different reservoirs, as evident by the large range, a dominant trend can be gleaned, of higher volume at the beginning of the dry season, followed by decline into the wet season. At the end of the wet season, the reservoirs fill again. This trend is highlighted by the average normalized operations curve, which is shown in black.

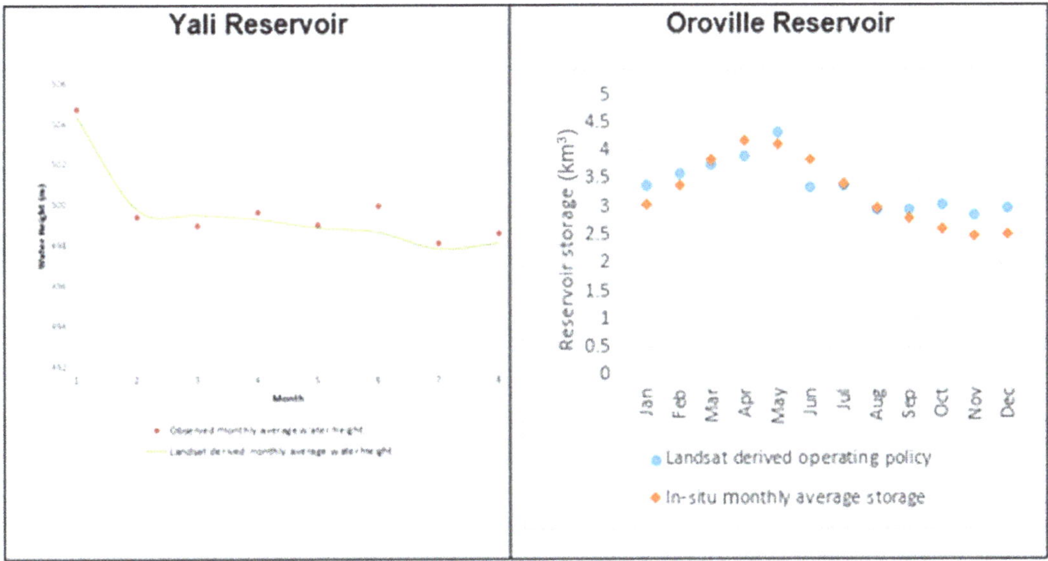

Figure 8.8 Rule curves based on reservoir storage change estimated from satellite data – known more appropriately as "inferred" rule curves. Left – Yali reservoir in Vietnam; right – Oroville reservoir in California, USA. These inferred rule curves were generated using Landsat data. [Reprinted with kind permission from American Geophysical Union; based on Bonnema and Hossain, 2017]

8.5   Tracking Reservoir Residence Time from Satellite Data

Once we set up a reservoir monitoring system for a river basin, such as RAT for the Mekong River, there are various things we can do with this capability. One such water management application is the tracking of residence time, or the average age of water behind the dam before it evaporates or is released from the reservoir. This average length of time for which water stays behind a reservoir can be an important parameter for water management applications. Residence time can be a controlling factor for sedimentation in the reservoir, thermal stratification (cooling at bottom layers during warm season), and poor water quality conditions such as harmful algal blooms. Satellite data can be used to derive an approximation of reservoir residence time. The key ingredients for estimating residence time are inflow (I), storage (S), and outflow (O). The driving concept behind the residence time calculation is that these terms obey a mass balance for every reservoir, given by Equation 8.4, where ΔS is the change in storage over a duration of time (let's ignore evaporation for now).

$$O = I - \Delta S \tag{8.4}$$

With the water in reservoirs obeying mass balance, several basic assumptions can be made in order to simplify the calculation of reservoir residence time:

1. A water parcel that enters the reservoir over a specific duration does not mix with other water parcels within the reservoir.
2. Water parcels exit the reservoir in order from oldest to newest.

Note that this is equivalent to assuming the reservoir behaves as a plug flow reactor. With these two assumptions, the length of time a parcel of water spends in the reservoir can be identified. Generally speaking, a monthly time step or shorter can be used, so the residence time of the inflow entering a reservoir during one month is the amount of time until all of the water is released that was in the reservoir just after the monthly inflow parcel arrived. This is expressed by Equation 8.5, where t_0 denotes the time step of interest, $\theta(t_0)$ is the residence time of the water entering the reservoir at time t_0, and t_R is the time step in which the water that entered at time t_0 is released from the reservoir. This time of release is calculated by summing the amount of water exiting the reservoir in each time step, $O(t)$, beginning at t_0 and ending when the sum is equal to the volume of the reservoir, $V(t_0)$ plus the amount of inflow, $I(t_0)$, at the time step of interest, t_0.

$$\theta(t_0) = t_R - t_0$$
$$I(t_0) + V(t_0) = \sum_{t}^{t_R} O(t) \tag{8.5}$$

To average the residence time of a single reservoir across time, an inflow weighted approach is used, as shown in Equation 8.6.

$$\theta_{t\text{-avg}} = \sum_{i=1}^{n} \frac{I_i \theta_i}{I_{avg}} \tag{8.6}$$

where θ_{avg} is the residence time averaged over some period of time, θ_i and I_i are the residence time and inflow, respectively, of time step i, and I_{avg} is the average inflow of the reservoir over the entire time period.

In order to estimate the collective effects of these reservoirs on the residence time of the basin as a whole, a similar averaging technique is used, shown in Equation 8.7.

$$\theta_{\text{b-avg}} = \sum_{j=1}^{n} \frac{I_j \theta_j}{I_{\text{avg}}} \tag{8.7}$$

where θ_{avg} is the basin-averaged residence time, θ_j is the time-averaged residence time of reservoir j, I_j is the time-averaged inflow of reservoir j, and I_{avg} is the average inflow of all reservoirs (i.e. the average of all I_j).

8.6 Flow Alteration by Reservoirs Using Satellite Data

Another thing we can track is that of downstream flow alteration by reservoirs upstream. If the dam does not act as a control structure, then there is no alteration of flow as water flows freely from upstream via the reservoir to downstream. Such a situation probably occurs during extreme cases when the reservoir is already full and the incoming flow is very high, so that the reservoir has no choice but to keep releasing what comes in. However, most of the time, reservoirs are actually altering the flow as water flows from upstream to downstream via the dam.

In essence, reservoirs alter flow via storage and release that do not follow natural physical rules but rather the man-made rules for water supply, flood control, hydropower, and other needs. We can use satellite data to track this flow alteration by reservoirs. Reservoir-imposed flow alteration (FA) can be defined here as the percent difference between I and O. Substituting the mass balance between O, I, and ΔS (Equation 8.1), shows that this is equivalent to the ratio between ΔS and I, shown in Equation 8.8.

$$FA = \frac{O - I}{I} = \frac{-\Delta S}{I} \tag{8.8}$$

Note that positive FA indicates that the streamflow was increased by reservoir operations and negative FA indicates streamflow was decreased by reservoir operations. When averaging FA across longer timescales, the result should be close to 0, owing to the principle of mass balance employed here. Thus, to derive meaningful information about a reservoir's overall effect on downstream flow, the absolute value of FA should be used when averaging. This averaging process is inflow-weighted, similarly to residence time, shown in Equations 8.9 (time average) and 8.10 (basin average).

$$FA_{\text{t-avg}} = \sum_{i=1}^{n} \frac{I_i FA_i}{I_{\text{avg}}} \tag{8.9}$$

$$FA_{\text{b-avg}} = \sum_{j=1}^{n} \frac{I_j FA_j}{I_{\text{avg}}} \tag{8.10}$$

where $FA_{t\text{-avg}}$ is the time-averaged flow alteration for a single reservoir, FA_i is the flow alteration of the reservoir at time step i, $FA_{b\text{-avg}}$ is the flow alteration averaged across the entire basin, FA_j is the time-averaged flow alteration of reservoir j, and all other quantities are as previously defined.

As an example for the heavily dammed Mekong River basin, Figure 8.9 displays how the resulting residence times, as derived entirely using satellite data and assumptions of a plug flow reactor, are distributed throughout the basin. Figure 8.9 also shows the resulting average absolute flow alterations for each reservoir in the Mekong. In general, using satellite data, a water manager can see that the residence times can range from 0.09 years to 4.04 years while the flow alteration ranges from 11% to 131%. Such information can be powerful in places where the information is absent or not measured using ground methods. For example – the reservoir with a higher residence time is likely a strong candidate for losing active storage through sedimentation management and can experience greater thermal stratification. In ecosensitive aquatic habitats during particular times when flows should remain natural and bountiful (such as during fish reproduction), reservoirs resulting in higher flow alteration could be candidates for minimizing this disruption at times critical for fish survival.

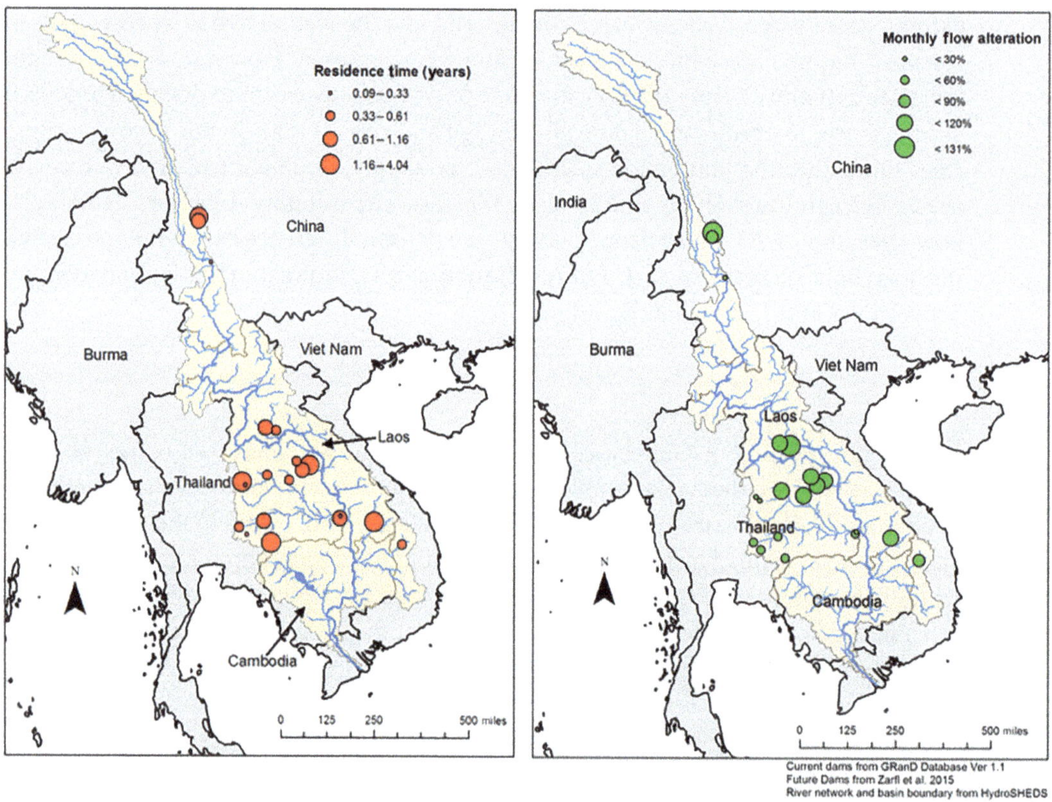

Figure 8.9 Residence time (left panel) and flow alteration (right panel) of major reservoirs of the Mekong River basin as derived from satellite data. [After Bonnema and Hossain, 2017]

8.7   Does Reservoir Shape Matter?

For what type of reservoirs do the estimates of surface area from current satellite missions work best? Large ones, small ones? Does the shape of reservoir relative to total areal size play a role in accurately detecting surface area and estimating storage change? It turns out that the size and shape of the shoreline of reservoir can play a role in careful estimates of storage change with areal estimation methods using visible, NIR, and SAR from satellites. The inaccuracy in tracking storage change trickles down to estimation accuracy of reservoir outflow. So size and shape of reservoir do matter, and it is good to be aware of this issue.

Biswas et al. (2021) used a test sample of 77 reservoirs in India to investigate the impact of reservoir size and shape on accuracy of storage change estimation. For shape, reservoirs were classified according to the ratio of nominal surface area (A) to the perimeter (P), here termed the irregularity index (A/P). Based on area–perimeter ratio, reservoirs were classified as highly irregular, very irregular, irregular, regular, very regular, and highly regular for this sample of 77 reservoirs (Table 8.2). Three reservoirs, each with a distinctive irregularity index, are shown in Figure 8.10 to illustrate the differences in reservoir shapes based on the irregularity index.

Figure 8.11 summarizes how shape of a reservoir plays a role in accuracy of storage change estimation of reservoirs using areal estimation methods. The larger the nominal surface area of the reservoir, the lower is the normalized RMSE (i.e. higher the accuracy) of satellite-based areal estimation of storage change. However, the size dependency is not exactly linear, and this is because the shape of the nominal reservoir shoreline plays a more definitive role in accuracy of surface area estimation and consequently storage change of reservoirs (see Figure 8.11 in relation to Figure 8.10). So we see that shape does matter in estimating storage change based on estimation of the reservoir surface area. The undermining role played by shape using areal estimation methods by visible, NIR, or SAR can be mitigated with the use of altimeters whenever they are available (note: the spatial coverage of altimeters for the world's reservoirs is significantly lower than that of NIR or SAR satellites that measure area). Fortunately, this is where SWOT can come to our rescue, by providing comprehensive spatial coverage and simultaneous measurement of elevation and area of the reservoir

Table 8.2 **Reservoir classification according to irregularity index [After Biswas et al., 2021]**

Classification	Irregularity index A/P ($km^2\ km^{-1}$)	Reservoir count
Highly irregular	Less than 0.25	13
Very irregular	0.25–0.40	25
Irregular	0.4–0.5	18
Regular	0.5–0.75	9
Very regular	0.75–1.0	7
Highly regular	Greater than 1.0	5

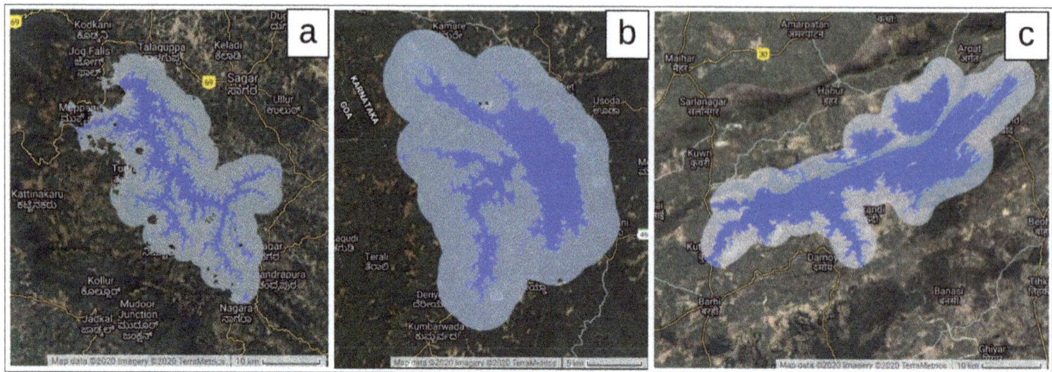

Figure 8.10 Example of reservoirs according to differing irregularity indices. **a**, Highly irregular: Linganamakki Reservoir over Sharavathi River, India, with *A/P* ratio 0.19 (surface area 146.49 km^2); **b**, regular: Supa Reservoir over Kalinadi River, India, with *A/P* ratio 0.51 (surface area 94 km^2); **c**, highly regular: Ban Sagar Reservoir over Sone River, India, with *A/P* ratio of 1.8 (surface area 384.3 km^2). The unit of A/P is in km [After Biswas et al., 2021]

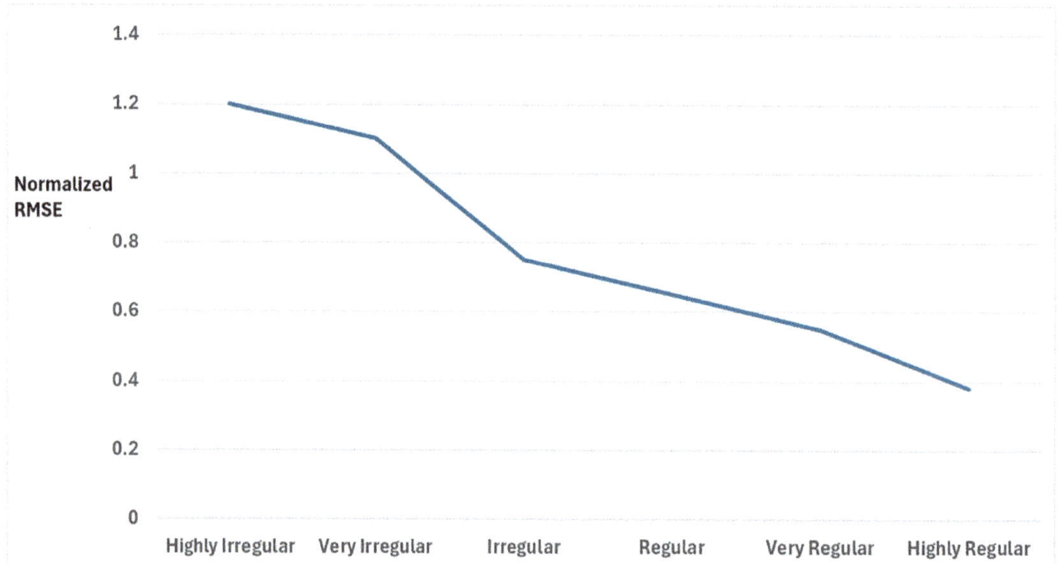

Figure 8.11 The role of reservoir shape in dictating the accuracy (in terms of root mean squared error, RMSE) of storage change estimates when compared against in-situ methods. Six area classification methods based on various remote sensing data were used to estimate reservoir surface area and then storage change. The y axis represents the normalized RMSE (normalized to the mean value) calculated for each reservoir shape category over these six area estimation methods. [For more details on the area estimation methods, see Biswas et al., 2021]

surface. This is as if SWOT is performing the combined role of all the altimeters and visible, NIR, and SAR satellites in one platform to afford global coverage of more reliable reservoir storage change estimates.

Case Study 8.1: Tigris–Euphrates Reservoir Tracking for Water Management

The Tigris–Euphrates River system is known as the birthplace of the Ancient Mesopotamian Civilization and the first place where organized irrigation began. Like most river systems, there is a distinct wet period that brings in high flows and sediments from upstream in eastern Türkiye to enrich the soils further downstream in Syria and notably Iraq. In more recent times (1950s through 1980s), Iraq and Syria built a series of dams in their countries to tame the flood pulse. For example, Iraq built a large reservoir near Baghdad to prevent the annual flooding of the city. The consequence of such actions of dam building and choking the river has been the steady decline of soil fertility of land that was once agriculturally productive and did not suffer from saline water intrusion from the Persian Gulf. Now enter Türkiye, the most upstream country controlling the lion's share of the source of water in the Tigris–Euphrates Rivers (Figure 8.12). From the 1980s, Türkiye has built a series of dams in the eastern region that were recently completed (known as the Southern Anatolya project). The result is that both Syria and Iraq are now even more parched than they used to be. With lack of access to water for each country of this basin, it has become fundamentally impossible to adapt and devise newer water management plans with a shared vision (Figure 8.12).

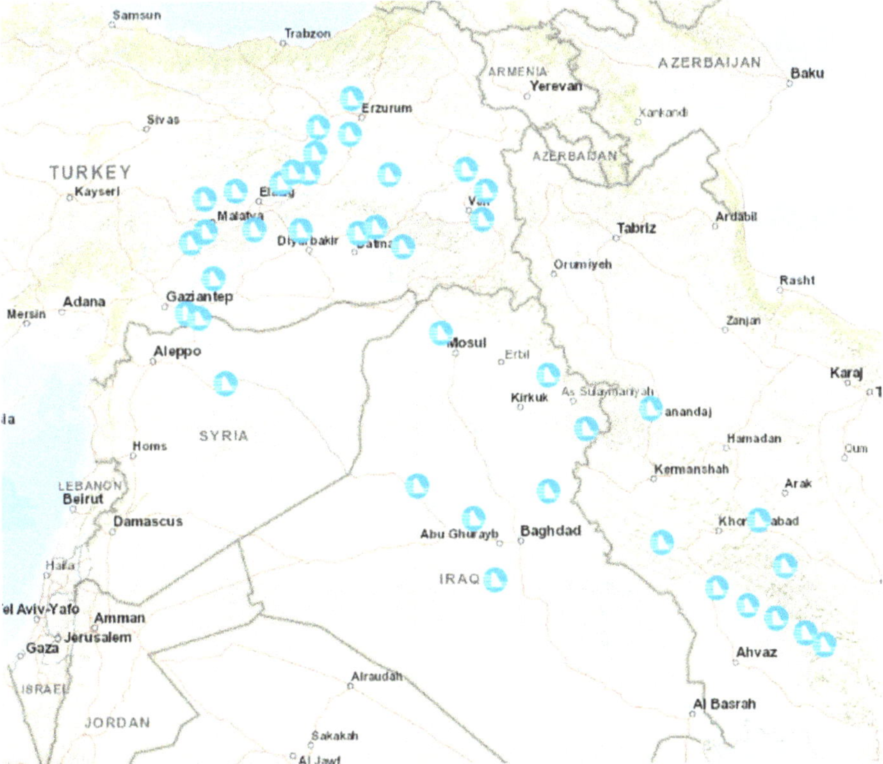

Figure 8.12 Location of major dams on the Tigris and Euphrates River system in Türkiye, Syria, and Iraq. [Dam locations were sourced from the GranD database by Lehner et al., 2011; © Openstreet Maps]

Case Study 8.1 (cont.)

This is where a satellite-based approach such as RAT and the SWOT mission can play a significant role in improving water management activities by agencies of these countries. Even if there is no cooperation or agreement, the most downstream country (like Iraq) that is vulnerable to climate and the more powerful upstream country controlling the water can at least make proactive decisions based on prompt information on Türkiye's withdrawals and releases. Other river basins that are faced with similar challenges to water management posed by lack of in-situ monitoring are the Nile River basin (the Blue Nile with Egypt, Sudan, and Ethiopia), the Mekong River basin (China, Vietnam, Laos, and Cambodia) and Indus (Pakistan and India). Readers can find customized RAT tools developed for stakeholders of these basins at http://depts.washington.edu/saswe/ indus, and http://depts.washington.edu/saswe/mekong. We encourage readers to also visit www .satellitedams.net and explore many of the dams in other places (such as the Tigris–Euphrates River basin) to understand how satellite data in the RAT framework can help monitor the reservoir states of inflow, outflow, and storage change. An example for Ataturk Dam in Türkiye is shown in Figure 8.13.

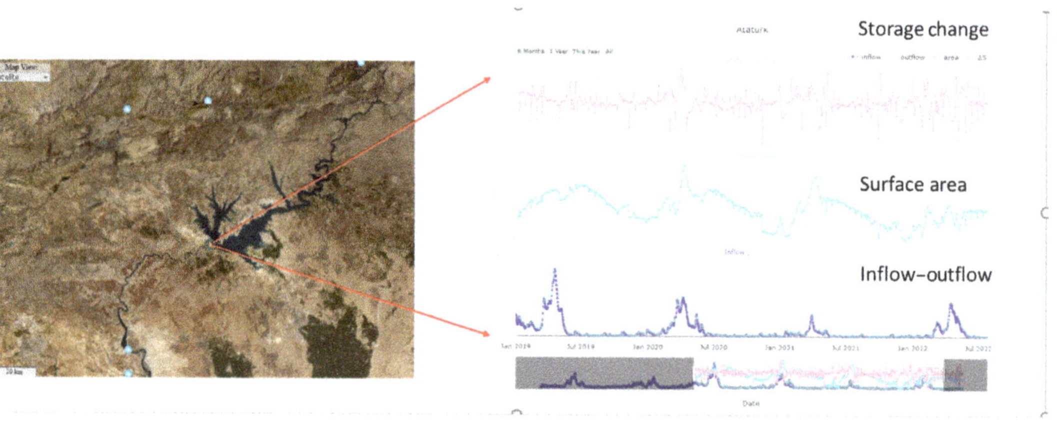

Figure 8.13 Reservoir state as monitored using satellite data and the RAT framework (on www.satellitedams.net) for Ataturk Dam in Türkiye in the upstream region of the Tigris–Euphrates River basin.

8.8   Conclusion

In this chapter, we learned how the various satellite remote sensing datasets on water, such as precipitation, reservoir surface area, elevation, bathymetry, and vegetation, can all be integrated in a hydrologic model and reservoir storage tracking system to completely profile a reservoir. This is a very new capability afforded by satellites that is increasingly levelling the playing field for water managers where data from traditional sources on reservoirs is fundamentally impossible to access. An example provided here was the reservoirs in the Tigris– Euphrates River basins spanning Iraq, Iran, Syria, and Türkiye. We introduced readers to the

Reservoir Assessment Tool (RAT) as an open-source tool for users to use the full power of satellites, cloud computing, and open-access software to track any reservoir. Such reservoir tracking capability is crucial for various water management applications such as irrigation and crop management that the reservoir's water may be used for. Therefore, in the next chapter, we will cover irrigation and crop management using satellites.

TUTORIAL ON SETTING UP RESERVOIR ASSESSMENT TOOL (RAT)[1]

[*For advanced users/readers only*]
In this tutorial we provide step-by-step instructions on how to set up the open-source Reservoir Assessment Tool (RAT) version 3.0 that we alluded to earlier in Sections 8.2 and 8.3. Because this tutorial requires advanced knowledge and experience with data science and programming in the appropriate computing environment, the setup of RAT 3.0 can be time-consuming and logistically challenging as an academic exercise. Therefore, this tutorial may be skipped by most readers if they do not have sufficient time to develop the computational and data science skill to master RAT 3.0. We expect a seasoned programmer with experience of coding in the Python environment to be able to follow and implement the instructions provided in this tutorial over a short period if they persevere and take advantage of the comprehensive resources we have already created for global and first-time users of RAT via www.satellitedams.net and www.ratdocs.io.

Background

In this tutorial, we will examine the impact of reservoir operations in the context of extreme precipitation events. Specifically, we will investigate whether these operations serve to mitigate or exacerbate the damage resulting from floods, using a real-world example described below.

In the initial week of August 2019, heavy monsoon rains led to severe flooding in the southern Indian state of Karnataka (Figure 8.14). Readers can search the internet to find more information on this devastating event. Here we provide a brief summary of the catastrophic flood.

Thousands of individuals were evacuated to safer locations and relief camps. The catastrophic event resulted in approximately 70 casualties and displaced around 700,000 people. On August 8, 2019, Karnataka experienced nearly five times its usual rainfall. This exacerbated an already ongoing flood in the Belagavi district of Karnataka where there is a dam known as Hidkal (Figure 8.14). To analyze the role of reservoir operations by Hidkal Dam, we will focus on how it operated.

We will use RAT 3.0 to analyze the reservoir operations of Hidkal Dam, which makes use of satellite remote sensing for tracking reservoir states (see Section 8.3). Because this tutorial is meant for advanced users with significant experience of coding in the Python environment, the reader is first encouraged to refer to the RAT documentation to learn more about

---

[1] Contributed by Sanchit Minocha, University of Washington.

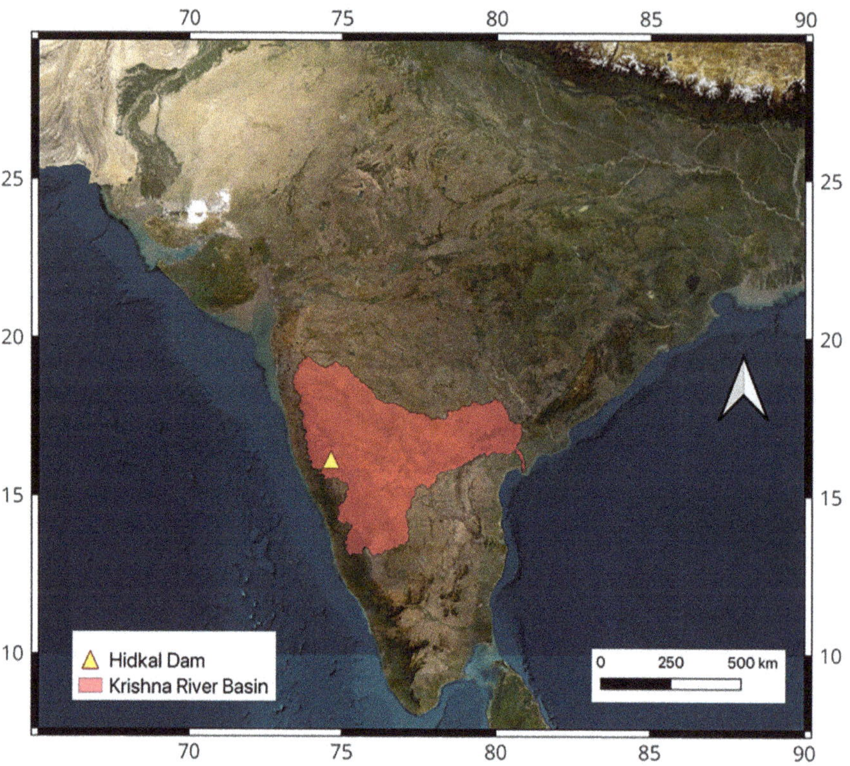

Figure 8.14 Location of the Hidkal Dam in the Krishna River Basin in India.

its installation, usage, and working (http://ratdocs.io or https://rat-satellitedams.readthedocs .io/en/latest/?badge=latest). Here we provide step-by-step instructions on how to set up RAT for Hidkal Dam in Karnataka (India) that should also be supplemented by the comprehensive documentation for RAT that the reader should refer to as needed.

Download

On this location in the cloud: www.cambridge.org/hossain, we have provided three important zipped files named 'codes', 'custom_data', and 'final_outputs'. Readers should download these three files and unzip them first before going on to the next section. Later in the tutorial, we will refer to them as needed to help readers complete the tutorial.

Requirements

1. For this tutorial, please use a Linux OS based laptop or workstation which has RAM of 8 GB or more and minimum hard disk of 512 GB.
2. Login credentials for AVISO user-account (for reservoir height data)
3. Login credentials for IMERG user-account (for accessing precipitation data for hydrologic model)
4. Login credentials for Earth Engine using service account (for reservoir storage change calculation)

To get the three sets of credentials (requirements 2, 3, and 4), follow the steps mentioned at https://rat-satellitedams.readthedocs.io/en/latest/QuickStart/GettingReady/ and create a secrets file accordingly.

Installation

Open a terminal and make sure your current working directory path has no spaces. Follow these steps in a terminal to get started:

Step 1. Create a new directory with name 'rat_tutorial'. We will refer to this directory as the RAT project directory.

```
mkdir rat_tutorial
```

Step 2. Change into the directory created.

```
cd rat_tutorial
```

Step 3. Create a new conda environment with name '.rat_env'. Note that '.' in a path means current directory whereas '.' in front of a folder/file name means it should be hidden. Type 'y' and press enter when asked to proceed in the terminal.

```
conda create --prefix ./.rat_env
```

Step 4. Activate the new environment.

```
conda activate ./.rat_env
```

Step 5. Install mamba using conda into the current environment from channel conda-forge. Type 'y' and press enter when asked to proceed in the terminal.

```
conda install mamba -c conda-forge
```

Step 6. Install RAT library using mamba. mamba is also a package manager just like conda, but it is more efficient in resolving dependencies version requirements. Type 'y' and press enter when asked to proceed in the terminal.

```
mamba install rat -c conda-forge
```

Activating Environment

To use RAT or its commands, you will always need to activate its Python environment once in a terminal. Open a new terminal and make sure your current working directory is the project directory. Type the following command and press Enter.

```
conda activate ./.rat_env
```

Initializing RAT

Initializing RAT takes care of setting up all the hydrological models and provides a set of default parameter files. Therefore, RAT requires only one-time initialization after installation. You do not need to initialize RAT every time before using.

In terminal, make sure your current working directory is the project directory and type the following command and press Enter. Again, please note that '.' means the path of the current working directory, which is the project directory's path. You can use either of the two mentioned commands, where the first uses Dropbox drive to download parameter files, the second uses Google drive.

```
rat init -d . -dr dropbox
```

OR

```
rat init -d . -dr google
```

Testing Installation and Initialization of RAT

Since RAT has now been installed and initialized, let us test whether RAT is working properly and whether your credentials are working fine. The default installation package of RAT comes with two test basins to test and verify whether your RAT has been installed and initialized successfully. The two basins are 'NUECES' in Texas and 'GUNNISON' in Colorado (USA). Open a new terminal and activate the environment as explained in the "Activating Environment" section. Let us test it on NUECES for this exercise with the following command in the terminal:

```
rat test -d <PATH_OF_RAT_PROJECT_DIRECTORY> -b NUECES -s
<PATH_TO_SECRETS_FILE> -dr dropbox
```

Replace **<PATH_OF_RAT_PROJECT_DIRECTORY>** with the path of the RAT project directory and the **<PATH_TO_SECRETS_FILE>** with the path of the secrets file (created in Requirements section). Also, you can replace 'dropbox' with 'google' in the above command if 'dropbox' does not work. You can check RAT produced outputs inside RAT project directory at data/test_output/Texas/basins/Nueces/final_outputs/ and compare them with the expected outputs provided in data/test_data/Nueces/expected_outputs/. RAT by default provides the complete comparison and will tell you if the files were matched or not. A successful test completion should have the output as shown below in the terminal:

```
Verifying Test Results for basin - NUECES ....
############################ Test-1 (AEC) ############################
Matching AEC files ...
Tested AEC files successfully. All files matched. Test Passed-100%.
############################ Test-2 (ΔS) ############################
Matching ΔS files ...
Tested ΔS files successfully. All files matched. Test Passed-100%.
############################ Test-3 (Evaporation) ############################
Matching Evaporation files ...
Tested Evaporation files successfully. All files matched. Test Passed-100%.
############################ Test-4 (Inflow) ############################
Matching Inflow files ...
Tested Inflow files successfully. All files matched. Test Passed-100%.
############################ Test-5 (Outflow) ############################
Matching Outflow files ...
Tested Outflow files successfully. All files matched. Test Passed-100%.
############################ Test-6 (Surface Area) ############################
Matching Surface Area files ...
Tested Surface Area files successfully. All files matched. Test Passed-100%.
```

Logging

Upon executing the test command provided above, two directories will be generated within the <PATH_OF_RAT_PROJECT_DIRECTORY>/data/. 'test_data' will house the files utilized for running RAT on the test basin, predominantly comprising input files and their corresponding expected output files. Meanwhile, 'test_output' will store all data generated by RAT, including final outputs and logs. RAT produces two log files, detailed below, which can be utilized for monitoring RAT's progress and troubleshooting any encountered errors.

1. Level-1 Log (concise log):
   Access the file in project directory at data/test_output/runs/logs/RAT_run-YYYYMMDD-*.log using a text editor that supports real-time updates. This enables us to observe the ongoing writing of the log file, providing insights into the step numbers executed by RAT and any encountered errors during the execution process.

2. Level-2 Log (detailed log):
   Open the file in project directory at data/test_output/Texas/logs/Nueces/RAT-NuecesYYYYMMDD-*.log (once again using an editor that permits real-time updates). Here, you can scrutinize every process taking place within each step of RAT. This detailed log is primarily employed for debugging errors.

You can further read about the entire directory structure that usually exists inside the project directory at https://rat-satellitedams.readthedocs.io/en/latest/RAT_Data/DirectoryStructure/.

Configuration File

Navigate to params/rat_config.yml in project directory. This file is a partially complete configuration file where you can observe the paths for model (or Python environment) and parameter files in the Metsim, VIC, and Routing sections. These paths were configured during RAT initialization, with the initialization command populating these paths. It's important to note that global data has not been downloaded, which is why these paths remain unpopulated. It is worth mentioning that within the params directory, there are two additional configuration files: 'rat_config_template.yml' functions as a template, while 'test_config.yml' was employed by RAT for conducting RAT tests. *[Side note: If you are curious what these terms "Metsim, VIC" or "routing" mean, you should refer to the papers on RAT and the documentation provided for RAT. However, here, we can safely treat these terms and components as a "black box" as the goal here is to set up RAT for Hidkal Dam and focus on the reservoir operations.]*

Information regarding each parameter in the configuration file is provided at https://rat-satellitedams.readthedocs.io/en/latest/Configuration/rat_config/.

Now, you are ready to utilize RAT for Hidkal Dam. **To begin with, download the 'custom_ data' folder inside the project directory**. Note that you should have already downloaded this 'custom_ data' earlier before starting the tutorial from this cloud location. Let us configure the file to analyze the reservoir operations of Hidkal Dam in the year 2019. In this instance, we are using a Python code to update the configuration file. Manual updates are also possible, albeit time-consuming.

```python
from rat.toolbox import config #to update/create config
from rat.toolbox.MiscUtil import rel2abs #to convert relative paths to absolute
from datetime import date #to create datetime.date objects
```

The above section of the Python script involves importing modules, and comments have been included to delineate the specific purpose of each module.

```python
# Define the parameters to update in the configuration file of rat
## Dam
station_global =True #True if file is shapefile otherwise False for csv file.
station_file = "custom_data/Hidkal_Dam_RAT_Tutorial/Dam/Hidkal_Dam.json"
station_file_col_dict ={
                'id_column':"GRAND_ID",
                'name_column':"DAM_NAME",
                'lon_column':"LONG_DD",
                'lat_column':"LAT_DD"
            }

## Reservoir
reservoir_file = "custom_data/Hidkal_Dam_RAT_Tutorial/Dam/Hidkal_Reservoir.json"
reservoir_file_col_dict ={
                'id_column':"GRAND_ID",
                'dam_name_column':"DAM_NAME",
                'area_column':"AREA_SKM",
                'dam_lat':"LAT_DD",
                'dam_lon':"LONG_DD",
                'dam_height':"DAM_HGT_M"
            }
## River Basin
basin_file ="custom_data/Hidkal_Dam_RAT_Tutorial/River_Basin/Krishna_River_Basin.json"
basin_file_col_dict ={
                'id':"MRBID",
            }
region_name ="India" # or Karnataka
basin_name ="Krishna"
basin_id =2312
```

```
## Grid Flow Direction
flow_direction_file = "custom_data/Hidkal_Dam_RAT_Tutorial/Flow_Direction_File/
Krishna_Basin_Flow_Direction.asc"

## Metsim Parameter Files
metsim_domain_file = "custom_data/Hidkal_Dam_RAT_Tutorial/Metsim_Parameter_Files/
Metsim_Domain.nc"

## VIC Parameter Files
vic_global = False #False if providing specific parameter files for the basin
vic_soil_param_file = "custom_data/Hidkal_Dam_RAT_Tutorial/VIC_Parameter_Files/Vic_
Soil_Param_Krishna_Basin.nc"
vic_domain_file = "custom_data/Hidkal_Dam_RAT_Tutorial/VIC_Parameter_Files/Vic_
Domain_Krishna_Basin.nc"

## Specifying Run Parameters
steps = [1,2,3,4,5,6,7,8,9,10,12,13,14] #Not running altimeter
start_date = date(2019,1,1)
end_date = date(2019,12,31)
spin_up = False #True if running without state files otherwise True
vic_init_state = "custom_data/Hidkal_Dam_RAT_Tutorial/Initial_State_Files/Vic_Init_
State_Krishna_Basin_2019-01-01.nc"
rout_init_state = "custom_data/Hidkal_Dam_RAT_Tutorial/Initial_State_Files/Routing_
Init_State_Krishna_Basin_2019-01-01.nc"

## Secrets & Configuration File
secrets_file = "params/secrets.ini"
config_file = "params/rat_config.yaml"
```

In the code snippet above, we have defined variables with preassigned values intended for inclusion in the configuration file. For this exercise, there is no need to modify the values of any variables, as all the paths are relative to the RAT project directory and therefore the Python script should be executed from the project directory itself. However, if you are using RAT for any other reservoir or for a different time-period and adjustments are required, you can modify the variable values accordingly.

For example, if you want to run RAT for Hoover Dam, you need to change the station file, reservoir file, basin file, flow direction file, metsim domain file, VIC domain file, and VIC parameter files for the hydrologic model. You will also have to ensure that the formats of these files remain the same. You can also change 'start_date' and 'end_date' variables to alter the duration for which you want to execute RAT. You can exclude the vic_init_state and rout_state_files and instead change 'spin_up' variable's value to True. With spin_up true, the computational time will increase as RAT will be executed for an additional 800 days.

It's important to note that the single-quote strings, representing dictionary keys, should remain unchanged. **Another important thing to ensure is that your secrets file is inside params directory, otherwise change the path accordingly**.

Now, we will construct a Python dictionary to update the configuration file using the previously declared variables. It is essential to note that the following code can be used as it is, regardless of the scenario where you are using RAT.

```python
params_to_update={

'GLOBAL':{
    'steps': steps,
    'basin_shpfile': rel2abs(basin_file),
    'basin_shpfile_column_dict': basin_file_col_dict
    },

'BASIN':{
    'region_name': region_name,
    'basin_name': basin_name,
    'basin_id': basin_id,
    'spin_up': spin_up,
    'start': start_date,
    'end': end_date,
    'vic_init_state': rel2abs(vic_init_state),
    'rout_init_state': rel2abs(rout_init_state)
    },

'METSIM':{
    'metsim_domain_file': rel2abs(metsim_domain_file)
    },

'VIC':{
'vic_global_data': vic_global,
    'vic_soilm_file': rel2abs(vic_soilm_param_file),
    'vic_domain_file': rel2abs(vic_domain_file)
    },

'ROUTING':{
    'station_global_data': station_global,
    'stations_vector_file': rel2abs(station_file),
    'stations_vector_file_columns_dict': station_file_col_dict
    },

'ROUTING PARAMETERS':{
    'flow_direction_file': rel2abs(flow_direction_file)
    },

'GEE':{
    'reservoir_vector_file':rel2abs(reservoir_file),
    'reservoir_vector_file_columns_dict': reservoir_file_col_dict
    },

'CONFIDENTIAL':{
    'secrets': rel2abs(secrets_file)
    }
}
```

The last step here before running RAT is to update the configuration file using the Python dictionary created above.

```python
config.update_config(config_file, params_to_update)
```

The complete code is present in update_config.py in the codes folder. You should have already downloaded the codes folder from this location on the cloud. Run the Python script in the terminal after ensuring your RAT Python environment is activated. After executing this Python script, you will notice that the configuration file inside the project directory at params/rat_config.yml will be populated with values defined by you.

Execution

Use the below command to execute RAT using the updated configuration file. It should take around 1–2 hours depending on the computational power of the machine for complete RAT execution. Replace the **<PATH_OF_RAT_CONFIGURATION_FILE>** with the path of the configuration file (prepared above). For example: it should look like 'rat_tutorial/params/rat_config.yml'.

```
rat run -p <PATH_OF_RAT_CONFIGURATION_FILE>
```

Analysis of RAT Outputs

Navigate to data/India/basins/Krishna/final_outputs inside the project directory to view the RAT generated inflow, outflow, surface area, evaporation, storage change, and area elevation curve. 'final_outputs' folder has already been provided to you that you should have downloaded earlier from this location on the cloud. You can visualize the data using Excel or create interactive visualization plots using the RAT module itself in Python. Below is the code to create interactive plots and the snapshot of the figures created. *[Side Note: The 'final_outputs' folder generated by RAT may contain additional outputs beyond those listed here, such as NSSC (Normalized Suspended Sediment Concentration), which relate to water quality rather than reservoir operations. Since our focus is on investigating reservoir operations, these additional outputs can be disregarded.]*

First, we will import RAT_RESERVOIR class from the RAT module that represents a reservoir object in Python.

```python
from rat.toolbox.visualize import RAT_RESERVOIR
```

Then we will create a reservoir instance and specify the path of the 'final_outputs' generated by RAT and the filename being used for the reservoir.

```python
hidkal_reservoir = RAT_RESERVOIR(
final_outputs = 'data/India/basins/Krishna/final_outputs', file_name = '4773_Hidkal.csv')
```

1. Inflow

```python
inflow_figure = hidkal_reservoir.plot_var(
    var_to_observe = 'Inflow',
    title_for_plot = 'Inflow Timeseries',
    xlabel = 'Date',
```

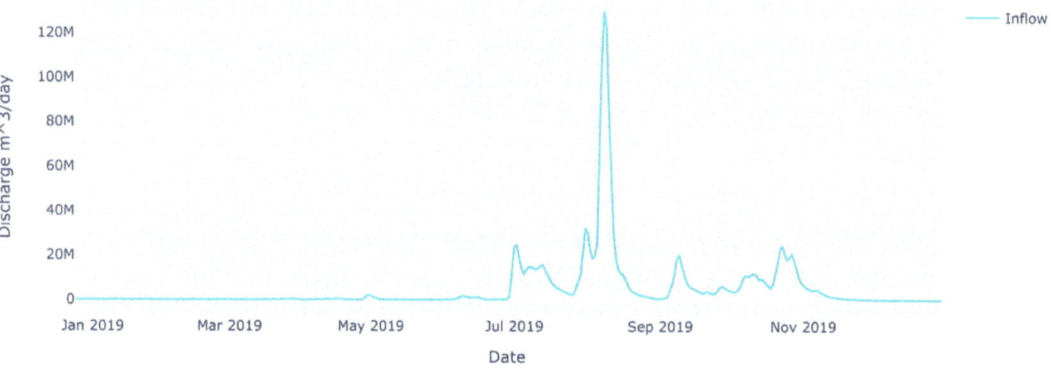

Figure 8.15 Time series of reservoir inflow for the Hidkal Dam for the year 2019. This inflow is estimated on the basis of a hydrologic model called VIC that is embedded as part of the RAT package. VIC runs using hydrometeorological data, notably precipitation, temperature, and windspeed, which are automatically extracted by RAT using credentials in the secrets file.

```python
    ylabel = 'Discharge',
    x_axis_units = '',
    y_axis_units = 'm^3/day'
)
inflow_figure.show()
```

Upon observing the Inflow time series interactive plot, zooming into the year 2019 reveals the graph below. Hovering over the peak inflow value indicates that it occurred on August 8, 2019. Consequently, RAT accurately identified the peak flood, aligning with our earlier mention in the introduction that Karnataka experienced rainfall five times higher than usual on this particular day (Figure 8.15).

2. Outflow

```python
Outflow_figure = hidkal_reservoir.plot_var(
    var_to_observe = 'Outflow',
    title_for_plot = 'Outflow Timeseries',
    xlabel = 'Date',
    ylabel = 'Discharge',
    x_axis_units = '',
    y_axis_units = 'm^3/day'
)
Outflow_figure.show()
```

We can observe that during the flood period, the reservoir's outflow reached its peak. Zooming in on Figure 8.16, we can see that the peak outflow occurred on August 13, 2019. The prior observation was on August 8. This is because RAT provides outflow estimations at a frequency of 2–5 days limited by the aggregate sampling frequency of the satellites it uses

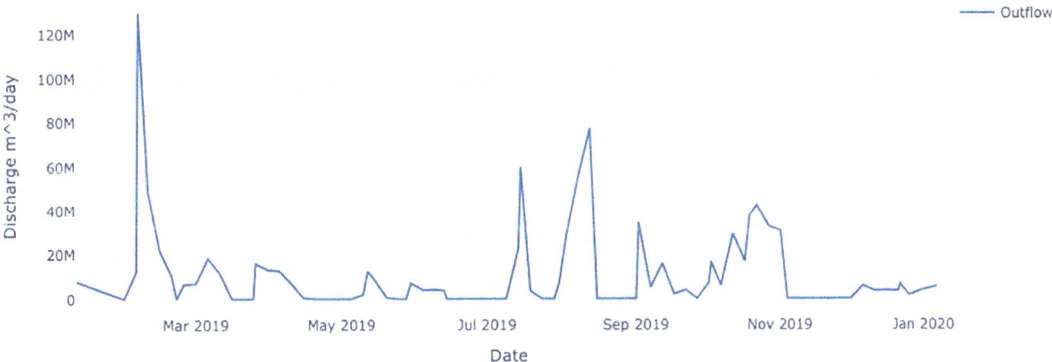

Figure 8.16 Outflow from Hidkal Dam using RAT. Here RAT calculates the outflow based on a mass balance of inflow, reservoir storage change, and evaporative losses mentioned earlier in Section 8.2.1 (see Figure 8.2)

to compute reservoir storage change. This can produce uncertainties about the occurrences in between satellite overpasses. Therefore, we can safely assume that the peak outflow was observed somewhere between the 8th and 13th of August. Despite this uncertainty, it is plausible to suggest that if the reservoir's outflow had been lowered beforehand, the damage caused by the incoming flood could have been mitigated by storing that excess flood water and releasing it more gradually over a longer time.

3. Surface Area of Reservoir

```
surface_area_figure = hidkal_reservoir.plot_var(
  var_to_observe = 'Surface Area',
  title_for_plot = 'Surface Area Timeseries',
  xlabel = 'Date',
  ylabel = 'Surface Area',
  x_axis_units = '',
  y_axis_units = 'Km^2'
)
surface_area_figure.show()
```

The surface area estimates were estimated using the Tiered Multi Sensor-Optical and SAR (TMS-OS; Das et al., 2022) algorithm at a frequency of 2–5 days (Figure 8.17; see Section 8.2). As satellite remote sensing data, despite its quantitative bias and uncertainty, are quite powerful for its qualitative trends, the trend of reservoir level is found to be accurately detected by RAT. The reservoir level was lowered from March to June 2019 which is the pre-monsoon period in South Asia. This standard practice is typically employed to accommodate the anticipated high inflows during the monsoon, allowing the reservoir to reach its full capacity by the end of the monsoon season. However, in this particular case, the reservoir was almost full at maximum capacity by early August, which is only halfway through the monsoon.

Figure 8.17 Time series of reservoir surface area as estimated by RAT

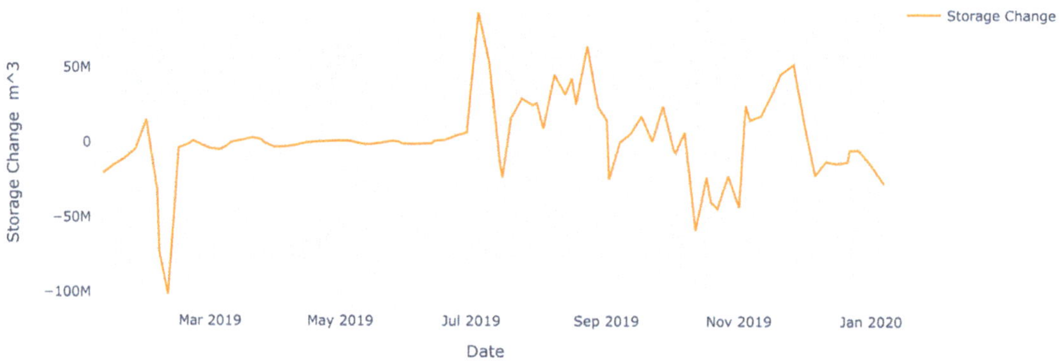

Figure 8.18 Time series of reservoir storage change by RAT using satellites and the TMS-OS algorithm.

4. Reservoir Storage Change

```python
storage_change_figure = hidkal_reservoir.plot_var(
    var_to_observe = 'Storage Change',
    title_for_plot = 'Storage Change Timeseries',
    xlabel = 'Date',
    ylabel = ' Storage Change ',
    x_axis_units = '',
    y_axis_units = 'm^3'
)
storage_change_figure.show()
```

The positive storage change observed during the months of July indicates that the reservoir commenced filling up in July and early August of 2019 (Figure 8.18). This is consistent with the surface area trends shown in Figure 8.17 earlier.

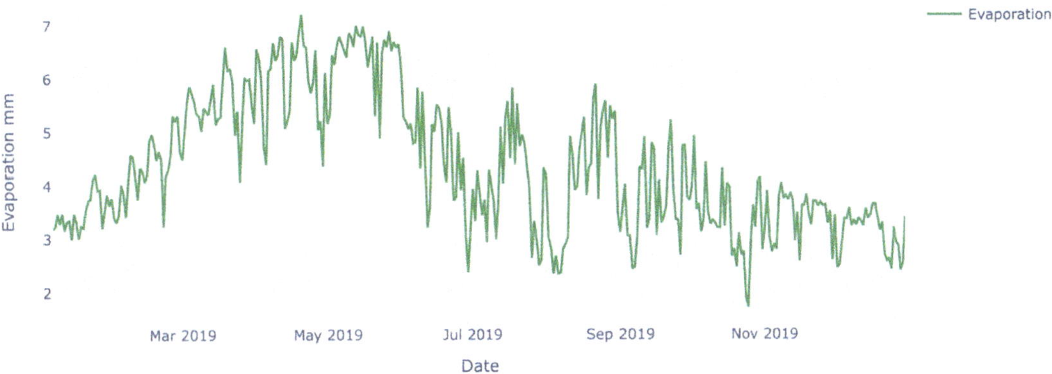

Figure 8.19 Time series of evaporation from the reservoir surface by RAT using the Penman method (see Das et al., 2022 for details).

5. Reservoir Evaporation

```python
evaporation_figure = hidkal_reservoir.plot_var(
    var_to_observe = 'Evaporation',
    title_for_plot = 'Evaporation Timeseries',
    xlabel = 'Date',
    ylabel = 'Evaporation',
    x_axis_units = '',
    y_axis_units = 'mm'
)

evaporation_figure.show()
```

The amount of evaporation from the reservoir was estimated by RAT at a daily frequency (Figure 8.19). At the time of flood peak, the evaporation is very low, which is probably due to high humidity in the air in that region.

6. Reservoir Area–Elevation Curve

```python
aec_figure = hidkal_reservoir.plot_var(
    var_to_observe = 'A-E Curve',
    title_for_plot = 'Area-Elevation Curve',
    xlabel = 'Area',
    ylabel = 'Elevation',
    x_axis_units = 'Km^2',
    y_axis_units = 'm'
)

aec_figure.show()
```

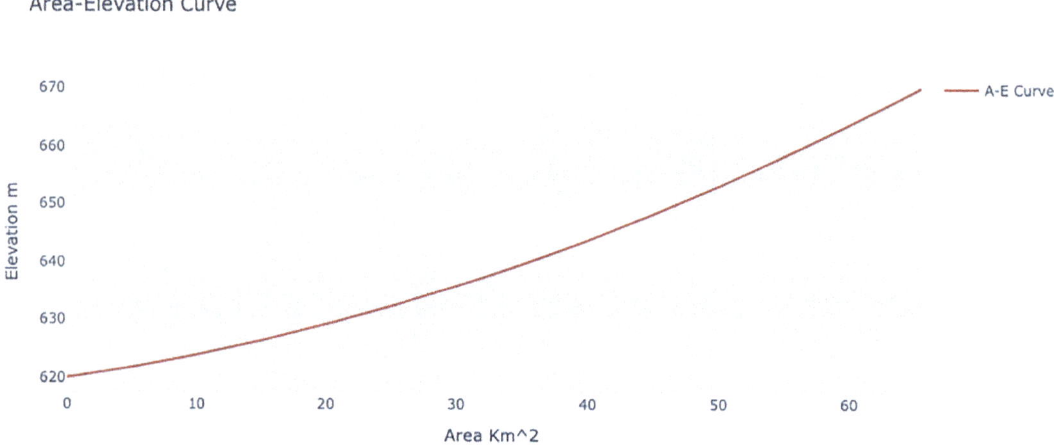

Figure 8.20 Area–elevation curve for Hidkal Reservoir of Karnataka as derived by RAT using the SRTM digital elevation data.

RAT can construct the area–elevation curve (AEC) for the Hidkal Reservoir by leveraging the SRTM digital elevation model available in Google Earth Engine (GEE), as no survey-observed AEC was available here (Figure 8.20). Since SRTM data was recorded in February 2000 when the reservoir already existed, the elevation data above the reservoir level (that existed during February 2000) can be mapped to give its bathymetry. Consequently, RAT extrapolates the AEC for lower reservoir levels. Where possible, users should apply their own judgement about the extrapolated region of the AEC due to potential uncertainty, and use in-situ or higher-quality reservoir topographic maps if available.

EXERCISES: CHAPTER 8

Q8.1   What are the flux components that comprise a reservoir state?

Q8.2   Describe the key steps to using a mass balance approach to track reservoir state from space.

Q8.3   What are the limitations of this mass balance approach and how can it be improved?

Q8.4   How far back do you think it is possible to track reservoir operations around the world? For which regions and reservoir types do you think the multi-decadal historical tracking of reservoir operations would work better and why?

Q8.5   Why do you think the shape of the reservoir shoreline plays a role in estimation accuracy of surface area and storage change? Use your understanding from previous chapters to explain, noting that the water surface is the target and the surrounding is the background.

Q8.6   How do you think SWOT will improve tracking of reservoir operations around the world?

REFERENCES

Biswas, N. K., F. Hossain, M. Bonnema, H. Lee, and F. Chishtie (2021). Towards a global Reservoir Assessment Tool for predicting hydrologic impacts and operating patterns of existing and planned reservoirs. *Environmental Modelling & Software*, vol. 140, 105043. https://doi.org/10.1016/j.envsoft.2021.105043

Das, P., F. Hossain, S. Khan, et al. (2022). Reservoir Assessment Tool 2.0: stakeholder driven improvements to satellite remote sensing based reservoir monitoring. *Environmental Modeling and Software*, vol. 157. https://doi.org/10.1016/j.envsoft.2022.105533

Das, P., F. Hossain, S. Minocha, et al. (2024). ResORR: a globally scalable and satellite data-driven algorithm for river flow regulation due to reservoir operations. *Environmental Modeling and Software*, vol. 176, 106026. https://doi.org/10.1016/j.envsoft.2024.106026

Hamman, J. J., B. Nijssen, T. J. Bohn, D. R. Gergel, and Y. Mao (2018). The Variable Infiltration Capacity model version 5 (VIC-5): infrastructure improvements for new applications and reproducibility. *Geoscientific Model Development*, vol. 11, pp. 3481–3496, https://doi.org/10.5194/gmd-11-3481-2018

Huffman, G. J., D. T. Bolvin, D. Braithwaite, et al. (2020). Integrated Multi-satellite Retrievals for the Global Precipitation Measurement (GPM) Mission (IMERG). In Levizzani, V., Kidd, C., Kirschbaum, D. B., et al. (eds.) *Satellite Precipitation Measurement. Advances in Global Change Research*, vol. 67. Springer. https://doi.org/10.1007/978-3-030-24568-9_19

Lehner, B., C. R. Liermann, C. Revenga, et al. (2011). High-resolution mapping of the world's reservoirs and dams for sustainable river-flow management. *Frontiers in Ecology and the Environment*, vol. 9, 494–502.

Loucks, D. P., E. van Beek, J. R. Stedinger, et al. (2005). *Water Resources Systems Planning and Management: An Introduction to Methods, Models and Applications*. UNESCO.

Minocha, S., F. Hossain, P. Das, et al. (2023). Reservoir Assessment Tool 3.0: a scalable satellite-based reservoir monitoring tool to mobilize the global water management community. *Geosci Model Dev*. https://doi.org/10.5194/gmd-2023-130

Pekel, J.-F., A. Cottam, N. Gorelick, and A. S. Belward (2016). High-resolution mapping of global surface water and its long-term changes. *Nature*, vol. 540, 418–422. https://doi.org/10.1038/nature20584

Penman, H. L. (1948). Natural evaporation from open water, bare soil and grass. *Proceedings of the Royal Society of London. Series A. Mathematical and Physical Sciences*, vol. 193, 120–145. https://doi.org/10.1098/rspa.1948.0037

SUGGESTED READING

Gao, H. (2015). Satellite remote sensing of large lakes and reservoirs: from elevation and area to storage. *WIREs Water*, 2, 147–157. https://doi.org/10.1002/wat2.1065

Gao, H., C. Birkett, and D. P. Lettenmaier (2012). Global monitoring of large reservoir storage from satellite remote sensing. *Water Resources Research*, vol. 48, W09504. https://doi.org/10.1029/2012WR012063

9   Crop and Irrigation Management from Space

9.1   Chapter Overview

In the previous chapter, we learned how satellite data to estimate various water targets such as precipitation and surface water can be combined in a model-reservoir system to track a reservoir's dynamic state and understand river regulation. In this chapter, we will cover how satellite data can be used to manage crops and irrigation. We will learn how the data can be used to estimate an area under a specific crop using classification techniques, which then helps us to understand the water need for that area. Next, we will learn methods to estimate crop water demand and actual crop water consumption.

9.2   Introduction

For many decades, we have used satellite data in the visible and near-infrared (NIR) wavelengths from orbiting platforms to track the time-varying spatial extent of vegetation and its type, its biomass, chlorophyll content, and health condition. For water management in agriculture (or agricultural water management), the two specific uses of satellite data are for irrigation and for tracking crop types that require specific irrigation schedules and a water management plan. In this chapter, we will learn how satellite data can be used for tracking crop type, which can inform a water manager of the crop water need and can be used to create a water allocation plan for an irrigation scheme. We will also learn how satellite data can be used to manage irrigation schedules and optimize the use of precious water to avoid waste.

9.2.1   Tracking Vegetation

First, let us review the basic concepts of how tracking of crops using satellites works in the visible/NIR wavelengths (note: the method is passive and will work only during daytime in clear sky conditions). This is again based on the basic principles of remote sensing that in Chapter 2 were referred to as the "golden rules" (see Section 2.7). Just like we learned in Chapter 6 for surface water using normalized difference indices, we can apply the same idea using vegetation as our target. Recall from Chapter 2 that NIR is highly reflective when there is vegetation present. One of the most commonly used indices for tracking vegetation is therefore the Normalized Difference Vegetation Index (NDVI) in Equation 9.1. Essentially, NDVI quantifies vegetation by measuring the difference between NIR (which vegetation strongly reflects) and red light (which vegetation absorbs).

$$NDVI = \frac{(NIR - Red)}{(NIR + Red)}.$$ (9.1)

In the formula above, the terms are the reflected energy amount (or reflectance) in the specific wavelength. NDVI always ranges from −1 to +1. But there is not a distinct boundary for each type of land cover. A negative value may indicate the likely presence of water. An NDVI value close to +1 indicates the very high likelihood of vegetation, possibly containing dense green leaves. An NDVI value close to zero may indicate a highly urbanized or bare region. Satellite sensors such as Landsat and Sentinel-2 both have the necessary bands with NIR and red (Figure 9.1).

If we have low reflectance (or low values) in the red channel and high reflectance in the NIR channel, this will yield a high NDVI value. Overall, NDVI is a standardized way to measure healthy vegetation. When you have high NDVI values, you have healthier vegetation. Figure 9.2 shows an example of a typical Landsat surface reflectance map in the visible bands (true color, cloud-free over Sacramento, California, USA) and the corresponding NDVI classified map with a color scheme varying from −1 to 1. Notice how the negative 1 values (blue) match so closely with the true-color surface water bodies on the left scene.

Nowadays, by leveraging ultra-high-resolution and multi-spectral imagery in the visible/NIR bands from drones and commercial platforms such as Planet, IKONOS, and QuickBird, with existing data from Landsat, MODIS, and Sentinel-1 (SAR) and Sentinel-2, we can also track

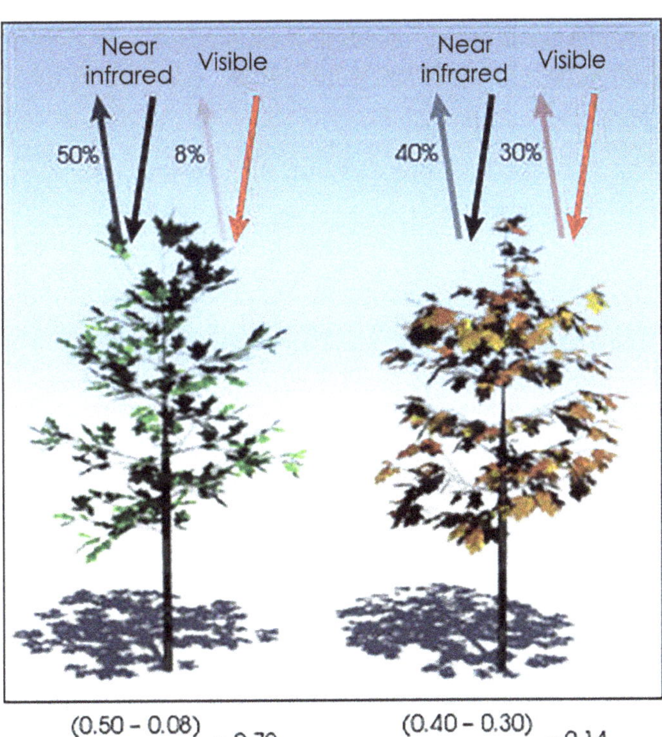

Figure 9.1 NDVI values for two different types of vegetation. [Image courtesy of European Space Agency]

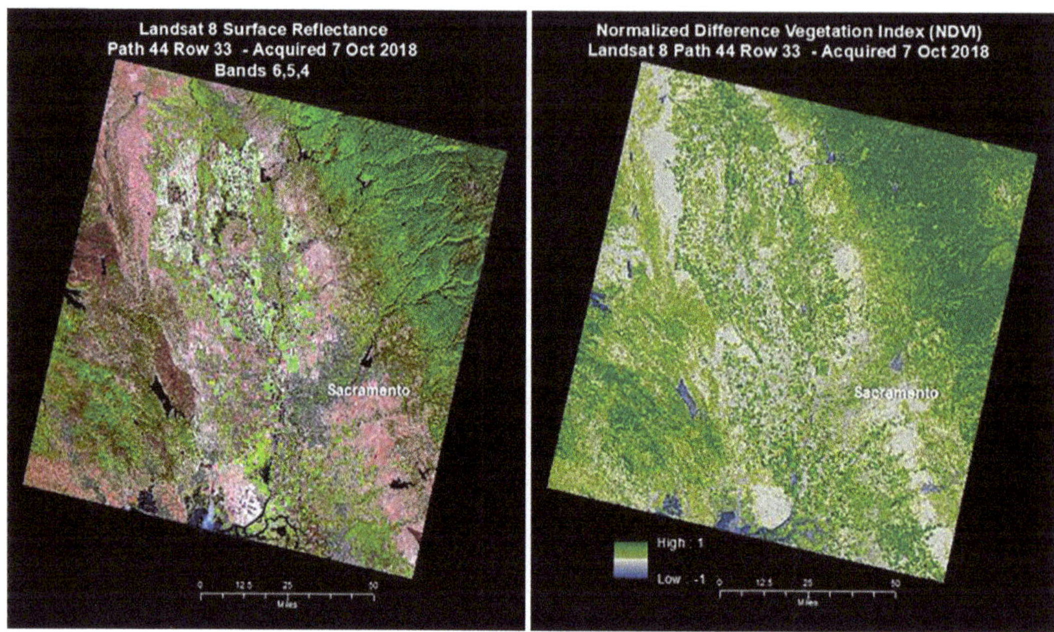

Figure 9.2 Left, Landsat 8 Surface Reflectance (SR); right, the SR-derived Landsat Surface Reflectance Normalized Difference Vegetation Index (NDVI). [Image: US Geological Survey]

biomass, vegetation health, vegetation stress, and so on. There is a vast body of literature on this topic. Readers can refer to the book *Remote Sensing of Vegetation: Principles, Techniques and Applications*, by Jones and Vaughan (2010).

9.2.2 Why is Crop Type and Its Area of Cultivation Important for Water Management?

Today, we grow many types of food, and many of these are staples that must be grown in large quantities to meet the growing food demand of the world. None of this can be grown without water. So, each major crop we cultivate has a water footprint, or a crop water demand needed for its growth from sowing to harvest. Knowledge of the crop type and the area under cultivation informs the water manager about the water needs for that region throughout the growing season and helps in planning optimal water allocation strategies. The water footprint of a crop is obviously dependent on climate, geographic location, and growth stage, as well as other factors such as soils, seed type, and use of fertilizer. Therefore, knowledge of what we are growing, where, and to what spatial extent informs us of the total water need to grow that type of food for a given period during the growing season, and allows us to make more informed decisions on irrigation schedules and water allocation from reservoirs or irrigation canals. In the past, this information was supplied from in-situ-based crop maps, cropping patterns, and ground information maintained by agricultural agencies and communities. Today, cropping is much more dynamic in terms of type and location. It is also expanding to new areas to keep up with demand for food. At the same time, agriculture is becoming more mechanized, and the land is getting consolidated by the few as the world marches towards urbanized settlements (see Chapter 1). For all these reasons we face today, the vantage of space using satellites has become necessary for agricultural water management.

9.3 Crop Water Demand (or Crop Water Need)

There are many methods or techniques to estimate the amount of water a crop needs throughout its growing season until harvest. Perhaps the most accurate benchmark measurement comes from the lysimeter, which is the cornerstone for developing other indirect and more feasible methods. A lysimeter is a direct way of measuring the amount of water a crop is consuming via evapotranspiration (Figure 9.3). It is a very expensive instrument that is time-consuming and costly to maintain. There are not as many lysimeters around the world as we would like that are maintained in a fashion similar to, say, rainfall gauges or river gauges. Thus, we do not have direct crop water information of sufficient geographical coverage and crop type around the world. Nevertheless, lysimeter-based readings form the cornerstone of all other indirect methods of estimating crop water demand. Figure 9.4 shows an actual lysimeter set up under the soil level by the Pakistan Council of Research in Water Resources (www.pcrwr.gov.pk) in the southern region of Pakistan, visited by the author during May 2022.

Of the various indirect methods available to estimate crop water demand, the one that is more relevant to water management applications and feasible to track using satellite data is the Food and Agriculture Organization (FAO) method based on Penman–Monteith evapotranspiration (ET). This method is also known as FAO56, as it was published in a report numbered 56. Here we will overview the FAO56 method of estimating crop water demand mainly to show (a) that crop water demand is specific to crop, location and growth stage, and (b) how it can be used in the latter sections of this chapter for irrigation management using satellite data.

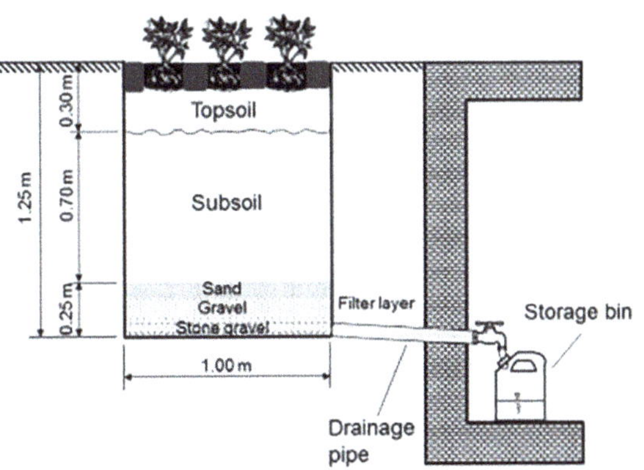

Figure 9.3 Schematic of a lysimeter. [After Meissner et al., 2020; image courtesy of Elsevier Sciences]

Figure 9.4 The top part of the lysimeter where the crop is grown, watered, and water use tracked. [Photo taken by the author during a visit to Pakistan on invitation from the Pakistan Council of Research in Water Resources; May 2022]

9.3.1   The FAO56 Method for Crop Water Demand Tracking

The crop water need (ET crop) is defined as the depth (or amount) of water needed to meet the water loss through evapotranspiration. In other words, it is the amount of water needed by the various crops to grow optimally.

The crop water need always refers to a crop grown under optimal conditions: that is, a uniform crop, actively growing, completely shading the ground, free of diseases, and with favorable soil conditions (including fertility and water). The crop thus reaches its full production potential under the given environment. The crop water need mainly depends on:

- Climate: in a sunny and hot climate, crops need more water per day than in a cloudy and cool climate
- Crop type: crops such as maize, rice, or sugarcane need more water than crops like millet or sorghum
- Growth stage of the crop: fully grown crops need more water than crops that have just been planted.

The FAO has recommended using the Penman–Monteith equation to estimate what we call reference evapotranspiration (ET_o). This is basically the amount of water a reference crop (typically tall grass or alfalfa) consumed through transpiration when water supply is not limiting. So, the conditions that would limit the ET_o would be the ambient weather variables defined by geographic location, elevation, climate, and soil wetness.

Penman–Monteith ET_o, which is a proxy for potential water demand for a reference crop, can be calculated over a 24-hour period following the steps outlined in the FAO56 report by Allen et al. (1998). The equation for ET_o is as follows:

$$ET_o = \frac{0.408\Delta\left(R_n - G\right) + \gamma\dfrac{900}{T+273}u_2\left(e_s - e_a\right)}{\Delta + \gamma\left(1 + 0.34u_2\right)} \tag{9.2}$$

where ET_o is reference evapotranspiration (mm day^{-1})
R_n is net radiation (MJ m^{-2} day^{-1})
G is ground heat flux (MJ m^{-2} day^{-1}), considered negligible (i.e. 0) here
T is mean air temperature at 2 m height (°C)
u_2 is wind speed at 2 m height (m s^{-1})
e_s is saturation vapor pressure (kPa)
e_a actual vapor pressure (kPa)
Δ is slope of saturation vapor pressure (kPa °C^{-1})
γ is psychrometric constant (kPa °C^{-1})

Fortunately, the input terms on the right-hand side of the above equation for ET_o can be rewritten with various parameterizations or computed using the following commonly measured variables:

1. Daily maximum temperature
2. Daily minimum temperature
3. Daily maximum relative humidity
4. Daily minimum relative humidity
5. Daily mean wind speed
6. Latitude of the location
7. Elevation of the location
8. Julian day of the year

If we look carefully at all the eight inputs above, we can conclude that satellite data can be used with modeled weather data to estimate ET_o without the need for in-situ ground information. Hence ET_o can be estimated from space (with the help of some weather data from numerical models). We will discuss more about the tracking of ET from space later in tutorials and case studies.

But first, let us make sure how we go from ET_o to actual crop water demand for a specific crop. ET_o calculated using Equation 9.2 is for a reference crop (well-watered grass of 0.12 m height) which is then converted for the actual crop by using two factors: Kc (crop coefficient) and Ks (soil stress index). Kc is basically a conversion factor that "adjusts" the crop water requirement for the specific crop in question, and it depends on growth stage and agro-climatic zones. FAO provides some standard Kc values for major crops for four distinct growth stages: initial stage (just planted); developmental stage (growing); mid-season or mature stage (flowering); and late-season stage (harvest) (Figure 9.5). After multiplying ET_o with Kc for the specific crop and its growth stage, we need to multiply by Ks, which is

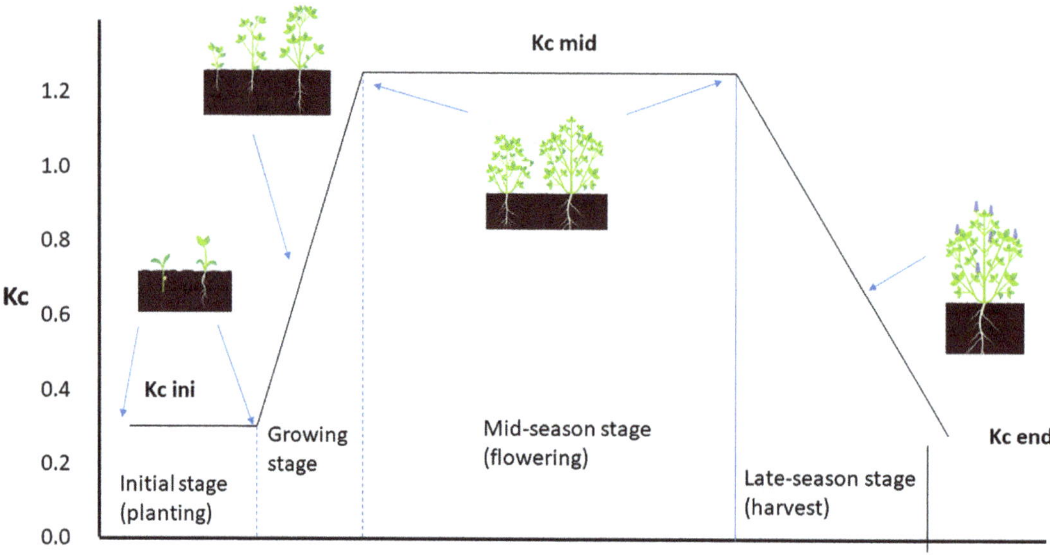

Figure 9.5 Example of how Kc value can change as a function of crop growth stage.
[After Allen et al., 1998]

basically a measure of the soil's saturation status (1 if it's saturated and water supply is not limiting; 0 if it's dry). Often, information on soil moisture or soil water content is not available and has to be assumed. For example, if a crop needs standing water or very wet soil for development, then Ks can be assumed as 1.0. Rice in Asia, for example, is often grown in the early stages with a standing pool of water. Although the flooding is for weed control, Ks can be assumed to be 1.0 during the early stages for the Asian variety of rice grown in monsoon climates. The whole process of converting the Penman–Monteith FAO56 ET_o to crop-specific ET_c for a given growth stage is summarized in Figure 9.6. Table 9.1 summarizes the Kc values for some of the major crops published by FAO that are specific to agro-climate zones.

9.3.2 Water Demand for Cultivating Rice

Let us talk about rice – the king of water-intensive crops. This will help us understand how satellite-based crop and irrigation management is necessary for water sustainability wherever rice is grown.

Rice is grown in large parts of the world and is the staple food source for almost half of the world's population (India, China, Southeast Asia, and major parts of Africa) (see Figure 9.7).

Agriculture accounts from anywhere from 50% to 90% of total freshwater demand, where rice occupies the first place (Figure 9.8) in terms of total freshwater consumption. In fact, rice accounts for almost 40% of the world's total irrigation water need.

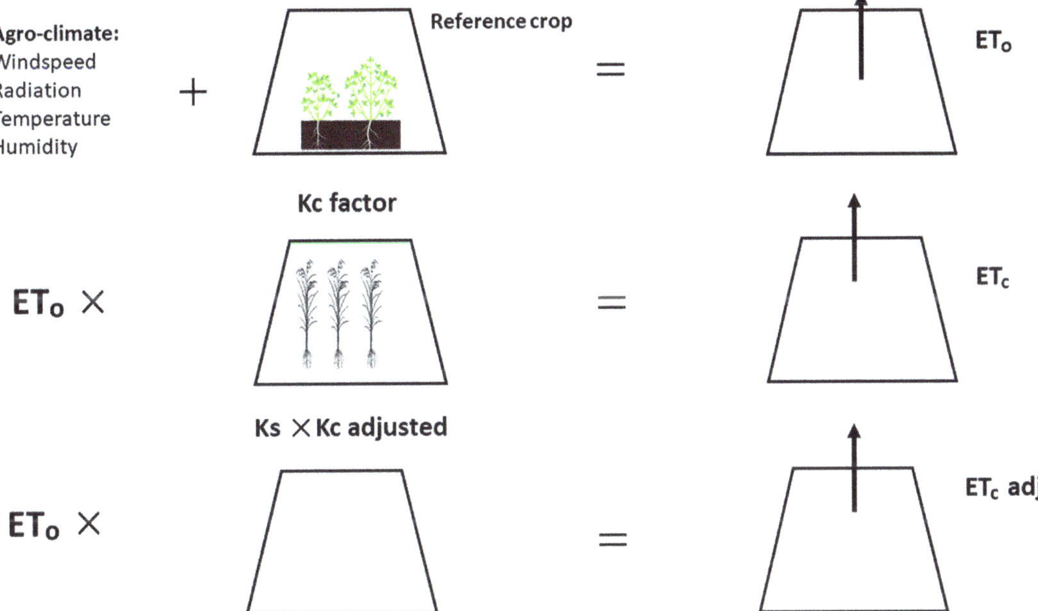

Figure 9.6 Applying Kc and Ks for estimating ET_c (crop water demand) from FAO56 Penman–Monteith reference evapotranspiration. [After Food and Agriculture Organization, Allen et al., 1998]

There is a lot of debate on whether rice is really that water-intensive. For example, the International Rice Research Institute has been recommending Asian farmers to practice alternate wetting and drying (AWD) to increase yield and optimize water consumption. Similarly, there are now better varieties of seeds for rice that are more water-efficient. Cultivation technology has also evolved in many places where the use of raised bed rather than the more common flood irrigation can reduce the water need for rice. However, no matter how we look at it, the Kc values for rice and the data in Figures 9.7 and 9.8 show that rice will continue to demand the lion's share of freshwater globally. Any freshwater management strategy must therefore be able to track the water demand from space, given the ubiquity and water-hungry nature of this particular crop.

Most rice farmers, particularly in Asia, do not have access to the latest knowledge and information. They continue to practice inefficient irrigation methods such as flood irrigation inherited from the previous generation, when water availability was plentiful compared with water demand. In fact, wasteful and water-intensive irrigation practices in many parts of the world have led to dramatic negative changes in groundwater storage (by over-pumping or groundwater mining). Many now forecast that critical breadbasket regions will become dry, with no groundwater to grow food (Figure 9.9). A recent study by the Central Groundwater Board of India has reported that western India is likely to run out of its groundwater in another 15 years (Singh, 2020).

We are also not really making optimum use of water and other cultivation methods to get the most out of our freshwater resources. Many studies show there is room for improvement with proper irrigation, use of fertilizers, or both. If we are going to solve the freshwater

Table 9.1 Kc values for various crops published by FAO

Note: depending on the planting date, climate and geographic location, these values will vary. These crop coefficient values assume the soil is not water-stressed. [Taken from Allen et al., 1998; www.fao.org/3/x0490e/x0490e00.htm]

Crop	Initial stage	Crop dev. stage	Mid-season stage	Late-season stage
Barley/oats/wheat	0.35	0.75	1.15	0.45
Bean, green	0.35	0.70	1.10	0.90
Bean, dry	0.35	0.70	1.10	0.30
Cabbage/carrot	0.45	0.75	1.05	0.90
Cotton/flax	0.45	0.75	1.15	0.75
Cucumber/squash	0.45	0.70	0.90	0.75
Eggplant/tomato	0.45	0.75	1.15	0.80
Grain/small	0.35	0.75	1.10	0.65
Lentil/pulses	0.45	0.75	1.10	0.50
Lettuce/spinach	0.45	0.60	1.00	0.90
Maize, sweet	0.40	0.80	1.15	1.00
Maize, grain	0.40	0.80	1.15	0.70
Melon	0.45	0.75	1.00	0.75
Millet	0.35	0.70	1.10	0.65
Onion, green	0.50	0.70	1.00	1.00
Onion, dry	0.50	0.75	1.05	0.85
Peanut/groundnut	0.45	0.75	1.05	0.70
Pea, fresh	0.45	0.80	1.15	1.05
Pepper, fresh	0.35	0.70	1.05	0.90
Potato	0.45	0.75	1.15	0.85
Radish	0.45	0.60	0.90	0.90
Sorghum	0.35	0.75	1.10	0.65
Soybean	0.35	0.75	1.10	0.60
Sugarbeet	0.45	0.80	1.15	0.80
Sunflower	0.35	0.75	1.15	0.55
Rice	0.35	1.05	1.20	0.90

sustainability problem and develop more robust policies for managing our finite freshwater resources, we must have a space-based approach for tracking the water demand for water-intensive crops and be able to track actual use globally, at actionable scales. Again, to get an accurate handle on the water need for crops, we need to know the crop type and the area of cultivation so that the total amount of water to be allocated for irrigation can be derived by water managers.

9.4   Tracking Area of Cultivation under Crop Type

Earlier, we had shown how visible and NIR reflectance imagery can be used to detect vegetation. For water management applications, this is usually not enough, as we also need to know

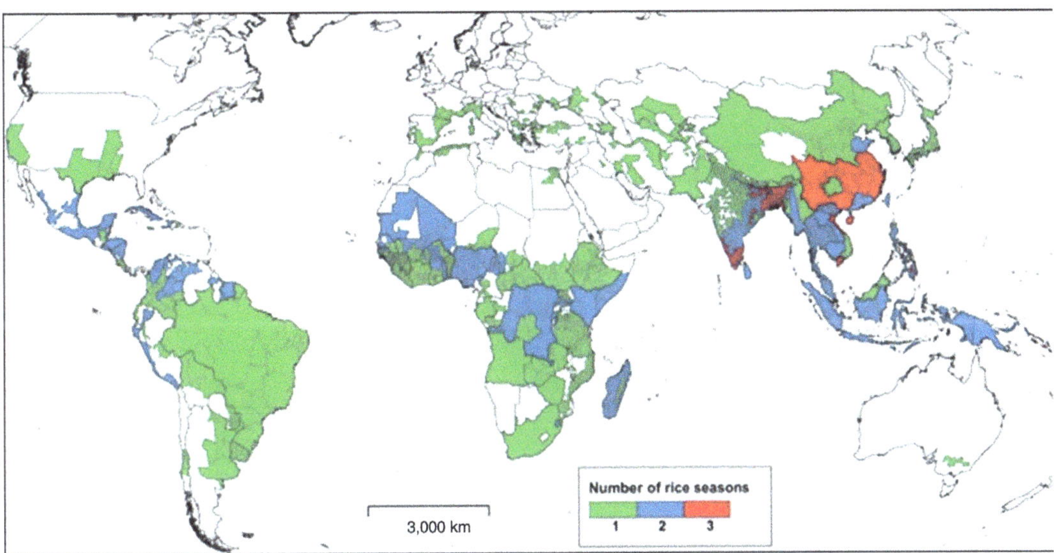

Figure 9.7 The widespread practice of rice cultivation. [After Laborte et al., 2017; image credit: Nature and Creative Commons License: http://creativecommons.org/licenses/by/4.0/]

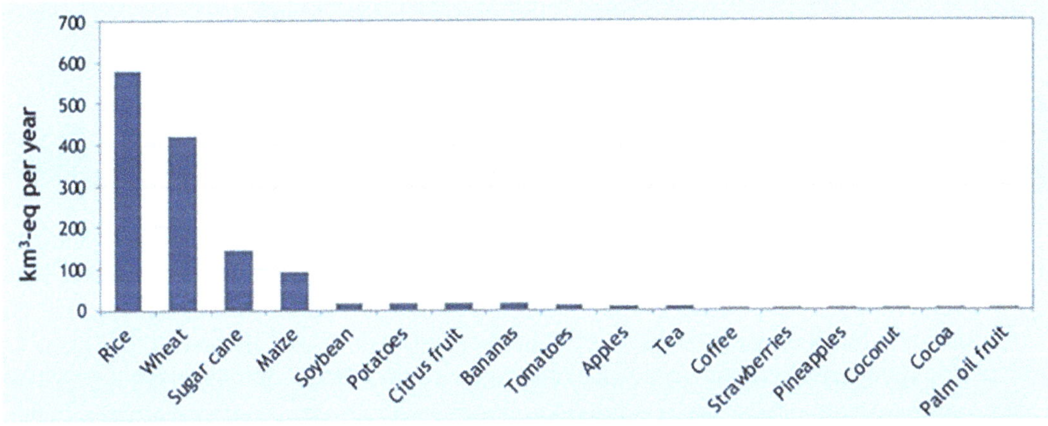

Figure 9.8 The total water (in km^3) used for major crops around the world showing the water-intensive nature of rice. For perspective, Lake Mead has about 30 km^3 of water when at capacity. So we use almost 20 Lake Meads to grow the world's rice today. [Image credit: taken from "Feeding Climate Change", Oxfam, June 2016, figure 6a, and adapted with permission of Oxfam, Oxfam House, John Smith Drive, Cowley, Oxford OX4 2JY, UK; www.oxfam.org.uk. Oxfam does not necessarily endorse any text or activities that accompany the materials, nor has it approved the adapted text.]

the crop type in the area classified as vegetated. Unfortunately, we do not have any normalized crop indices such as Normalized Difference Rice Index or Normalized Difference Corn Index that would work well using such satellite data. In essence, the NDVI would be of similar quantitative value for rice, corn, or wheat if the color was relatively green due to chlorophyll.

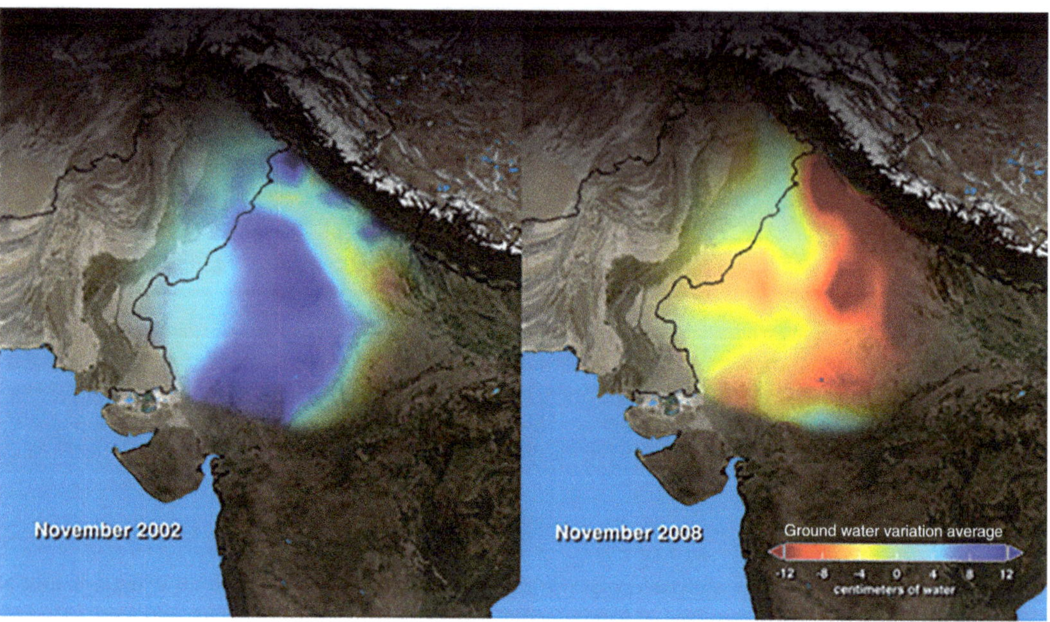

Figure 9.9 A satellite called GRACE (Gravity Recovery and Climate Experiment) shows a net negative change in water storage in the western Indian and eastern Pakistan regions (the breadbaskets for these two nations) through over-irrigation and over-pumping of the ground aquifers. [Image credit: NASA Scientific Visualization Studio]

We may be able to track the specific color or vegetation stress, but there is no crop-specific index or numerical methodology based on satellite data.

So how do we track from space what crop type is there and the areal extent of it? One way to do this nowadays for water management applications is with supervised classification and using machine learning of satellite data. Basically, if we know the exact location, time, growth stage, and crop type from ground information that is highly trustworthy, then we can "train" satellite imagery (say from Landsat, or Sentinel) to classify the rest of the scene's pixels as belonging to that crop type or not. We would also need to supply the "training" using machine learning with a list of locations that are the converse of the crop ("non-crop" – such as urbanized location, water, barren soil, dense forest) so that the training can learn and "supervise" the satellite imagery to filter out those types of pixels that are definitely not the crop type. The more data we feed from ground sources for training with crop and non-crop data, the more accurately the method works.

In this section, we show an example for rice classification for Bangladesh, which is the world's fourth largest rice growing nation. We have picked rice as we have already discussed its importance in water management, given the water-intensive nature of its production, and we use Sentinel-2 visible imagery (Figure 9.10).

While one could apply visible imagery such as Sentinel-2 or Landsat, such methods may underperform because of the widespread cloud cover during the monsoon season and the presence of fog during winter. However, in other places, further west, or more arid, such as

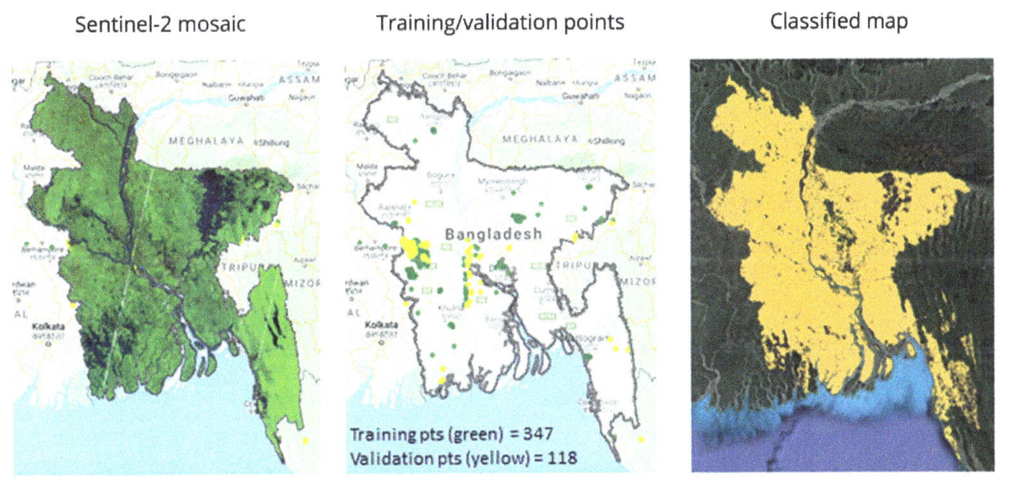

Figure 9.10 Example of land classified as rice growing in Bangladesh, using Sentinel-2 visible imagery. The total area under rice cultivation for the July–November season in 2021 is found to be 10.82 million hectares, which is an important piece of information for the country's food security planning. The middle panel shows location of training (validation) points where the location of rice cultivation was verified directly.

in Pakistan or Egypt, we could use visible imagery (Waleed et al., 2022). One way to overcome this challenge is to use the active microwave remote sensing from SAR imagery such as Sentinel-1 and perhaps combine it with visible imagery when and where available.

The SAR-based crop mapping was initially limited to small study areas because of the intensive data processing required. However, with the advent of high-performance cloud computing resources (see Chapter 4), we can now process very large geospatial data analyses such as those from Sentinel-1 SAR imagery. A point to note is that Sentinel-1 is angle-looking SAR, and it is the same active remote sensing instrument that can track surface water by measuring backscatter. Here, we can use the same principle as we did for Sentinel-2 using some type of data classification. First, we need to train the backscatter values to known locations of rice.

Today there are many such training methods in literature that one can use and train satellite data against ground observations to create a more online tool for classifying cropped area for a crop. We have created one such online tool based on cloud computing in Google Earth Engine (GEE) and on Sentinel-1 SAR data for rice area classification for Bangladesh (Figure 9.11). The goal here was to generate a 10-m paddy rice map (i.e. at the spatial resolution of Sentinel-1) for three distinct seasons: Boro (December/January–April), Aus (April/May–June/July), and Aman (July/August–November) for the entire country of Bangladesh, using GEE and time-series Sentinel-1 SAR imagery.

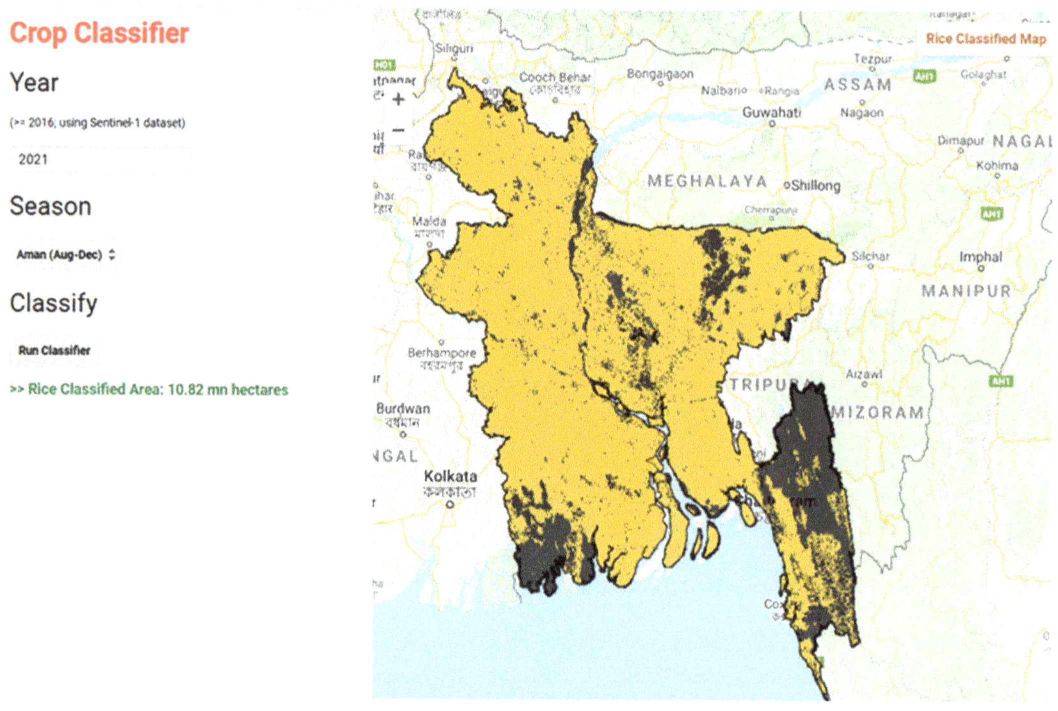

Figure 9.11 Example of a cloud-based platform for classifying microwave SAR imagery from the Sentinel-1 mission to identify rice-growing areas in Bangladesh.

Case Study 9.1: An Operational Satellite-Based Irrigation Advisory System (IAS) for Pakistan[1]

The Case of the Indus Basin Irrigation System (IBIS)

The Indus Valley is home to the world's largest irrigation system based on surface water. This system is known as the Indus Basin Irrigation System (IBIS; Figure 9.12). When IBIS was designed 80 years ago, the motivation was to bring more area under cultivation at low cropping intensity of one crop per year. However, IBIS is now being used to support cultivation of crops two to three times a year. While the amount of water that is typically available in any given year has remained the same, there is now more competition and demand for water among different sectors of the economy (such as energy, food, and industry) and also with neighboring India, which is home to the Indus River headwaters and shares groundwater aquifers. To address the increased demand for water, groundwater pumping is now required to supplement the surface water of IBIS.

[1] After Hossain et al., 2017

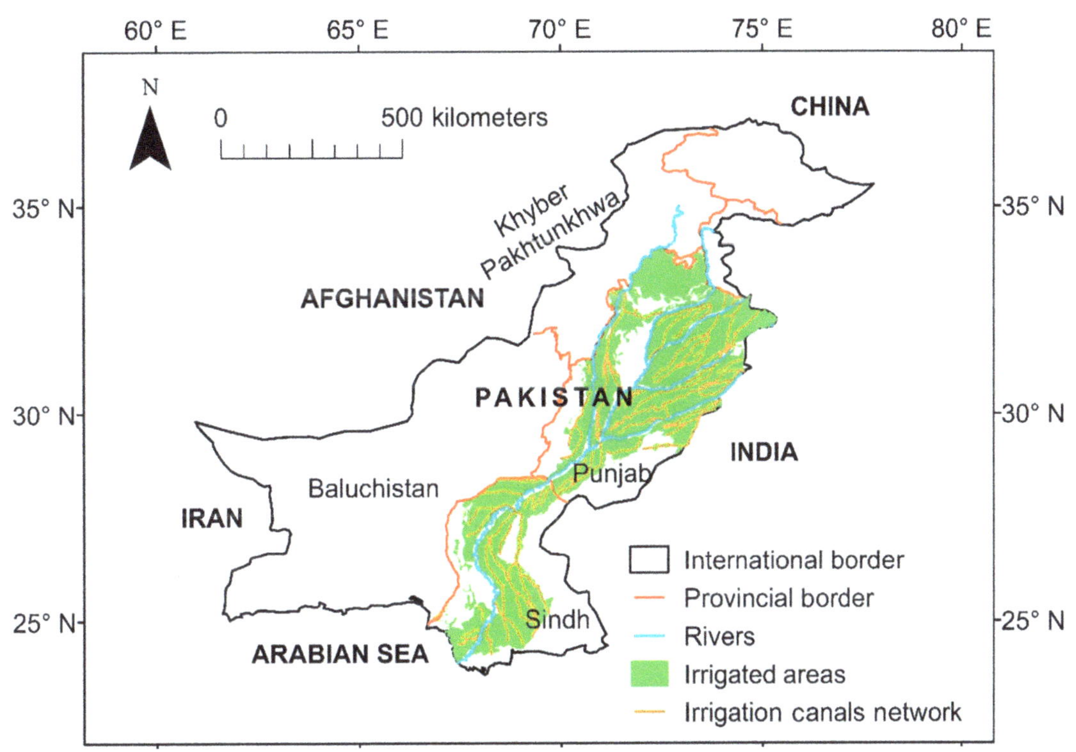

Figure 9.12 Location of the Indus Basin Irrigation System in Pakistan (green areas). [Reprinted with kind permission from Nature Publishing; after Muzammil et al., 2021]

However, there exists no water pricing system for farmers to irrigate their lands using groundwater, while a modest pricing scheme exists for the archaic IBIS surface water irrigation system. Farmers typically supplement the surface water irrigation with groundwater, which now meets more than 60% of total annual water demand. The more groundwater that is pumped for an inefficient irrigation approach, the more rapid is the decline in the water table. Pumping water from a greater depth also increases the fuel cost. This is further complicated by the fact that modern farming knowledge for crop water management is lacking, as mentioned earlier, because farming knowledge has been handed down from previous generations. For instance, the crop water requirements for rice, which consumes more than 60 percent of irrigation water in Pakistan, are 600 mm in Punjab province and 1,400 mm in Sindh province based on field measurements (lysimetric studies; source: Pakistan Council of Research in Water Resources). In contrast, the farmers apply almost 2,200 mm, resulting not only in a substantial loss of water but also an increase in fuel cost to pump water, and a lower crop yield. The water use efficiency of rice averages 0.45 kg of rice per m^3 of water in Pakistan compared with the world average of 0.71 kg m^{-3} of water. In a few irrigation districts of the Indus region, this efficiency is as low as 0.08 kg m^{-3} of water. Since overwatering reduces crop yield as well as increasing

Case Study 9.1 **(cont.)**

the cost of maintaining the supply of groundwater, it is no surprise that many farmers do not find farming profitable enough to sustain their livelihood.

Using Satellite Data to Track Water Requirement

It is obvious now that to mitigate the water crisis and develop a more robust water management strategy for IBIS, guidance on crop water requirements based on environmental conditions and location would be needed for the entire Pakistan region. This is the satellite-based Irrigation Advisory System (IAS). If farmers could be told specifically how much to irrigate (i.e. pump less when possible), given their tendency to overwater, then traditional mindsets could begin to change. The groundwater pumping component of irrigation could be driven by actual crop water demand and not by age-old convention dating back to when cultivation was one crop a year in IBIS. For consistent and data-driven messaging, farmers could be sent the information on crop water demand, along with weather forecast and suggestions for skipping irrigation as a Short Messaging System (SMS) pushed to farmers' cellphones. The whole system is now the cornerstone of IAS.

Advising Farmers on Irrigation Based on Demand Versus Supply

To advise farmers on how much to irrigate according to actual requirements (and reduce reliance on groundwater when possible), satellite precipitation data is useful. In Chapter 5, we have learned that we can now track precipitation from space globally at sub-daily timescales and at fairly high spatial resolution. Thus, if farmers could be told whether the precipitation that occurred recently was adequate to meet crop water demand, or if they could be informed of the net irrigation required in the coming week (demand minus amount already supplied via precipitation), then they would have a chance to use surface or groundwater irrigation more optimally. Not only that, if rain was forecasted and if that message could be relayed in the context of crop water demand and water supplied via recent precipitation, farmers could identify opportunities to skip irrigation (Figures 9.13a–b). The rationale is therefore based on comparing the demand with supply (precipitation). The demand for water is based on the crop- and location-specific ET data (as crop water requirement; Figures 9.14a and Figure 9.14b). The supply is precipitation and groundwater pumping. One can obtain precipitation data from the Global Precipitation Measurement (GPM) data product called IMERG, available in 10×10 km grids. Whenever supply (from precipitation) exceeds demand (i.e. crop water demand estimated from ET), farmers are sent messages reassuring them that they can pump less or no groundwater. Similarly, the deficit water (i.e. demand minus supply) is communicated to farmers as an irrigation amount they are encouraged to comply with by making sure the groundwater supplements the surface water from IBIS.

Using all the satellite data, ET information for a specific crop grown by a farmer (which the local agency would know a priori), combined with forecast information, a typical IAS message on a farmer's cellphone could read as follows:

Dear farmer friend, we would like to inform you that your wheat crop does not need irrigation due to sufficient rainfall during the past week.

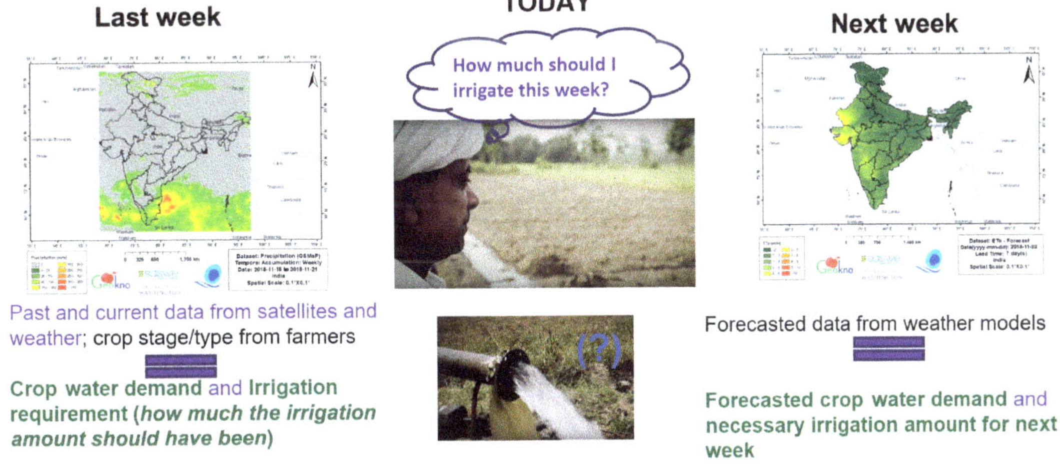

Figure 9.13a From farmer's perspective, showing how irrigation decisions need to be made based on past and near-future needs and conditions.

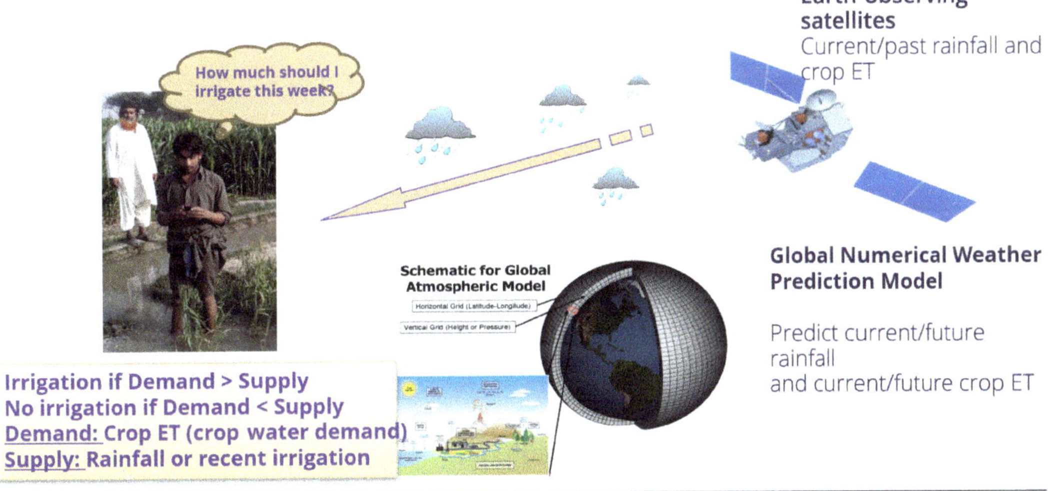

Figure 9.13b The overall schematic for the satellite-based Irrigation Advisory System (IAS) that is now operational in Pakistan.

Case Study 9.1 (cont.)

Or,

Dear farmer friend, we would like to inform you that the irrigation need for your banana crop was 2 inches during the past week.

These messages can be customized according to location, and crop type (note: crop type will let us convert ET_o to ET_c and do the demand versus supply calculations). Such an IAS has been operational in Pakistan, by the Pakistan Council of Research in Water Resources, serving anywhere from 20,000 to 100,000 farmers.

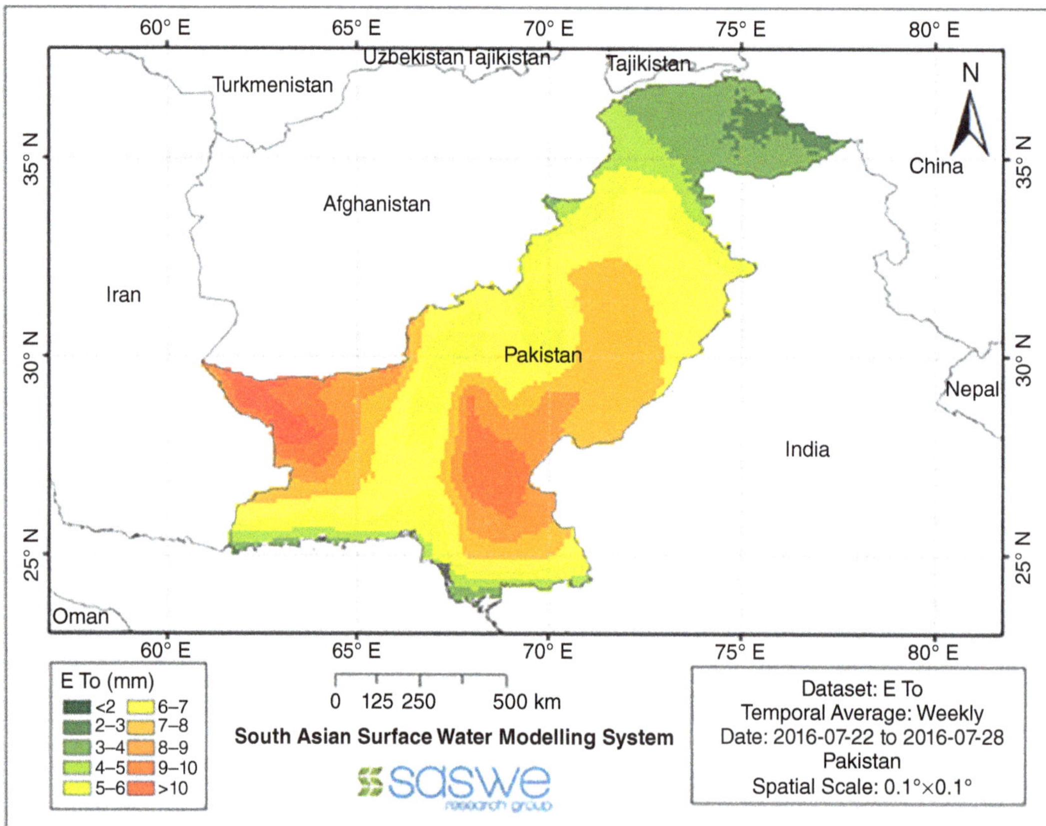

Figure 9.14a A map of ETo (weekly) based on FAO56 method (Allen et al., 1998) using NASA satellite data produced by the author's research group (SASWE; www.saswe.net).

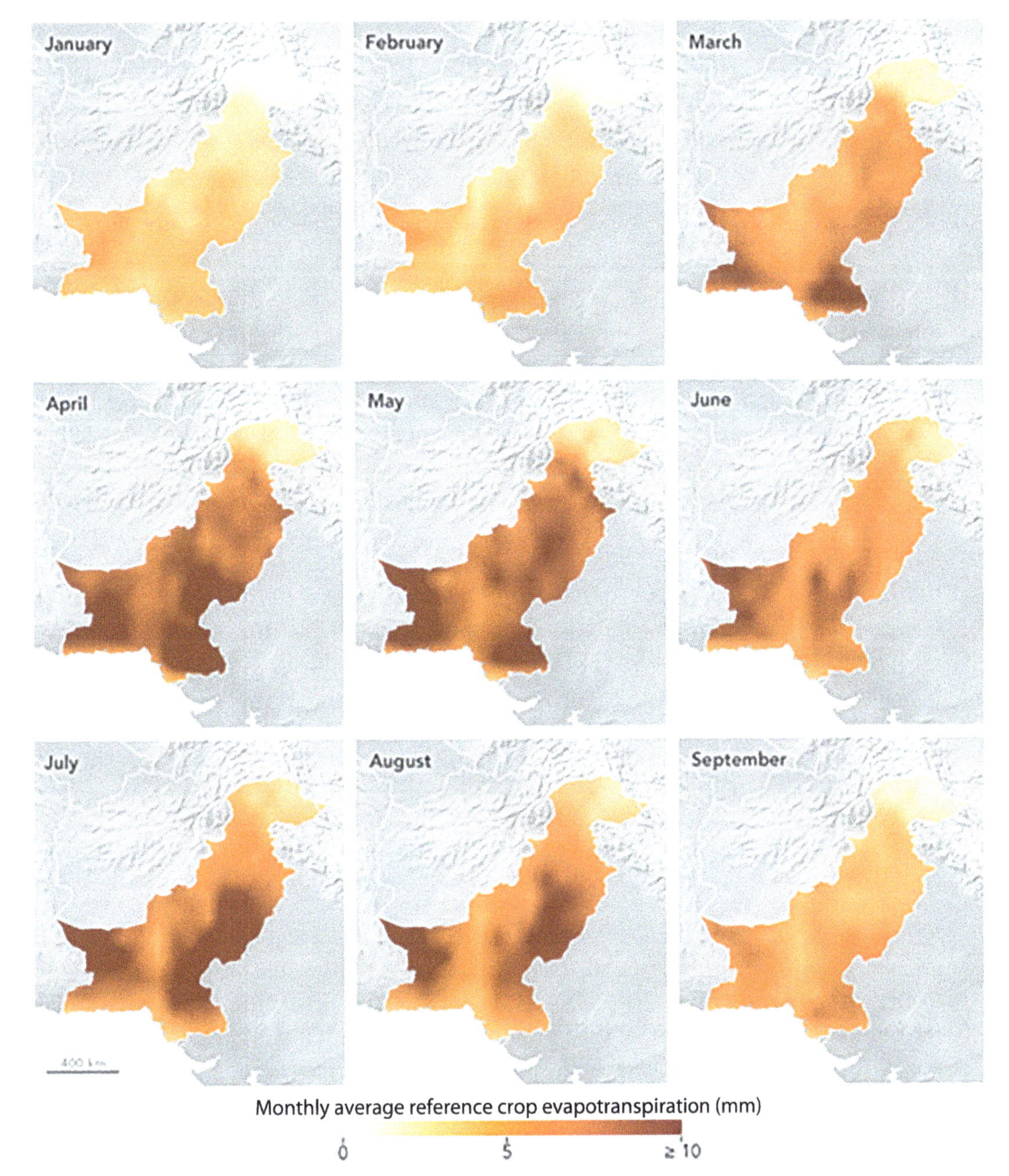

Figure 9.14b Monthly variations in ETo calculated by IAS using satellite data. [Image courtesy of NASA]

9.5   Tracking Actual Evapotranspiration Using Thermal IR Data

The previous section was all about using satellite and model data to track crop water demand using the FAO56 method and integrating that with precipitation data from satellites to provide farmers with an advisory that is designed for optimal irrigation. Alternately, the same approach could be used for a surface-water canal irrigation system to calculate the minimum amount of water that would need to be supplied to each command area and delivered via a canal regulatory system. Such an approach could also ensure that no area of an irrigation scheme is supplying more water than needed (with, of course, losses accounted for, along with some factor of safety).

We can also use satellite data, notably the visible/NIR and thermal infrared (TIR) bands to estimate the actual water consumption (ET) variability where availability of ground-based data is a constraint. SEBAL (Surface Energy Balance Algorithm for Land) and METRIC (Mapping Evapotranspiration at High Resolution and with Internalized Calibration) are such two widely used techniques (Liou and Kar, 2014). These methods work on the principle of tracking the energy balance for a crop. Whatever energy is unaccounted for is essentially assumed to be the latent heat source for driving evaporation (Figure 9.15). Discussing SEBAL is sufficient to cover both methods in this chapter to gain an understanding of how satellite data can be used to track actual ET. The SEBAL model solves the surface energy balance to compute ET using satellite images and weather data. Since the satellite image provides information for the overpass time only, SEBAL computes an instantaneous ET flux for the image time. A series of equations are incorporated in SEBAL model that compute net surface radiation, soil heat flux, and sensible heat flux to the air. The residual energy flux is then calculated by subtracting the soil and sensible heat fluxes from the net radiation at the surface. This residual energy (latent heat) enables liquid water to phase transition to water vapor, i.e. evapotranspiration, ET. Thus, for each pixel of the image, the ET flux (in units of $[L/T]$) is calculated as a residual of the surface energy budget equation:

$$\lambda ET = R_n - G - H \tag{9.3}$$

where λET is the latent heat flux, λ is the specific latent heat of vaporization, R_n is the net radiation flux at the surface, G is the soil heat flux, and H is the sensible heat flux (all in W m^{-2}).

The net radiation flux at the surface (R_n) represents the actual radiant energy available at the surface. It is computed by subtracting all outgoing radiant fluxes from all incoming radiant fluxes (Figure 9.16). This is given in the surface radiation balance equation:

$$R_n = R_S \downarrow - \alpha R_S \downarrow + R_L \downarrow - R_L \uparrow - (1 - \varepsilon_0) R_L \downarrow \tag{9.4}$$

Here σ is the solar albedo (the reflectivity of solar radiation for planet Earth); ε is the longwave thermal emissivity (hence $1 - \varepsilon$ is reflected); and S and L indicate shortwave and longwave.

In Equation 9.3, the soil heat flux (G) and sensible heat flux (H) are subtracted from the net radiation flux at the surface (R_n) to compute the "residual" energy available for evapotranspiration (λET). Soil heat flux is empirically calculated using vegetation indices, surface temperature, and surface albedo. Sensible heat flux is computed using wind speed observations, estimated

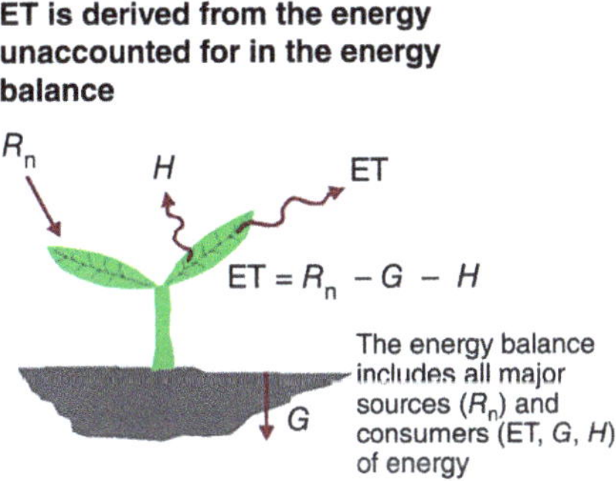

Figure 9.15 Energy balance in the SEBAL method for estimating actual ET. Note: here the term ET in the figure represents an energy flux that is obtained by multiplying the ET volumetric flux (in units of L/T) by the specific latent heat of vaporization λ.

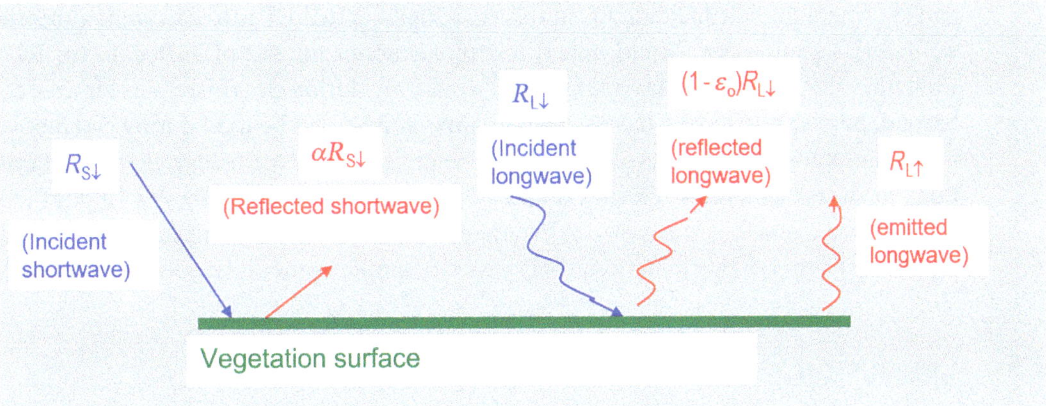

Figure 9.16 The outgoing and incoming radiation fluxes for a vegetative surface

surface roughness, and surface to air temperature differences. SEBAL uses an iterative process to correct for atmospheric instability due to the buoyancy effects of surface heating.

Once the λET is computed for each pixel, an equivalent amount of instantaneous ET (mm hr^{-1}) is readily calculated by dividing by the latent heat of vaporization (λ). These values are then extrapolated using a ratio of ET to reference crop ET to obtain daily or seasonal levels of ET. Reference crop ET, termed ET$_r$, is the ET rate expected from a well-defined surface of full-cover alfalfa or clipped grass and is computed using ground weather data. We can compute daily or 24-hour ET by assuming that the variations in instantaneous ET are not significant over the 24-hour period.

The SEBAL algorithm basically requires visible/NIR imagery to compute NDVI to identify cropped area and TIR imagery to compute land surface temperature. Recall from Chapter 3 that, using brightness temperature and knowledge of emissivity, one can estimate kinetic temperature relatively easily at longer wavelengths, albeit with atmospheric corrections needed for humidity, vapor, and stray light. We will cover surface temperature remote sensing in greater detail for water bodies in the next chapter.

In summary, SEBAL needs the following inputs:

- Reflectance of light energy (from satellite visible imagery such as Landsat)
- Vegetation indices (from satellite visible imagery such as Landsat)

- Surface temperature (using TIR data from Landsat)
- Relative variation in surface temperature (using TIR data from Landsat)
- General wind speed (from weather station or global weather models
- Humidity (from weather station or global weather models)
- Solar radiation (from parameterizations based on season, geographic location, Julian day)
- Air temperature (from weather station or global weather models)

The SEBAL algorithm as a software based on satellite TIR data and NDVI data from Landsat is actually more complex and involved, as it involves a series of processing steps. Reader can refer to the detailed training manual developed by SEBAL developers at: https://posmet.ufv.br/wp-content/uploads/2017/04/MET-479-Waters-et-al-SEBAL.pdf. Also, a better way to grasp the steps and see where the satellite data fits in this tracking of actual ET is to carry out a tutorial. Before we end this section, we should note a few things about the use of TIR data for ET tracking. First, this TIR-based SEBAL tracking will work only during cloud-free and fog-free conditions. Second, passive microwave-based surface temperature could be used during cloud cover, but its more than kilometer-scale resolution would be of little value for field-scale agricultural applications for water managers. Third, Landsat's current sampling frequency is 16 days (collectively for missions 8 and 9, it is 8 days). While there are plans to reduce this interval, one will have to assume uniform ET rate in between overpasses or employ other interpolation methods.

9.6   Conclusion

In this chapter, we have covered how satellite data can be used to manage crops and irrigation. We learned how satellite data can be used to estimate an area under a specific crop using classification techniques. Such techniques help us to understand the water need for that area. Next we will learn methods to estimate crop water demand and actual crop water consumption. In the next chapter, we will cover the remote sensing of surface temperature using thermal infrared sensing. This capability is important in tracking the surface water temperature of rivers and understanding how reservoirs may be regulating it. Water temperature is a water quality parameter. Because temperature controls many biological processes, water management needs to consider this parameter to make sure that water is being managed in a way that maintains both the required quantity and quality.

TUTORIAL ON USING SEBAL ET AND PENMAN ET TO TRACK IRRIGATION NEED[2]

Note: All relevant codes/scripts, data, and accompanying shape files for the geographic domain are available on this link: www.cambridge.org/hossain (copy and paste in the browser if the link does not work directly).

Before starting this tutorial, it will help if we read the following two reference articles to build relevant context:

---

[2] Contributed by Shahzaib Khan, University of Washington

A. Bose, I., Hossain, F., Eldardiry, H., et al. (2021). Integrating gravimetry data with thermal infra-red data from satellites to improve efficiency of operational irrigation advisory in South Asia. *Water Resources Research*, 57, e2020WR028654. https://doi.org/10.1029/2020WR028654

B. www.nasa.gov/missions/gpm/nasa-data-helps-bangladeshi-farmers-save-water-money-energy/

Brief Introduction to SEBAL ET and Penman ET

The SEBAL ET model was developed for estimating actual evapotranspiration (ET_a) and other surface energy balance components. The model uses satellite imagery and ancillary meteorological data to derive ET_a. SEBAL calculates the energy balance components by considering the differences between incoming and outgoing shortwave and longwave radiation and the energy used in evaporation and soil heat flux. By determining these energy fluxes, SEBAL allows for assessing water consumption in agricultural fields, natural vegetation, and other landscapes, making it a valuable tool for water resource management.

The Penman equation is used to estimate potential evapotranspiration (PET). PET calculation combines both energy (radiative) and aerodynamic (wind and vapor pressure deficit) components. PET, after it is adjusted for the specific crop and soil saturation conditions, represents the crop water demand. In other words, SEBAL ET can be considered a good proxy for how much water has been consumed by crops through transpiration, while PET adjusted for crop and soil conditions indicates how much water the crops needed for optimal growth. If SEBAL ET is greater than PET (crop and soil adjusted), then there is likely overirrigation in the region (assuming the region is being irrigated). Vice versa, if PET (adjusted for crop and soil conditions) is greater than SEBAL ET, the region is likely being underirrigated. Bose et al. (2021) provides more details to understand this idea.

The Penman formula is given in Equation 9.2. More details about this formulation to calculate PET can be found at www.fao.org/3/x0490e/x0490e06.htm#penman%20monteith%20equation

Region of Interest (ROI): The tutorial and exercise are based on the Rupasi union in northeast Bangladesh. Here a "union" is the smallest administrative unit in Bangladesh. Below is the map of the union's location (Figure 9.17).

In this tutorial, we will estimate if the region of interest is under or over-irrigation using the SEBAL and Penman equation. We will be using the Google Earth Engine (GEE)-based Python API. In essence, this is local computing but with a cloud option to access the datasets in GEE. This is very similar to the Chapter 6 tutorial using Python to calculate area of surface water before and after Belo Monte Dam project.

Step 0: Prerequisite step. We need to install the following packages to run the script in Python.

a. Matplotlib [conda install matplotlib]
b. Rasterio [conda install -c conda-forge rasterio]
c. Earth Engine Python API [conda install -c conda-forge earthengine-api]

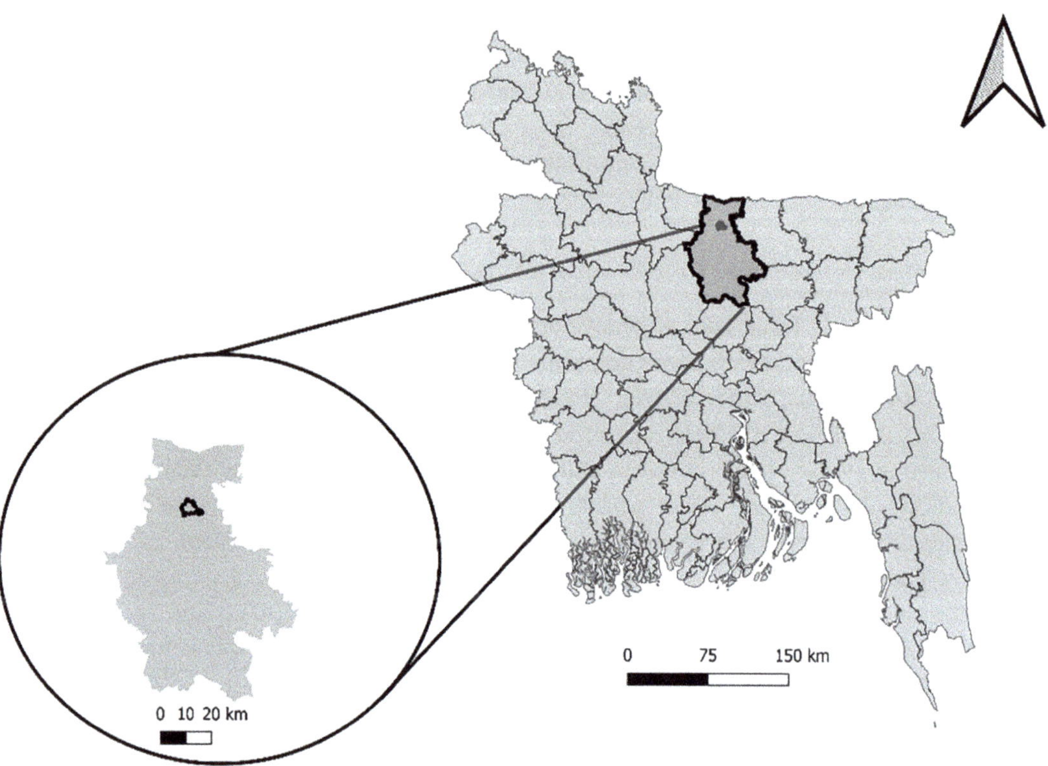

Figure 9.17 Rupasi Union in Bangladesh.

Step 1: Importing the required libraries.

```python
# Importing required libraries
import ee
import os, datetime, requests, zipfile, time
import urllib.request
import datetime,math
import numpy as np
import subprocess as sub
import rasterio as rio
import matplotlib.pyplot as plt
import warnings
warnings.filterwarnings("ignore")
```

Step 2: Initializing GEE Python API. This step is a must; otherwise, the local machine will not be connected to the cloud server

```python
# Initialise Earth Engine
# ee.Authenticate() # If you are running the earth engine first time in your
machine, you need to run the ee.Authenticate() command first.
ee.Initialize()
```

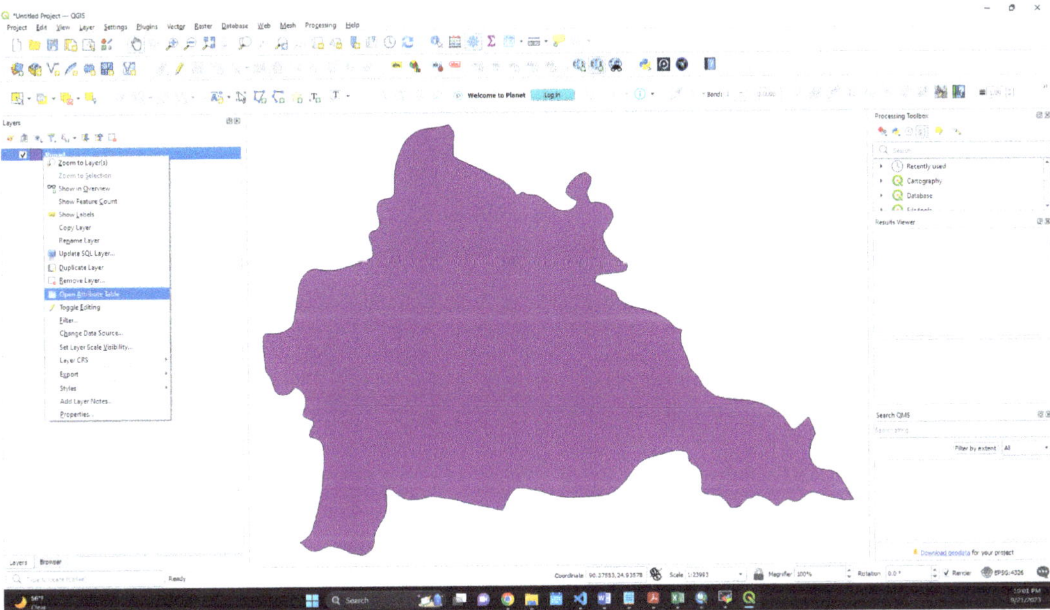

Figure 9.18 QGIS dashboard for Rupasi Union, Bangladesh.

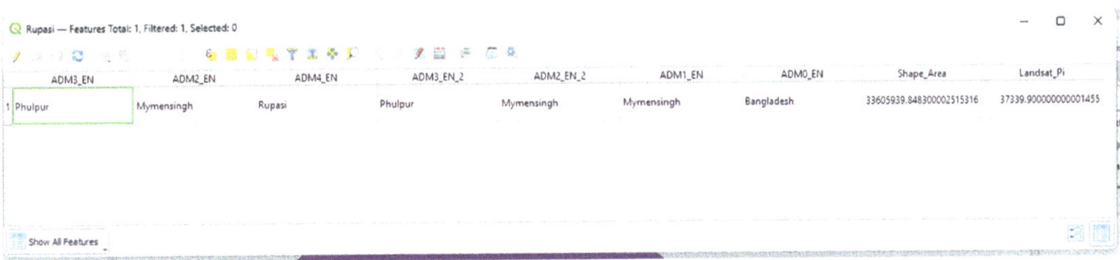

Figure 9.19 Opening attribute table for Rupasi Union in QGIS.

Step 3: Understanding the shapefile features. For the tutorial, Rupasi (the fourth administrative level of Bangladesh) is used as a region of study. For ease, the shapefile of Rupasi has been provided. Refer to the link provided at the beginning of this tutorial. On this same link the shapefile can be accessed. Readers can explore the attributes of the shapefile as mentioned above using QGIS or Python. Here a snapshot of the attributes of the given shapefile is shown using QGIS software (Figure 9.18), which can be obtained by right clicking on the shapefile in QGIS (or any GIS software) and clicking on 'open attribute table' option (Figure 9.19).

To upload the shapefile to GEE, click on the "New" button in the Assets Tab and select "Shape files" as shown below (Figures 9.20 and 9.21).

Click on 'Select' and upload the shp, dbf, prj, shx, cpg files and give the name of the uploaded shapefile in the "Asset ID" tab. After uploading is finished copy the Asset ID. For example, in the tutorial, the Asset ID is "users/skhan7/PhD/TextBook/Rupasi".

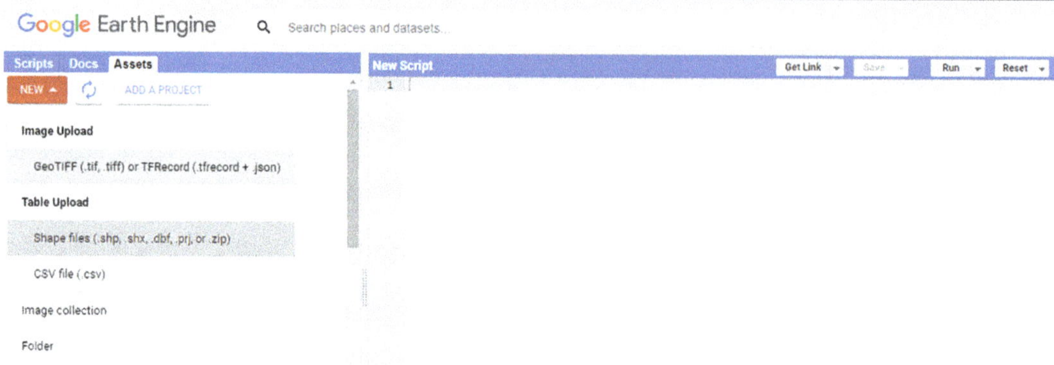

Figure 9.20 Uploading shapefile to GEE. [Credits: Google Earth Engine]

Figure 9.21
Uploading
shapefile to GEE.

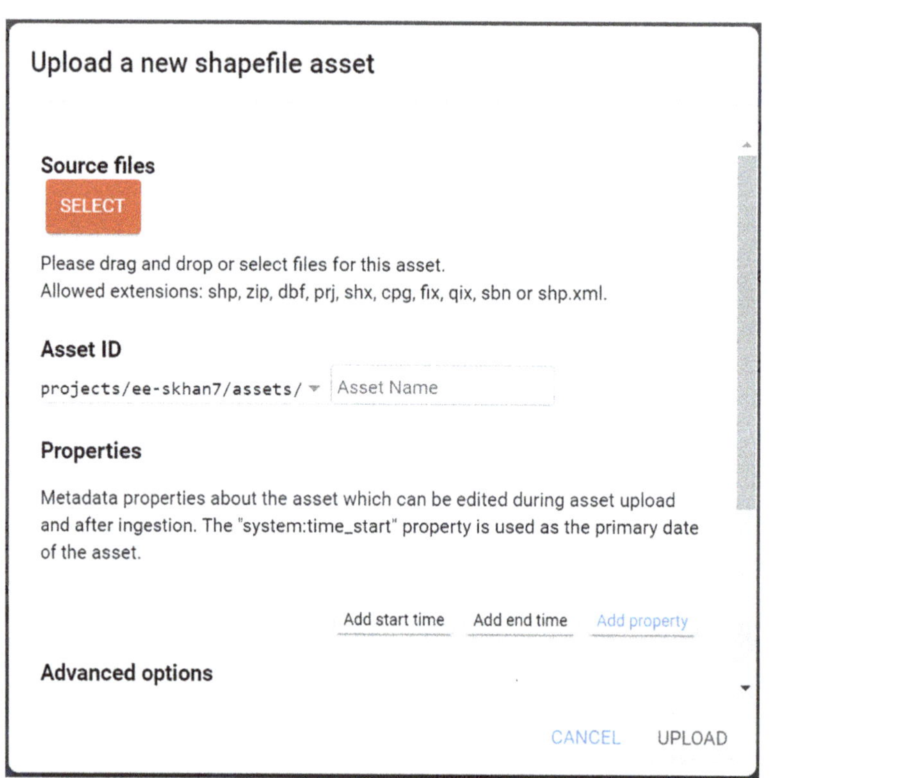


```
###########*******Module (1) USER INPUT (ROI and Intereset Dates and Planting
Date)********##########/
alldistsbd = ee.FeatureCollection('users/skhan7/PhD/TextBook/Rupasi'); # Users
need to change the Asset ID as per their requirement.
upazilalist = alldistsbd.reduceColumns(ee.Reducer.toList(1), ['ADM3_
EN']).get('list').getInfo()
glcc = ee.Image("COPERNICUS/Landcover/100m/Proba-V-C3/Global/2019");
wktimes = ["lastweek",'currentweek']
```

Serial no.	Attribute name	Meaning
1	ADM0_EN	Country
2	ADM1_EN	Division
3	ADM2_EN	District
4	ADM3_EN	Upazila
5	ADM4_EN	Union

Step 4: Understanding the data. Below is the table showing all the datasets used in the algorithm.

Dataset	Dataset ID	Band name	Description	Unit	Spatial resolution	Temporal resolution	Temporal aggregation
GLDAS	NASA/GLDAS/ V021/NOAH/ G025/T3H	Tair_f_inst	Air temperature at 2 m	K	25 km	3 hourly	Daily
		Wind_f_inst	Wind speed at 10 m	m/s	25 km	3 hourly	Daily
		Qair_f_inst	Specific humidity	Kg/kg	25 km	3 hourly	Daily
		Psurf_f_inst	Pressure	Pa	25 km	3 hourly	Daily
Landsat 8	LANDSAT/LC08/ C02/T1_TOA	B2	Blue		30 m	16 days	
		B4	Red		30 m	16 days	
		B5	NIR		30 m	16 days	
		B6	Shortwave IR		30 m	16 days	
		B10	Low-gain Thermal Infrared 1		Resampled from 60 m to 30 m	16 days	
		B11	High-gain Thermal Infrared 1		Resampled from 60 m to 30 m	16 days	
SRTM	USGS/ SRTMGL1_003		Digital Elevation Data				
GLCC		discrete_ classification	Land Use Land Cover Data	%	100 m		

Step 5: Overview of the script. Readers are **strictly** advised to understand SEBAL and the Penman–Monteith equation. The script cannot be understood without proper knowledge of the two methods. In the script, we have estimated the evapotranspiration through Penman–Monteith and SEBAL, downloaded the respective TIF files, and then estimated the irrigation needs based on observed (SEBAL) and expected (Penman–Monteith) evapotranspiration.

Step 6: Inputting the desired dates in the function. The tutorial will estimate the SEBAL ET and Penman ET for two weeks for the rice crop. We want to visualize the over/underirrigation from Feb 7, 2023 to Feb 21, 2023. The following screenshot shows the same. Line 22 and line 23 show how dates are given as input in the GEE.

```python
def sebal_eto_Landsat():
  for wktime in wktimes:
    if wktime == "lastweek":
      startdate ='2023-02-07'
      dayVal = 14
    if wktime == "currentweek":
      startdate = '2023-02-14'
      dayVal = 7
    print('startdate = ',startdate)
    enddate = '2023-02-21'
    print('enddate = ',enddate)
    startDate=ee.Date(startdate)
    endDate=ee.Date(enddate)
```

Step 7: Setting the planting date and crop coefficient. For rice, the planting date in Bangladesh is mid-January. Hence, a planting date of "2023-01-15" has been selected. The planting date depends on the type of crop. Based on the crop and its planting date, the crop coefficient is empirically derived through years of studies and observations. The behavior of the crop coefficient with the number of days it has been sown is shown from line 67 to line 77 in the snapshot below. For a crop other than rice, lines 67 to 77 must be changed accordingly.

```python
planting_YY_MM_DD = '2023-01-15' ## Change the planting date as per you crop
planting_date = datetime.datetime.strptime(planting_YY_MM_DD, '%Y-%m-%d').date()
print('Difference between planting date:',abs((my_datetime - planting_date)).days)
num_days = abs((my_datetime - planting_date).days)
print('Number of days = ',abs((my_datetime - planting_date).days))
growth_kc = 1.23

if num_days<=20:
    growth_kc=1.1
elif 20<num_days<=45:
    growth_kc=1.1+(0.3/25)*(num_days-20)
elif 45<num_days<=105:
    growth_kc=1.4
elif 105<num_days<=115:
    growth_kc=1.4+(-0.4/10)*(num_days-105)
elif 115<num_days<=140:
    growth_kc=1
else:
    print('Number of days is greater than 140')
    pass
print('growth_kc=',growth_kc)
```

Step 8: Running the function. The following lines of script execute a multiple function, i.e. creating the directory where the TIF files will be stored, clearing the previously run data, running the SEBAL and Penman function, unzipping the TIF files downloaded by running the SEBAL and Penman function, and finally converting the unzipped TIF to usable TIF files.

```python
if __name__ == '__main__':
  print('Making Directory if it does not exist')
  makeDirectory()
  print('Finished making directory if they were not present')
  print('Clear Data Started')
  clearData()
  print('Clear Data Finished')
  print('sebal_eto_Landsat Started')
  sebal_eto_Landsat()
  print('sebal_eto_Landsat Finished')
  print('unziptiffs Started')
  unziptiffs()
  print('unziptiffs Finished')
  print('convertTiffs Started')
  convertTiffs()
  print('convertTiffs Finished')
```

Step 9: Estimating the irrigation over the region. To estimate if the region is under/over-irrigated, we need to check the difference between SEBAL ET and Penman ET. If the difference between the two is positive, the observed water availability is more than the crop's actual water demand, hence overirrigation. There are some assumptions here that we will not go into the details of. The calculation of ET is based on the assumption that the Kc (crop coefficient value) and scale of weather data as input to PET calculation is representative. In addition, the soil stress factor Ks is assumed to be 0.5 in this case, which implies that the average long-term soil saturation is 50%. With these assumptions, if the difference is negative, the crop water demand exceeds the available water supply. Hence this is underirrigation. More details about over- and underirrigation status can be found in Bose et al. (2021). In the following snapshot of the script, we are trying to visualize the SEBAL ET and Penman ET over the region of interest.

```python
# Open the TIFF file using Rasterio
penman = './Data/uploads/landsat_lastweek_penman_Phulpur.tif'
sebal = './Data/uploads/landsat_lastweek_sebal_Phulpur.tif'
with rio.open(penman) as dataset:
  penman = dataset.read(1) # Assuming you want to display the first band

with rio.open(sebal) as dataset:
  sebal = dataset.read(1) # Assuming you want to display the first band

# Create a Matplotlib figure and axis
fig, ax = plt.subplots(nrows = 1, ncols = 2,figsize=(20, 6))

# Display the image with a colorbar
cax = ax[0].imshow(penman, cmap='viridis') # You can choose a different colormap
if desired
# plt.colorbar(cax, ax=ax[0], label="Penman",fontsize = 14)
cbar = plt.colorbar(cax, ax = ax[0])
cbar.set_label(label="Penman ET (mm)",fontsize =15)
cax1 = ax[1].imshow(sebal, cmap='viridis') # You can choose a different colormap
if desired
cbar1 = plt.colorbar(cax1, ax = ax[1])
cbar1.set_label(label="SEBAL ET (mm)",fontsize =15)
```

```python
ax[0].set_title('Penman Evapotranspiration', fontsize = 21, fontweight = 'bold')
ax[1].set_title('SEBAL Evapotranspiration', fontsize = 21, fontweight = 'bold')
fig.suptitle('Evapotranspiration',fontsize = 24, fontweight = 'bold')
plt.show()
```

Step 10: Estimating mean irrigation percentage.

```python
# To estimate the mean irrigation percentage over the Rupasi area for rice in the
last week (2023-02-07 to 2023-02-14), we will estimate the mean while ignoring
the Nan values

mean_irr = np.nanmean(irrigation)

print(f'Mean Irrigation Percentage over Rupasi Area for Rice: {mean_irr:.2f} %')
```

Results

Below are the TIF files generated from the tutorial. Readers are advised to work on the exercise using the same code provided (Figures 9.22 and 9.23).

```
Mean Irrigation Percentage over Rupasi Area for Rice: -92.87 %
```

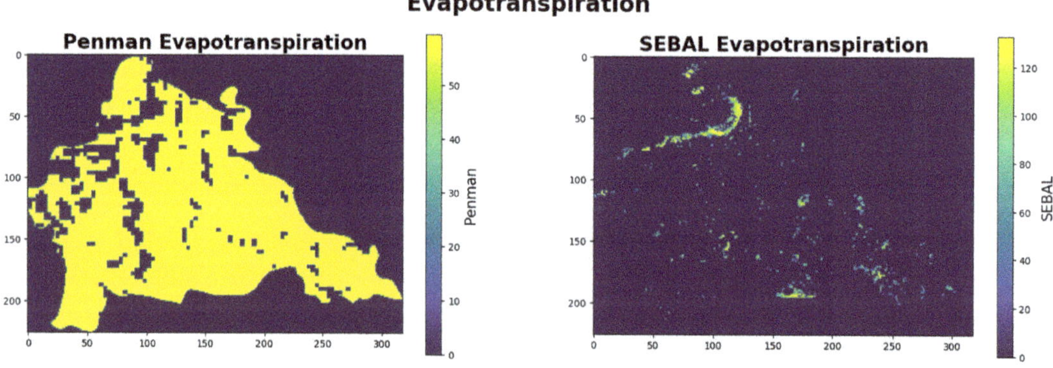

Figure 9.22 Map of Penman PET and SEBAL ET.

Figure 9.23 Map of over/underirrigation for Rupasi Union. Most of the union for rice cultivation appears to be underirrigated, subject to caveats and assumptions summarized in Step 9.

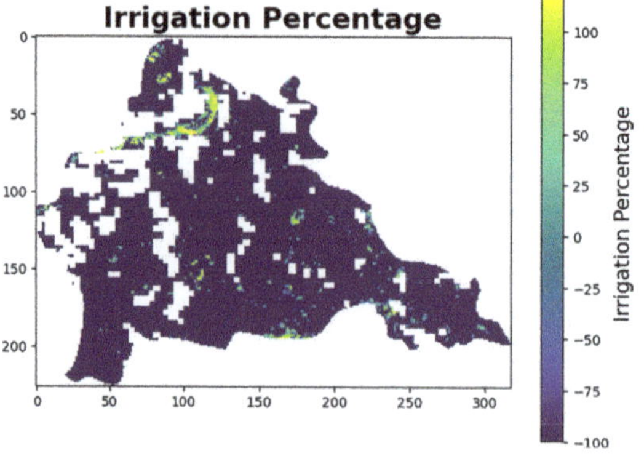

EXERCISES: CHAPTER 9

Q9.1   What is the normalized index used to track vegetation using satellite data? What are its limitations?

Q9.2   Why is tracking crop type and area under cultivation critical for water management applications?

Q9.3   How can crop type be identified using satellite data?

Q9.4   What is crop water demand? Name one of the key methods for estimating it, and list the inputs needed for its calculation.

Q9.5   Explain briefly the meaning of the terms "reference ET_o" and "crop ET". What coefficients are used for calculation of crop ET?

Q9.6   Discuss briefly how the tracking of crop type and crop water demand can help in better water management practices.

Q9.7   What is the SEBAL algorithm, and what inputs does it need? Highlight the inputs that are available from satellite data.

Q9.8   What are the limitations of the SEBAL algorithm based on Landsat data?

Q9.9   **Data-based exercise (Contributed by Shahzaib Khan, University of Washington):** Using your understanding from the tutorial, determine the over/underirrigation zones for Rupasi's wheat crop from March 1, 2023, to March 15, 2023, considering the following crop coefficient's variation based on the number of days after sowing. Assume the planting date for the wheat crop to be November 15, 2022.

Tip: Use the Crop Coefficient vs Days figure to change the if else condition for the Kc values.

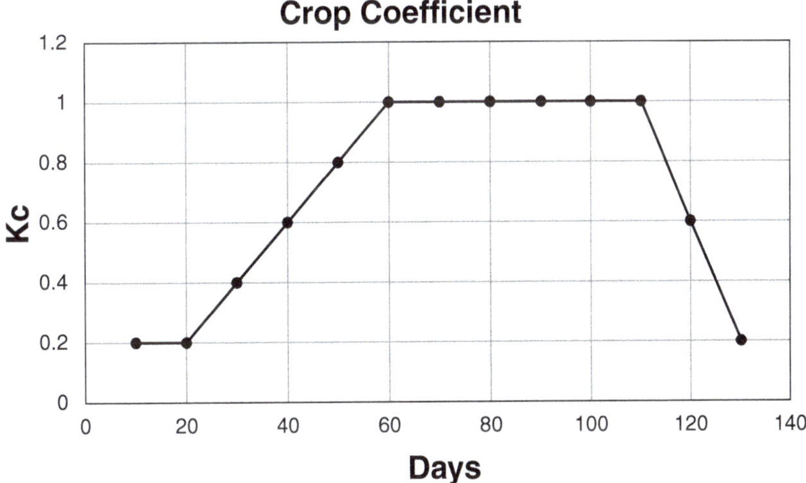

REFERENCES

Allen, R. G., L. S. Pereira, D. Raes, and M. Smith (1998). *Crop Evapotranspiration – Guidelines for Computing Crop Water Requirements*. FAO Irrigation and Drainage Paper 56. FAO. Available at: www.fao.org/3/x0490e/x0490e00.htm (last accessed June 3, 2023)

Bose, I., F. Hossain, H. Eldardiry, et al. (2021). Integrating gravimetry data with thermal infra-red data from satellites to improve efficiency of operational irrigation advisory in South Asia. *Water Resources Research*, 57, e2020WR028654. https://doi.org/10.1029/2020WR028654

Hossain, F., N. Biswas, M. Ashraf, and A. Z. Bhatti (2017). Growing more with less using cell phones and satellite data. *EOS*, vol. 98. https://doi.org/10.1029/2017EO075143

Jones, H. G. and R. A. Vaughan (2010). *Remote Sensing of Vegetation: Principles, Techniques and Applications*, Oxford University Press.

Laborte, A., M. Gutierrez, J. Balanza, et al. (2017). RiceAtlas, a spatial database of global rice calendars and production. *Scientific Data*, vol. 4, 170074. https://doi.org/10.1038/sdata.2017.74

Liou, Y. A. and S. K. Kar (2014). Evapotranspiration estimation with remote sensing and various surface energy balance algorithms – a review. *Energies*, vol. 7, 2821–2849.

Meissner, R., H. Rupp, and L. Haselow (2020). Chapter 7 – Use of lysimeters for monitoring soil water balance parameters and nutrient leaching. In Vara Prasad, M. N. and Pietrzykowski, M. (eds.) *Climate Change and Soil Interactions*, 171–205. Elsevier. https://doi.org/10.1016/B978-0-12-818032-7.00007-2

Muzammil, M., A. Zahid, and L. Breuer (2021). Economic and environmental impact assessment of sustainable future irrigation practices in the Indus Basin of Pakistan. *Scientific Reports* 11, 23466. https://doi.org/10.1038/s41598-021-02913-9

Singh, G. (2020). India's food bowl heads toward desertification. *EOS*, vol. 101. https://doi.org/10.1029/2020EO147581

Waleed, M., M. Mubeen, A. Ahmad, et al. (2022). Evaluating the efficiency of coarser to finer resolution multispectral satellites in mapping paddy rice fields using GEE implementation. *Scientific Reports*, vol. 12, 13210. https://doi.org/10.1038/s41598-022-17454-y

SUGGESTED READING

Bhattarai, N., and P. Wagle (2021). Recent advances in remote sensing of evapotranspiration. *Remote Sensing*, vol. 13, 4260. https://doi.org/10.3390/rs13214260

Ozdogan, M., Y. Yang, G. Allez, and C. Cervantes (2010). Remote sensing of irrigated agriculture: opportunities and challenges. *Remote Sensing*, vol. 2, 2274–2304. https://doi.org/10.3390/rs2092274

Zhang, K., J. S. Kimball, and S. W. Running (2016). A review of remote sensing based actual evapotranspiration estimation. *WIREs Water*, vol. 3, 834–853. https://doi.org/10.1002/wat2.1168

10 Satellite Remote Sensing of Surface Water Temperature

10.1 Chapter Overview

In the previous chapter we covered how satellite remote sensing can be used to classify areas under a crop, and estimate their crop water demand and actual crop water consumption. This information can be used for irrigation management using satellite data. In this chapter, we will cover how satellite data can be used to estimate the temperature of surface water. We will cover the basic principle behind the estimation technique, understand its limitations, and then build some data literacy to derive the surface temperature of water in regulated rivers ourselves.

10.2 Introduction

Satellite remote sensing can allow us to estimate the temperature of targets in question by recording the radiation emitted in the longer electromagnetic (E-M) wavelengths of infrared and microwave (IR and MW). For example, we learned in Chapter 3 that all bodies at a temperature above −273 °C naturally radiate E-M energy. Planck's equation tells us at which wavelength this radiation is maximum, based on the body's temperature. At relatively low and planetary temperatures in the ranges of 0–40 °C, most of this E-M energy is at longer wavelengths of thermal IR (TIR) and MW. At these longer wavelengths, we can use the Rayleigh–Jeans approximation, and the brightness (radiative) temperature is a product of emissivity and kinetic (surface) temperature of interest. Also, at the TIR wavelengths, there are many naturally occurring bodies such as snow and water with a very high emissivity that are essentially black bodies in those wavelengths. We have also seen from the tutorials in the previous chapter that the TIR-based land temperature plays a direct role in estimating actual evapotranspiration, according to the energy balance algorithm called SEBAL.

In this chapter, we will concern ourselves with temperature estimation for surface water to understand the practicality of the method and its relevance for water management.

10.2.1 Why Track Surface Water Temperature for Water Management?

Ever since TIR detectors of various bands were put on Landsat missions in the 1980s, we have had the ability to track surface temperature at scales of 100 m. This capability has been widely used to develop numerous land surface temperature products for assimilation into global land surface models, climate models, and crop water tracking models. For water management

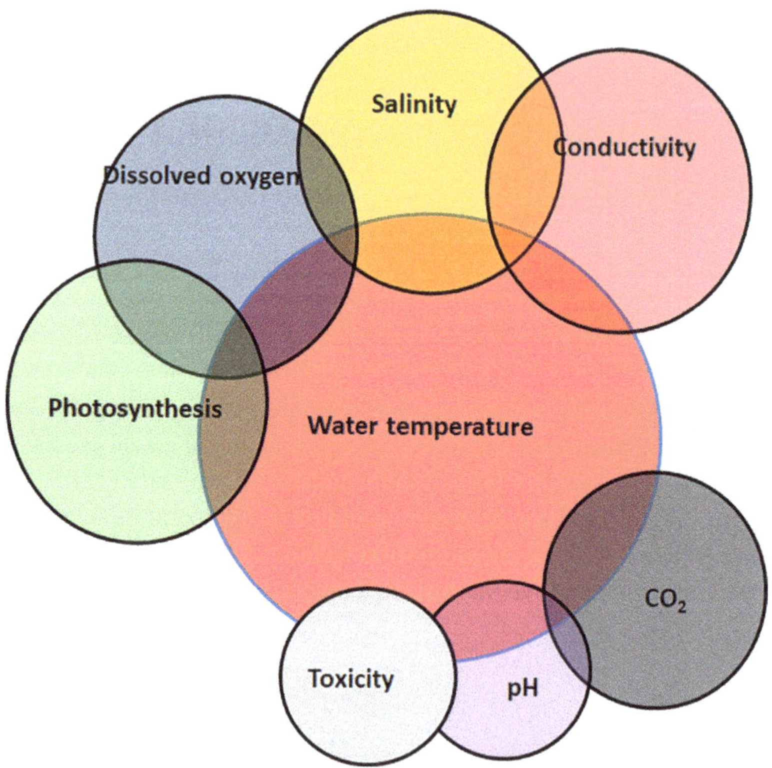

applications, this vast data source on water surface temperature has been relatively underutilized until recently. Here we will make a case as to why water managers should also track surface water temperature as an important water quality parameter.

Temperature is a water-quality metric controlling biological activity, potential for algal bloom, dissolved oxygen, pH, and aquatic habitat for fish and other marine animals (Figure 10.1). Thus, water management, and in particular integrated water management, should be guided by the water temperature that may be altered by the management decision. The case of reservoir management, which is a typical water management application, can make this point clear.

Any large body of water, such as a reservoir or lake, that is deep and located in a region that experiences seasonal temperature changes from cold (winter) to warm (summer) ambient temperature will often stratify into distinct thermal layers (Figure 10.2). This thermal stratification depends, however, on the bathymetry of the reservoir; deeper reservoirs will stratify, but shallow and wide ones will not do so as easily. The overall climate and location can also play a role. Typically, during summer time, the bottom layers of the reservoir are colder, owing to solar radiation being able to warm only the upper layers. During winter, the situation reverses, as the upper layers start cooling owing to ambient temperature dropping. This results in a "turnover" or mixing as warmer water will try to float to the top. We have long known this thermal effect for reservoir operations, which is particularly exacerbated for hydropower dams that generate power by pushing the colder waters at the bottom of the pool through the penstock (Figure 10.2).

The United States, like most developed nations, has tracked the temperature of rivers up and down a reservoir at many places using in-situ temperature probes. But what about the rest of the world where such in-situ data is mostly absent? It turns out there is a lot we can decipher in many of the developing regions of the world using satellite remote sensing of surface temperature. This is despite the fact the temperature we estimate is at best representative of the surface (skin) and not a depth-averaged temperature, and subject to requirements of clear sky, atmospheric corrections, and minimum water width.

When we do apply TIR data from satellites, we can see some stark thermal differences upstream, on, and downstream of a reservoir that can be attributed to reservoir operations. Take, for example, Figure 10.3, which was derived using Landsat TIR data for a reservoir in

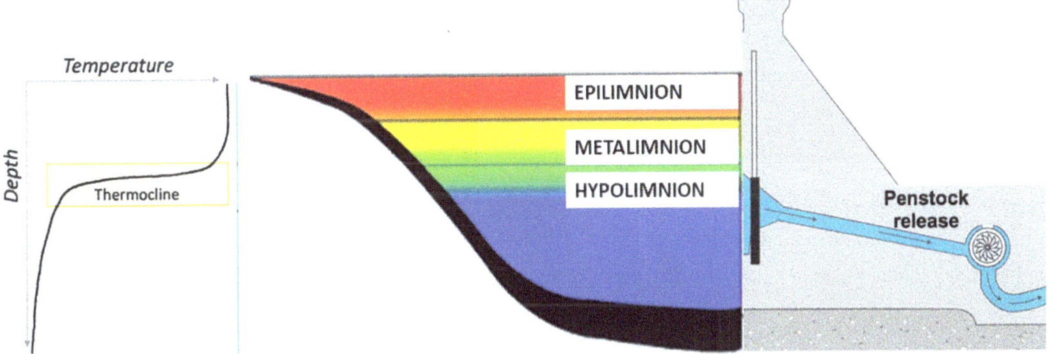

Figure 10.2 The thermal stratification of a hydropower reservoir during summer time.

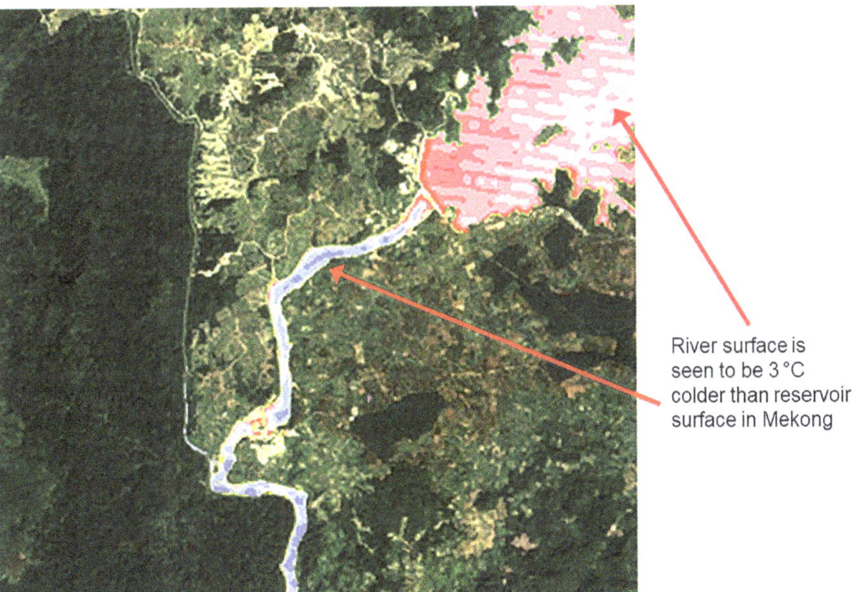

Figure 10.3a Qualitative temperature difference of surface water above and below a dam on the 3S river system of the Mekong River basin. Red indicates warmer temperatures, while blue is colder.

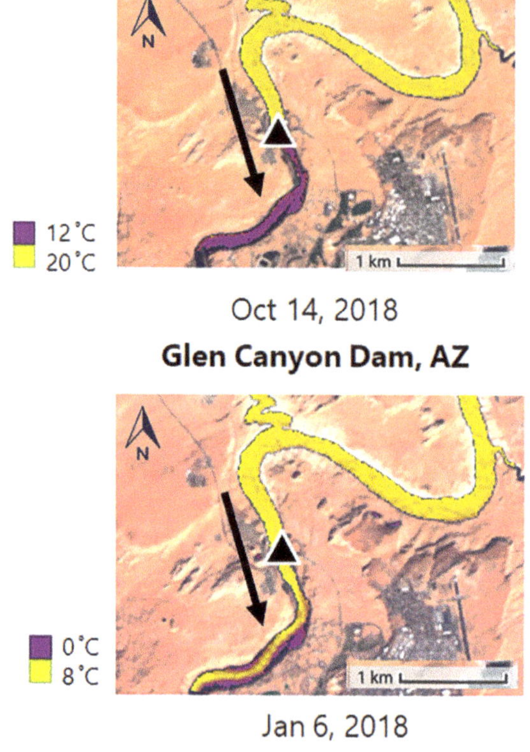

Figure 10.3b Landsat-TIR water surface temperature downstream and upstream of Glen Canyon Dam in Arizona, United States. There could be a variety of reasons why the surface water temperature is different (or similar) during those two particular times. What is noteworthy is that satellite remote sensing is able to detect the variability in terms of surface temperature.

the Mekong River basin in the 3S river system. The 3S rivers are Sekong, Srepok, and Sesan (hence "3S"), which mostly drain from Vietnam and meet with the main stem of the Mekong River in its southern stretch. We can see a distinct cooling pattern even using surface temperature estimates from satellites.

10.3 How Confidently Can Satellite-Based Surface Water Temperature Be Used in Water Management?[1]

To dig deeper and have confidence in the satellite-based surface water temperature data and that it is indeed attributable to dam operations, let us look at Figure 10.4, based on the work reported by Bonnema et al. (2020) on the 3S rivers of the Mekong.

By leveraging TIR data from 1,364 satellite TIR images from Landsat 4, 5, 7, and 8, we can construct the annual time series of average late dry season (January–April) water surface temperature of the three major 3S rivers, aggregated from the entire river reach downstream of all current dams (Figure 10.4, left panel – reach highlighted in green), from 1988 through 2018 (Figure 10.5). From these time series, a remarkable relationship emerges between the

---

[1] After Bonnema et al., 2020.

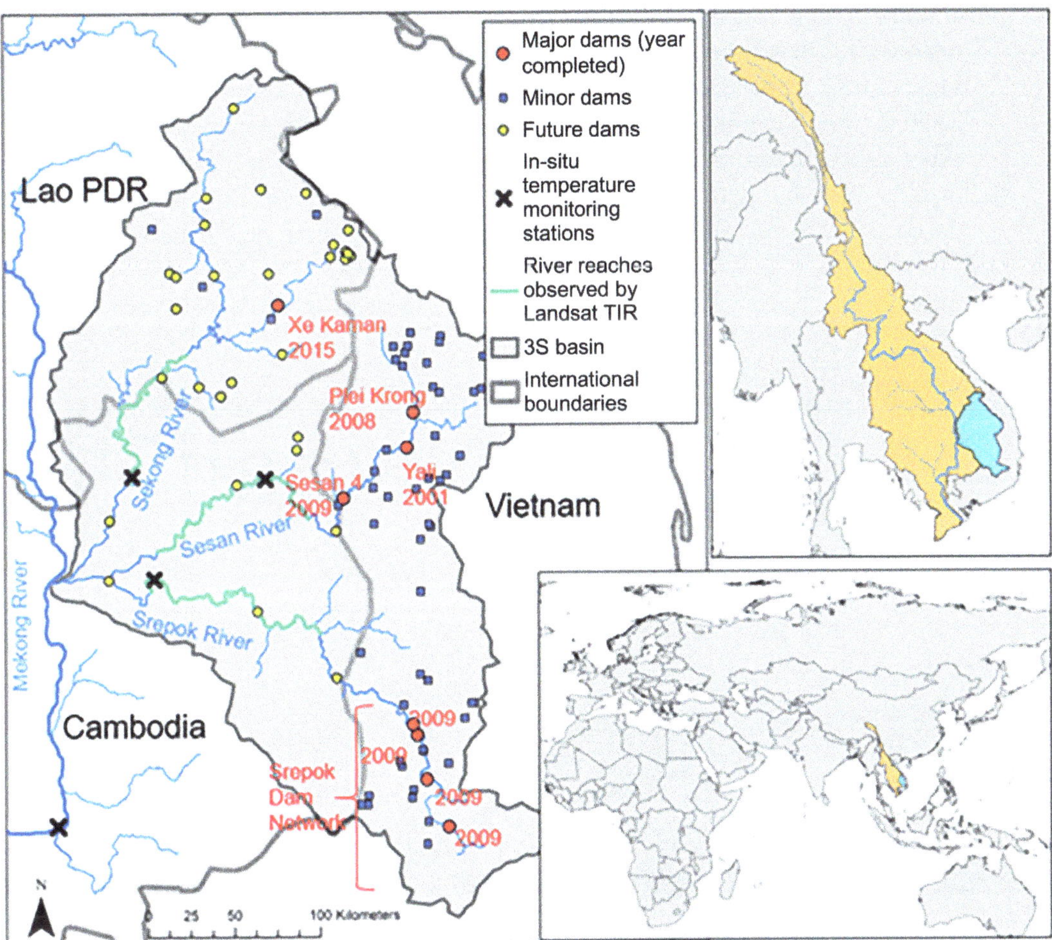

Figure 10.4 Map of 3S (Srepok, Sesan, and Sekong) river basin showing existing and future (under construction, planned, and proposed) dams as of 2017, as well as the locations of three water-quality monitoring stations. Major dams were defined as dams located on a major river or a tributary with a significant impounded reservoir (>25 km^2 surface area). Minor dams consisted of dams on smaller tributaries, and run-of-the-river style dams on major rivers with no impounded reservoir and unlikely thermal stratification. [After Bonnema et al., 2020]

construction and operation of major dams (dams on major rivers with >25 km^2 surface area) and downstream river temperature. Major dams were defined as dams that form significant reservoirs on the main stem or on a major tributary of one of the three major 3S rivers. On all three rivers of the 3S, significant temperature decreases were observed within one year of the beginning of operations of a major dam. These temperature changes were most clearly observed on the Sesan River, which experienced the construction of three major dams. In 2001, the dry-season water temperature of the Sesan River experienced a 1 °C decrease, corresponding with the commissioning of the Yali Dam, the first major dam on the river (Figure 10.5, middle panel). The water temperature dropped again between 2008 and 2009, by around 2 °C, and again, this temperature change corresponds with the timing of two major dams, Sesan 4 and Plei Krong, starting operations (Figure 10.5, middle panel). The Srepok River also experienced dam development in 2009, with a network of four dams coming online in 2009 (see Figure 10.4). Once again, a sharp decrease in temperature was observed. Between 2007 and 2009, the water temperature dropped by 1.4 °C (Figure 10.5, lower panel). The Sekong River experienced a temperature decrease corresponding with the commissioning of the Xe Kaman Dam in 2015, but with less severity than the other two rivers, only dropping by 0.7 °C over the course of three years (Figure 10.5, upper panel). This can be attributed to the fact that the

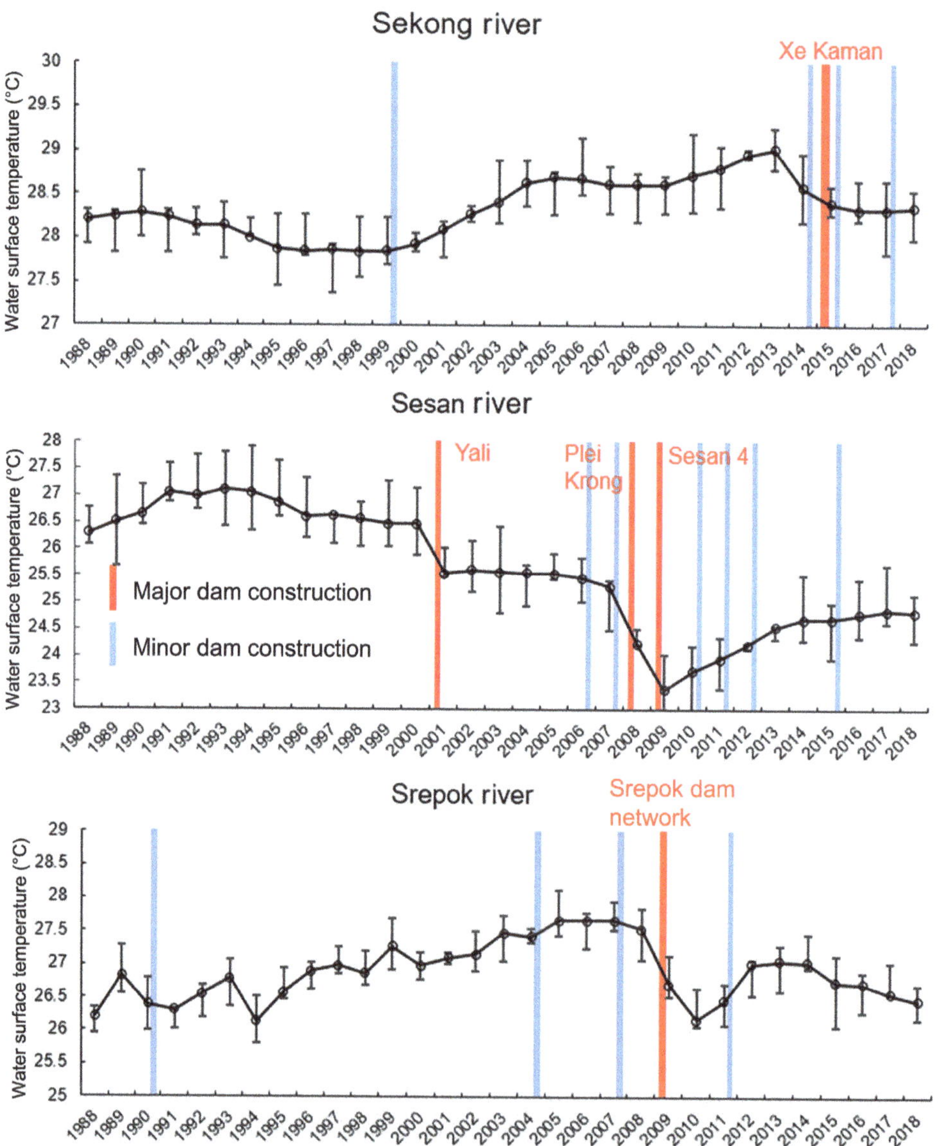

Figure 10.5 Timeline of dam development in each sub-basin and corresponding average annual Landsat-based dry-season river temperature downstream of dam development. Error bars represent 50th percentile bounds. [After Bonnema et al. 2020]

Xe Kaman Dam impounds a major tributary of the Sekong River and not the Sekong River itself, meaning that it influenced the temperature of only a portion of the water observed by the satellite data. Unfortunately, the tributary is not wide enough in the dry season to be accurately resolved by the thermal Landsat data, so the thermal impacts of this dam can only be deduced downstream after the impacted water has mixed with relatively free-flowing water. The 0.7–1.4 °C temperature decreases observed on all rivers appear larger than the observed uncertainty in the annual averages, which ranges from 0.1 to 0.3 °C.

Another characteristic of the Landsat temperature time series is that the river surface temperature appears unaffected by the influence of minor dams. Here, dams were classified as minor for one of two reasons. Several of the dams on the main stems of the major rivers are run-of-the-river style impoundments and do not create a sizeable impoundment. This means the water behind the dam does not thermally stratify as easily, resulting in little thermal alteration to the downstream river. Most of the minor dams, however, are located on smaller tributaries and impound only a fraction of the total basin discharge.

A third feature of the satellite-based temperature time series is that all three rivers experience temperature changes unrelated to dam development (Figure 10.5). Each river goes through periods of warming. For the Sekong and Srepok rivers, these warming periods last more than 10 years and overall equal or exceed the temperature decreases apparently caused by reservoirs, resulting in river temperatures ending around the same as they were in 1988, at the beginning of the time series. The Sesan River, however, goes through shorter periods of warming, which are not enough to offset the temperature drops. Clearly, there are other factors influencing river temperature, and some possible sources of warming are explored in the following sections.

Now let us pause for a moment and go back to the question we asked in this section: "How confidently can satellite-based surface water temperature be used in water management?" First we need to highlight the caveats or requirements that are associated with remote sensing of surface water temperature. These are: (1) need for clear sky conditions; (2) need for atmospheric corrections; (3) width of water body target must exceed the spatial resolution of TIR data; (4) the temperature estimated is only at the surface and not a depth-averaged temperature of water. We understand point 1 well by now, as we have covered this issue in earlier chapters. Point 2 basically refers to the fact that the satellite sensor's receiver up in space measures the emitted radiation as well as reflection from the top of the atmosphere. The radiation from the ground target has propagated through the atmosphere and undergone attenuation. This means that the satellite TIR sensor does not record all the radiation in the TIR wavelength coming from the target (the water body). Hence corrections for this need to be applied. In the above study, we applied some basic corrections that are available, and we have verified that for long-term climatologic studies based on differences from the baseline, the nature of correction does not affect the conclusions, as there is only a systematic bias in temperature derivations. The reader can refer to literature on correction algorithms that can be used to correct Landsat TIR data and make it more representative of the water surface. For example, the single channel algorithm (Jiménez-Muñoz et al., 2009) for temperature extraction can be used with atmospheric correction for top-of-atmosphere (TOA) reflectance. The third point addresses the spatial resolution of sensors like Landsat in the TIR wavelength – typically around 60–100 m (depending on the specific mission). So water bodies (rivers and reservoirs) narrower than 100 m will likely have land temperature contamination. For the last point, we need to remember that the surface water temperature can be modulated by the overlying air temperature as well as the underlying water column.

Obviously if the temperatures estimated by satellite had no discernible effect attributable to large hydropower dams and their operations, we could not have used satellite-based surface water temperature so confidently. So, the answer to the question in this section is, "It depends." As water managers, we need to first be very mindful of the above four caveats or requirements that go with satellite surface water temperature estimation. Water managers need to proceed with caution and verify that the observed changes or anomalies are not due to other factors unrelated to water management or dam operations. Overall, we are of the opinion that satellite surface water temperature can be a powerful dataset to empower water managers to make decisions on water allocation that are more sensitive to ecosystem services needs (see concluding section).

10.3.1 How Far Down Can the Effect of Thermal Alteration of Water Be Seen by Satellites?

Given that the rivers in the 3S basin are cooling and hydropower dam development is the most plausible driver, a follow-up question emerges: how far downstream does this effect propagate? To help to answer this, we can apply the same Landsat monitoring technique to three more locations, the 3S river outlet just upstream of where it meets the Mekong River, and the Mekong River just upstream and just downstream of the 3S confluence (Figure 10.4, left panel; and Figure 10.6, right panel). At the 3S outlet with the Mekong River, we see signs of cooling, but rather than the sharp, discrete decreases observed on the individual rivers, the temperature here slowly decreases over the course of the 30-year period. Given that hydropower dams appear to be the primary source of cooling in the major rivers, it is highly likely that this slow, steady temperature decline represents a cumulative cooling effect experienced by the major rivers upstream of the Mekong River confluence (Figure 10.6). While this cooling trend is fairly prominent at the 3S confluence, it becomes more ambiguous after mixing with the Mekong River. The Mekong River generally experienced warming at this location within the 30-year period, and none of the thermal characteristics appeared to correspond clearly with hydropower dam development in the 3S. However, comparing the temperature of the Mekong River before and after it has mixed with the 3S reveals the influence of the 3S on Mekong River temperature (Figure 10.5). Before 2000, the 3S generally appeared to have had a slight warming effect on the Mekong River when the two rivers mixed, with the downstream 0.2–0.4 °C ± 0.31 °C warmer than the upstream. Given the uncertainty, a longer record of historical data would be needed to establish this warming effect more definitively. After 2000, this trend shifts, with the 3S appearing to cool the Mekong River by 0.8 °C ± 0.42 °C downstream of the confluence. The timing of this shift roughly corresponded with the beginning of operations of the Yali Dam on the Sesan River in 2001. This is where signs of cooling end for now. The cooling signal disappears by the time the water reaches an in-situ temperature monitoring station approximately 100 km downstream of the 3S confluence (shown in Figure 10.4 left panel).

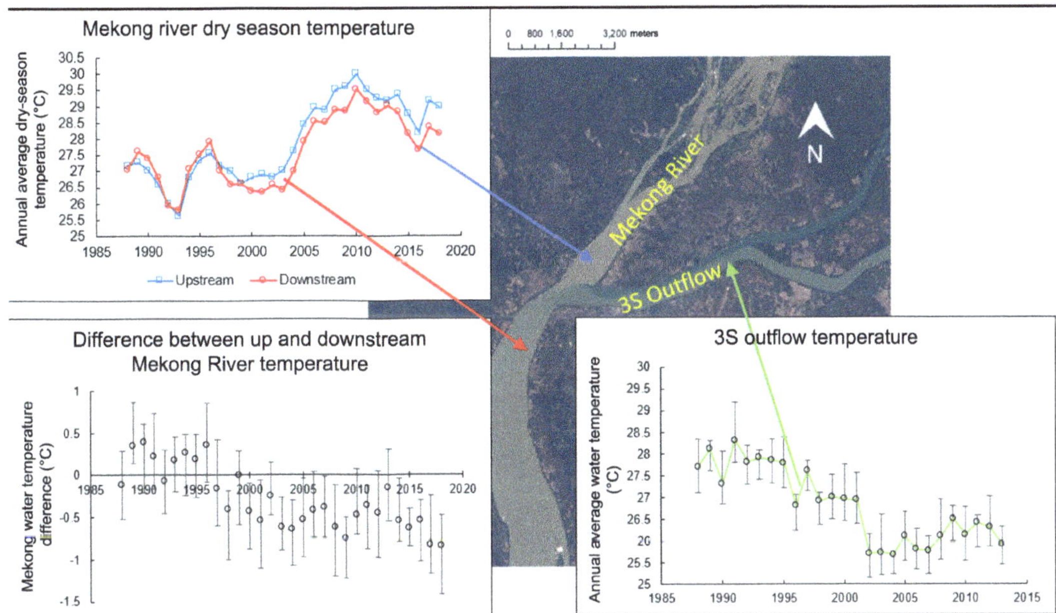

Figure 10.6 Landsat-based dry-season temperature trends of the 3S basin outflow, and the Mekong River upstream and downstream of the 3S–Mekong confluence. The difference between the upstream and downstream Mekong River temperatures is also shown. [After Bonnema et al., 2020]

So, it is quite clear from the above example, that satellite-based surface water temperature has value in data-limited regions where there is no in-situ monitoring of river or reservoir temperature. Given that temperature is an important parameter for aquatic habitat and biological activity, knowledge of remotely sensed surface water temperature can improve water management decisions that are traditionally driven by water quantity information alone. With knowledge of water quality parameters such as water temperature, water management can be made more integrated and compatible with the ecological and environmental needs of the river system. Now, let us see how temperature is estimated from Landsat TIR data.

10.4 How Is Surface Water Temperature Estimated from TIR Bands?

To remind ourselves, let us review the basics once again. Provided that the emissivity of a material is known, its absolute temperature (kinetic temperature, T_{kin}) can be derived from the radiation it emits. If the emissivity is not considered, only the brightness temperature (radiant temperature, T_{rad}) of the material can be determined. Figure 10.7 shows an example of thermal infrared radiation image over Sacramento (USA) city. The kinetic temperature is related to radiant temperature as follows:

$$T_{rad} = \varepsilon^{1/4} \, T_{kin} \qquad (10.1)$$

Figure 10.7 A snapshot of visible light (left) and thermal radiation (right) to show surface temperature in Sacramento. Surface temperatures vary on the scale of individual buildings. [Credit: NASA/Marshall Space Flight Center]

Here the T_{rad} (radiant temperature, known also as brightness temperature) is what is sensed by the satellite and the T_{kin} (kinetic temperature) is what we measure with a temperature probe.

The radiant temperature of a real material is always lower than its kinetic temperature. However, for a black body with $\varepsilon = 1$, it applies that:

$$T_{\text{rad}} = T_{\text{kin}}$$

Fortunately, emissivity for water in the TIR range is near-1 (see Table 10.1).

Scattering processes are negligible in the TIR region because of the long wavelength, but atmospheric absorption and emission by water vapor, CO_2, and O_3 are prevalent. As mentioned in a previous section, we need to remember that the Landsat receiver records radiation at the top of the atmosphere. By that time, the radiation has already propagated through the atmosphere and is not representative of the radiation emitted from the surface of the target. So, the signal recorded at the sensor consists of the radiation emitted from the terrain element modified by the transmission of the atmosphere, the atmospheric upwelling radiation, and the atmospheric reflected downwelling radiation. In short, the radiation recorded by the satellite needs to be "corrected" to make it representative of the target (such as a reservoir or river) on the ground. The corrections for surface temperature can be quite involved. In this chapter, we will try to learn, as potential water managers and practitioners, how temperature data from satellite TIR bands can be extracted and used practically with basic corrections already available. The best way to build such literacy is through hands-on tutorials and data-based exercises.

Table 10.1 Emissivity in the TIR wavelength range for various types of surfaces

Material	Average emissivity in the thermal infrared range (8–14 µm)
Clear water	0.95
Healthy green vegetation	0.96–0.99
Dry vegetation	0.88–0.94
Concrete	0.95
Granite (natural surface)	0.96
Ice	0.97
Quartz	0.93

Data sourced from www.thermoworks.com and NASA.

10.5 How Well Does Satellite-Based Water Temperature Compare Against In-Situ Records?

Using Landsat TIR-band observations, we can derive estimates of surface water temperature at three key locations of a reservoir where there also exists an in-situ record. These locations are: (a) right upstream of the reservoir near the inlet, averaged over a 1-km-long reach; (b) on reservoir, averaged over the reservoir surface extent; and (c) downstream of the reservoir at outlet, averaged over a 1-km-long reach. Comparison of all three temperature time series can allow a dam operator or a water manager to monitor potential thermal modification due to reservoir operations (Figure 10.8, right panel, for reservoirs in the US). Here, the TIR band from Landsat 7 and 8 was acquired at 60 m and 100 m spatial resolution, respectively, and that from Landsat 5 at 120 m. Because of this, we should remember that surface water temperature should be treated with skepticism for water extents narrower than 200 m. Fortunately, more than 20% of the world's rivers are likely to be wider than 200 m (Allen and Pavelsky, 2018). So, this implies that Landsat-based surface water temperature tracking for rivers near reservoirs should work for 20% of the world's rivers, and it should work for most reservoirs and lakes.

10.6 Can We Predict the Likely Thermal Impact of Current and Future (Planned) Dams Around the World?

Today, aquatic biodiversity is at very high risk in many large river systems such as the Congo, Amazon, and Mekong, where extensive hydropower dams are planned. In a recent study, Ahmad et al. (2021) predicted the thermal impact of 216 planned dams around the world in terms of their likely change to downstream river temperatures. Their predictions are based on a novel data-based framework for planned dams, called "FUture Temperatures Using River

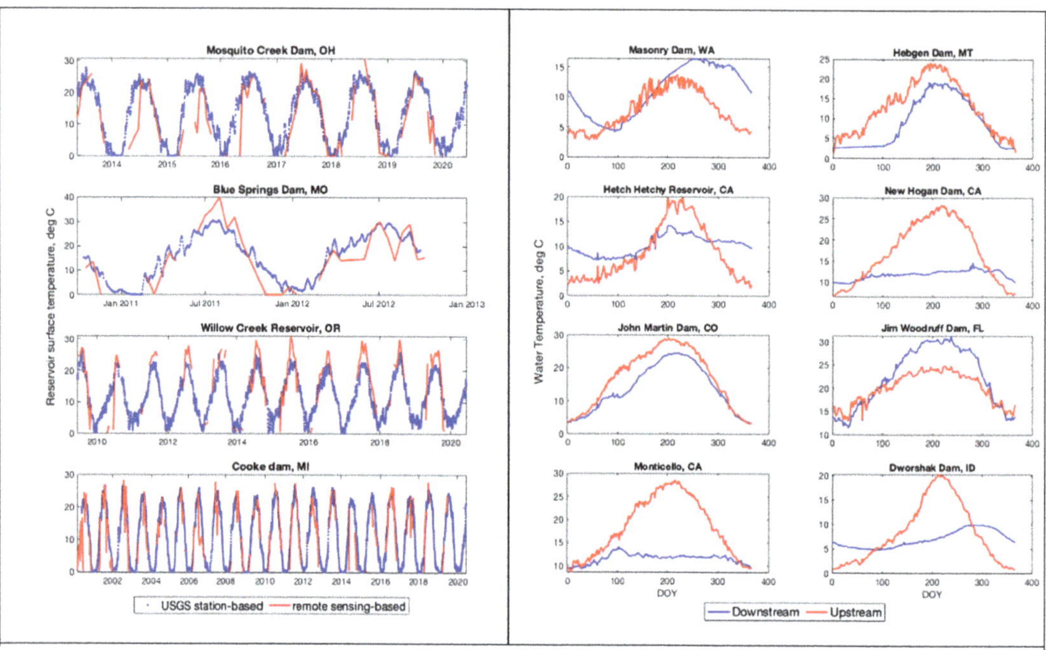

Figure 10.8 Left panel: comparison of estimated surface temperature from Landsat TIR data with US Geological Survey in-situ temperature gauges (sampled near top of surface) at selected dam locations in the United States. Right panel: comparison of river temperature estimated by satellite TIR data upstream and downstream of reservoir from US Geological Survey gauges, showing wide range of thermal modification due to reservoir management.

hISTory" (FUTURIST), using satellite TIR data. The FUTURIST framework is based on the key premise that a long record of the past thermal impact is a reasonable representation of the near-future impact due to planned dams.

The Ahmad et al. (2021) study employed a historical record of river temperature changes from in-situ gauges and remote sensing observations over a set of dams across the globe, capturing variability in climate and hydrology to train an artificial neural network (ANN) model. This data-based ANN approach predicted temperature change between upstream and downstream rivers. The ability of ANN model to capture nonlinearities in predicting a dam's thermal impacts makes the technique transferrable to other dams with unobserved conditions. Ahmad et al. (2021) reported high predictive skill over dams in varying climates across the continents during model validation. Also, for training, FUTURIST required widely available inputs that were either one of the dam's structural properties or variables derived from remote sensing products. TIR remote-sensing-based thermal change was used for model training and validation where in-situ monitored values were absent or scarce. The ANN model was then applied at planned hydropower sites worldwide to predict the likely thermal impacts.

Validation of FUTURIST-modeled impacts for dams worldwide showed promising results, with a root mean squared error of 2.5 °C (0.9 °C) and categorical accuracy of 63%

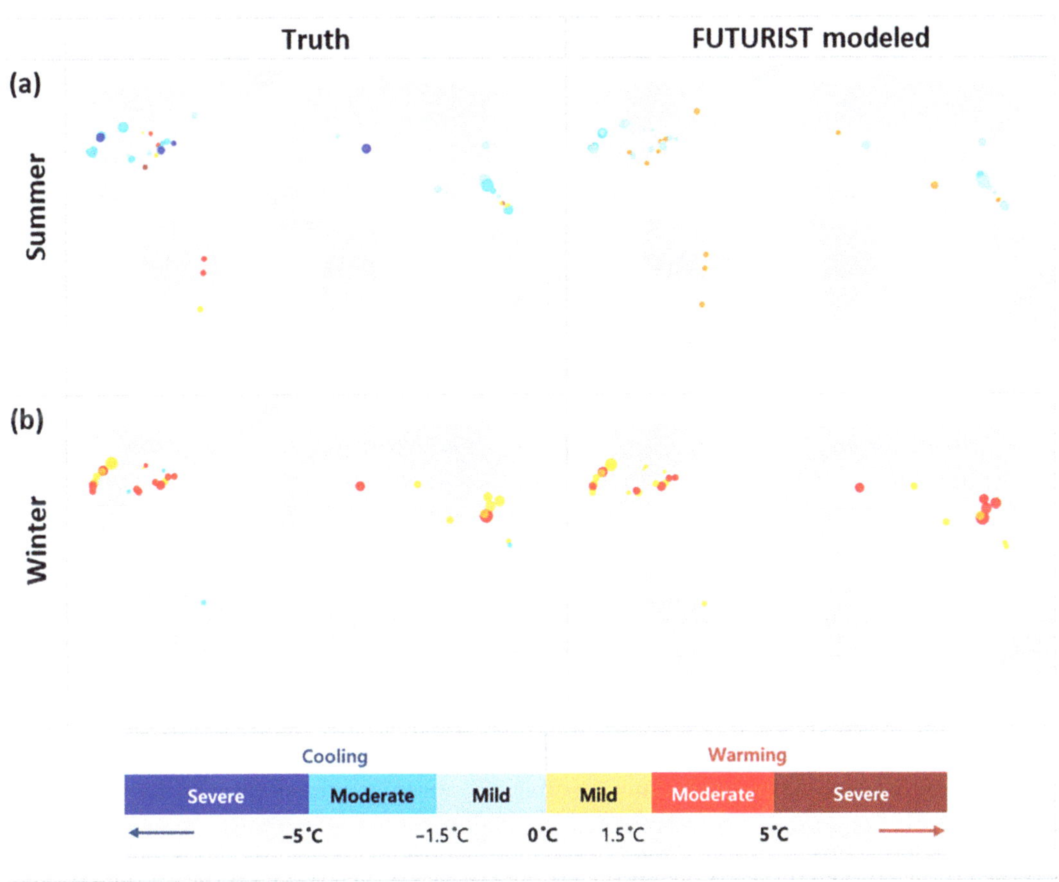

Figure 10.9 Validation of the Landsat TIR-based FUTURIST framework in predicting thermal effect of current dams. [After Ahmad et al., 2021]

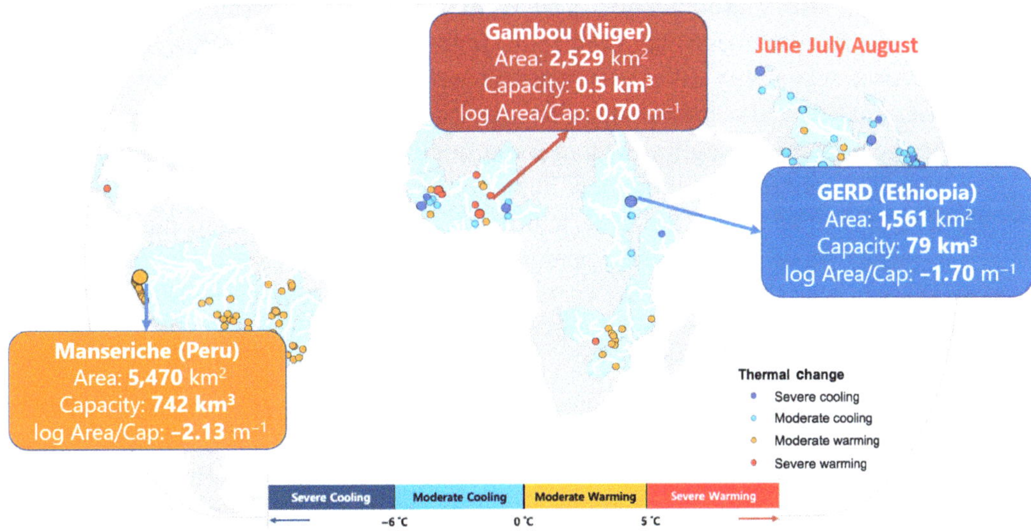

Figure 10.10 Potential thermal modification of future dams based on multi-decadal record of Landsat TIR water temperature data and machine learning. Cap., capacity. [After Ahmad et al., 2021]

(88%) during the summer (winter) season (Figure 10.9). The trained ANN model afforded prediction of the likely thermal impacts of 216 planned dams. Results suggest that during the summer season, 73% of future dams will potentially cool downstream rivers by up to 6.6 °C. Winter-season operations were predicted to consistently warm downstream rivers by temperatures of up to 2 °C (Figure 10.10). Reservoirs that experience strong stratification have the most potential to impact downstream pre-dam thermal regimes (Figure 10.10). For copious existing or planned dams worldwide for which the thermal impacts are yet to be mapped, a tool like FUTURIST trained on multi-decadal satellite TIR datasets for water temperature provides water managers with an efficient path forward to carry out a global thermal assessment. Such satellite-based tools can allow management decisions to investigate sustainable hydropower expansion plans so that planned dams can be redesigned and operated in a more ecosensitive manner than the existing ones, or not built at all if thermal impacts are likely to be too severe.

10.7 The Future of Remote Sensing of Surface Temperature of Water

We want to leave the reader with some thoughts on what the future might hold for satellite-based surface water temperature data, which is only recently gaining traction for water management applications thanks to advancements in cloud computing and data processing. As mentioned earlier, like all methods, satellite TIR has limitations, such as spatial (water bodies need to be wider than 200 m) and atmospheric (cloud-free and corrections) requirements. Passive microwave could be used for temperature remote sensing to overcome cloud conditions, but the spatial resolution would be too coarse for meaningful water management applications at rivers and reservoirs.

Setting all limitations aside, here is one reason why we feel that satellite estimation of water surface temperature can improve fundamentally water management. For decades, the physical availability of water (quantity) and the biochemical state of water (quality) have been studied and acted on separately by scientists and the management community. Yet it has become increasingly clear that water quality, which affects the health of humans and of aquatic ecosystems, is intricately linked to decisions made on management of water quantity. In the real world, however, making water quantity decisions that are also constrained by water quality conditions remains far from a mainstream water management practice. This can now change with the vantage of space and global observational capability, at multiple scales and sensors, that is unique to the satellites maintained by the space agencies of the world (notably NASA). Temperature is one such major water quality parameter that is relatively easier remotely sensed. Even where there are in-situ measurements, ground probes provide only a point-based view of the thermal condition of river systems, leaving the majority of river reaches and reservoir surfaces "dark". Hence, temperature remote sensing of managed water systems, which can provide a more informed view of the thermal state of surface water, should be embraced by the water management community around the world (Figure 10.11).

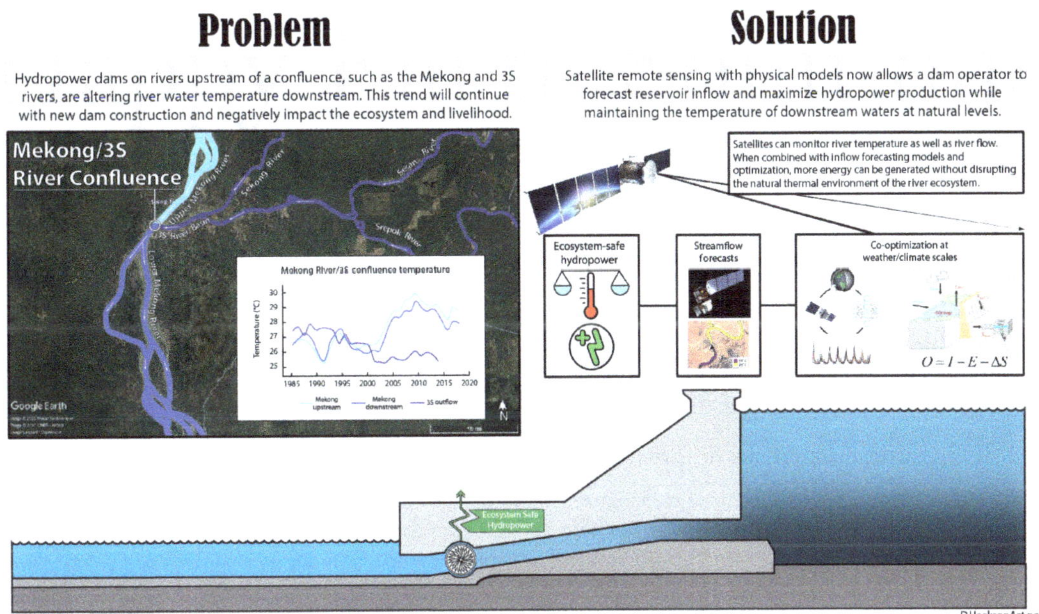

Figure 10.11 What is the future of hydropower generation and reservoir management using satellite-based TIR data?

10.8 Conclusion

In this chapter, we have covered how remote sensing of surface temperature of water works using passive sensing. We demonstrated a few real-world examples of how multi-decadal satellite estimates of water temperature can help us understand how reservoir regulation has altered river temperature downstream. Now that a reservoir can be tracked for its water quantity as well as quality (water temperature at the surface), water management can be potentially further improved with the joint consideration of both quantity and quality information on water. With this chapter, we conclude the technical part of this book. In the next two chapters, we will delve more into social and governance issues spanning social justice, equity (Chapter 11), and citizen science (Chapter 12).

TUTORIAL ON USING LANDSAT SATELLITE DATA TO TRACK SURFACE TEMPERATURE OF WATER[2]

Introduction

This tutorial will help us learn how to generate a time series of water surface temperature values for the Franklin D. Roosevelt Reservoir. The reservoir is created by the Grand Coulee Dam,

---

[2] Contributed by George Darkwah, University of Washington.

located along the Columbia River in the state of Washington, USA. The focus of this tutorial is on the reservoir water temperature, as well as the immediate downstream water temperature that may be dependent on how the dam is operated. As we have learned in this chapter by now, thermal infrared (TIR) sensors aboard satellites such as Landsat can be used to determine the temperature of the Earth's surface, including land and water. We will determine the reservoir's water temperature time series from 2019 to 2022. To do this, we will use data from the Landsat 8 satellite mission which has been in operation since 2013.

***Note**: The following sections cover some background information about the data and concepts used in the tutorial as well as a guide for Setting Up Your Environment. You may skip to the Tutorial section if you are comfortable with the background and setup.*

Landsat Data Products

Landsat data comes in three different processing levels (Levels 1–3). Level 1 products are the raw digital numbers (DN) from the satellite sensors. Additional processing is required to convert the DN to meaningful surface reflectance and temperature values. Level 1 products are organized into three different tiers: Real-Time (RT), Tier 1 (T1), and Tier 2 (T2). The real-time tier refers to newly acquired data. This data is made available for download within 12 hours of acquisition. The quality of the real-time data is assessed and then categorized into either Tier 1 or Tier 2. Tier 1 contains the highest-quality data while Tier 2 contains data that does not meet the quality standards.

Levels 2 and 3 are user-ready products derived from the Level 1 data. This reduces the amount of processing required before using the data. The Level 2 product contains surface reflectance and surface temperature that can be used for further analyses. The current collection of Landsat data is Collection 2, which utilizes improved reprocessing algorithms and data distribution capabilities (Landsat Collection 2, 2021).

In this tutorial, we will use data from Landsat 8, Collection 2, Level 2, Tier 1 which is accessible through the Google Earth Engine (GEE) platform (https://developers.google.com/earth-engine/datasets/catalog/LANDSAT_LC08_C02_T1_L2).

Identifying Water Pixels

We have already covered this topic of surface water classification in Chapter 6. We will cover it again here in the context of extracting temperature data only from pixels that are detected to be surface water. There are several ways to classify water pixels in a satellite image, but in this tutorial, we will explore two different approaches. One of the most popular ways is to use the Normalized Difference Water Index (NDWI) (McFeeters, 1996). The NDWI uses the green and near-infrared (NIR) bands to identify water pixels. NDWI ranges from −1 to 1 and is calculated using the formula:

$$NDWI = \frac{Green - NIR}{Green + NIR} \tag{10.2}$$

where Green and NIR are the green and NIR bands respectively.

To use the NDWI, one must establish a threshold index above which a pixel will be classified as either water or not. Although there is no fixed threshold for water, it is common to start with an initial threshold of 0.2 and adjust it until the desired discrimination between water and non-water pixels is achieved.

Another way to identify water pixels is to use information from the Quality Assessment (QA) band of the Landsat Level 2 products that will be used in this tutorial. One advantage of using information from the quality band to identify water pixels is that it is easily reproducible, and it does not require subjective decisions about the threshold for water pixels. The QA bands can also be used to identify pixels with cloud cover and high saturation. Hence, it will be important for identifying cloud-free, quality satellite observations too. See additional details about the QA band in Appendix A.1 of this chapter.

Thermal Infrared Temperature

Land and water temperature is obtained from the thermal infrared (TIR) band. The raw satellite observations of the TIR represent the top-of-atmosphere (TOA) measurements; however, the single-channel algorithm (Jiménez-Muñoz et al., 2009) is used to atmospherically correct the TOA measurements and convert them to surface temperature. The Landsat 8 Collection 2 Level 2 products contain the atmospherically corrected surface reflectance (SR) and surface temperature (ST). In this tutorial, we will use the atmospherically corrected dataset. If you are interested in deriving temperature using the single-channel algorithm, see Appendix A.2 of this chapter.

Setting Up Your Environment

This tutorial uses the GEE platform and is presented in both JavaScript and Python API. To refresh your knowledge of GEE, you can refer back to the Chapter 4 tutorial. JavaScript users can run the command in the GEE Code Editor (registration required). Python users can run the code either on Google Colab or their local computers in a Jupyter Notebook. The following steps will guide you to set up your preferred platform for this tutorial.

Required Data

1. Shapefiles of the region of interest (ROI). The shapefiles that will be used in this tutorial are provided in the supplementary files.
   a. Franklin D. Roosevelt Reservoir (Grand Coulee Dam) – obtainable from the Global Reservoir and Dam Database (GRanD) (Lehner et al., 2011).
   b. Downstream of the Grand Coulee Dam – manually delineated area along the downstream reach of the dam. Preferably, the shapefile should cover 10 km of the reach. The delineation does not have to be exact. Allow enough buffer around the reach. If you are working in the GEE Code Editor, there is the option to delineate this area directly in the Code Editor, but preparing or obtaining the shapefile externally and uploading it on GEE is recommended to stay consistent with the steps in this tutorial.
2. Landsat 8 Datasets (hosted on the GEE dataset catalog: https://developers.google.com/earth-engine/datasets/catalog). Data can be obtained for the GEE data catalog using the ID of the dataset as it is stored in the data catalog. The IDs of the following data are provided in parentheses.
   a. Landsat 8 OLI/TIRS Collection 2 atmospherically corrected surface reflectance, Tier 1 ("LANDSAT/LC08/C02/T1_L2")

Upload Data into GEE Assets

Prepare the shapefiles so that they contain all the required file extensions (set of *.shp*, *.shx*, *.dbf*, *.prj*, etc.) in one folder (one shapefile per folder). You can compress the folder into a single zip file for the upload. Upload the prepared shapefile into your Assets folder in the GEE Code Editor using the following steps (Figure 10.12):

1. Select the "Assets" tab in the left pane of the Code Editor to open the Asset manager.
2. Click the "New" dropdown button.
3. Select "Shape files (.shp, .shx, .dbf, .prj, or .zip)".
4. In the *Upload* dialogue, click on the "Select" button to choose the source files.
5. Select all the set of files in the shapefile's folder or the compressed zip file.
6. Navigate to the asset folder of choice and enter the name of the dataset. If you want to specify a sub-folder path, enter the sub-folders separated by a forward slash (/).
7. Click upload to upload the shape file into the asset folder.

After uploading, click on the uploaded shapefile and inspect the details of the asset (Figure 10.13).

1. Take note of the "Table ID" in the left side of the "Asset details" dialogue.
2. Also, check the Features of the uploaded features to understand how the dataset is organized. If you uploaded the files attached in this tutorial, take note of the "reach_id" column of the downstream shapefiles and the "RES_NAME" column of the reservoir shapefile. This information will be used later in the tutorial.

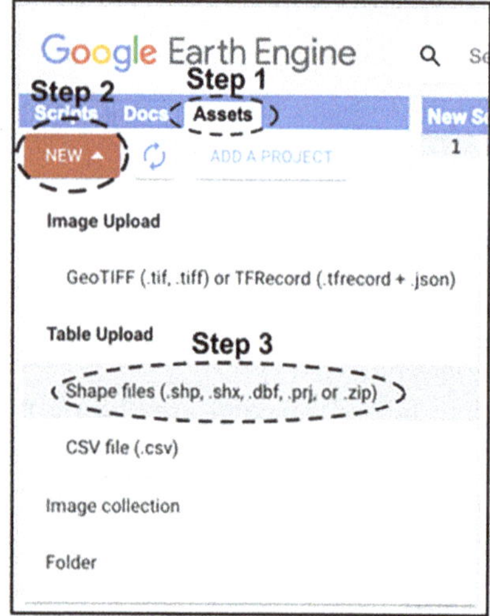

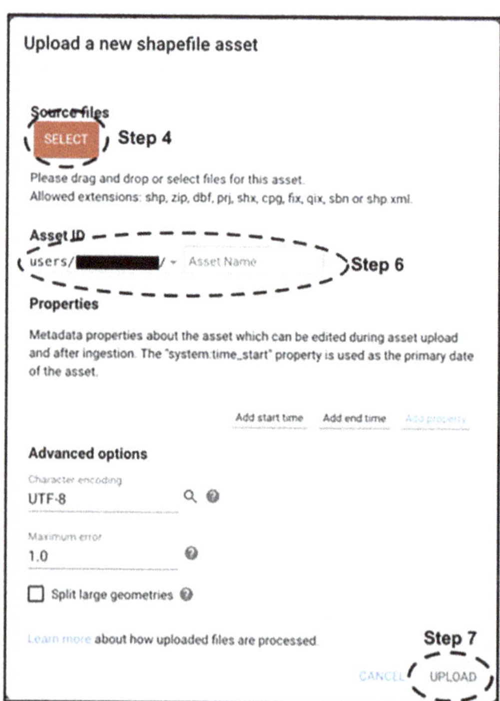

Figure 10.12 Steps in uploading shapefiles to the GEE Code Editor platform. [Credit: Google Earth Engine]

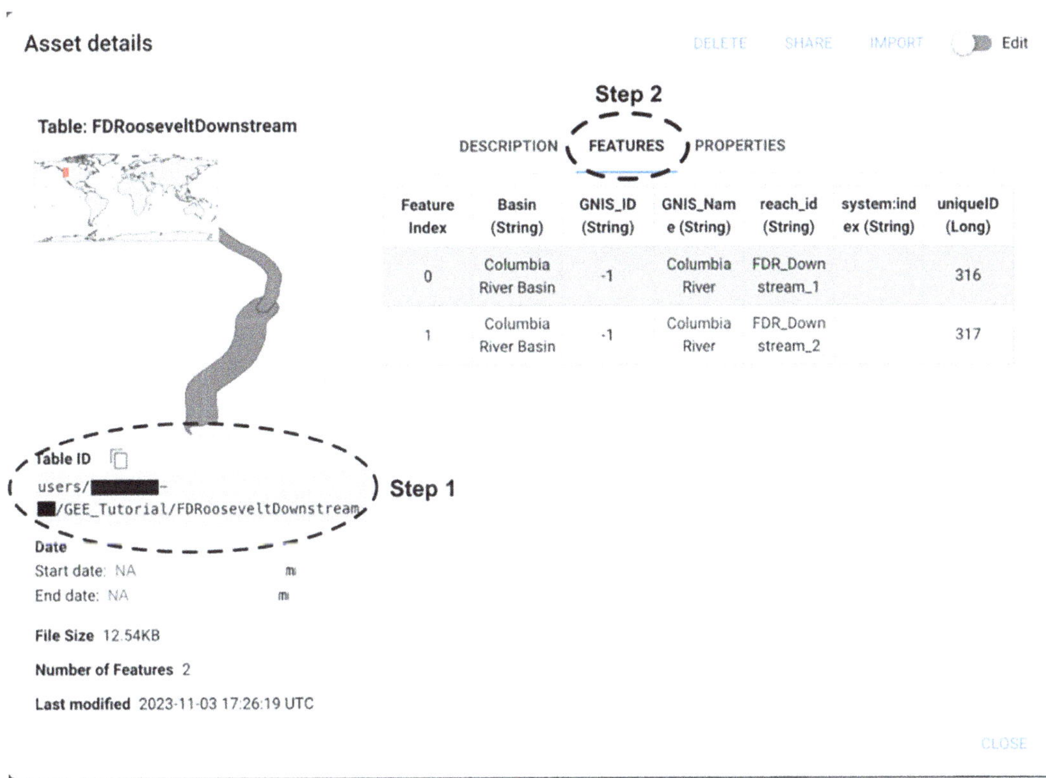

Feature Index	Basin (String)	GNIS_ID (String)	GNIS_Name (String)	reach_id (String)	system:index (String)	uniqueID (Long)
0	Columbia River Basin	-1	Columbia River	FDR_Down stream_1		316
1	Columbia River Basin	-1	Columbia River	FDR_Down stream_2		317

Figure 10.13 Inspection of uploaded shapefiles in the GEE Code Editor. [Credit: Google Earth Engine]

The "Table ID" will be used to import the dataset into our script and the specific attributes of the uploaded shapefiles will be used to filter features in the shapefile.

Python Only: Install and Import Required Packages

For Python users, if you are working on your local computer instead of Google Colab, you will need to install the GEE Python package and other auxiliary packages that will be used in this tutorial. Install the required packages by running the following commands in your Jupyter Notebook:

```
1. # Install the Earth Engine Python API (https://developers.google.com/earth-
   engine)
2. ! conda install -c conda-forge earthengine-api -y
3. ! conda install -c conda-forge geemap -y # Install Geemap package (https://
   geemap.org/)
4. ! conda install numpy -y # Install NumPy (https://numpy.org/)
5. ! conda install -c conda-forge pandas -y # Install Pandas (https://pandas
   .pydata.org/)
6. ! conda install -c conda-forge matplotlib -y # Install Matplotlib (https://
   matplotlib.org/)
7. ! conda install -c conda-forge geopandas -y # Install GeoPandas (https://
   geopandas.org/)
8. ! pip install ipython==8.15.0 # Install IPython (https://ipython.org/)
9.
```

If you are installing the packages directly from your terminal, you can ignore the preceding exclamation mark (!) and the training "–y" in every line of the code above. You can also skip the installation if the packages are already installed on your computer.

Now import the required packages into your Jupyter Notebook. Notice that there are some additional packages that are imported even though they were not installed in the previous step. These packages are pre-installed in the Python environment.

```
1. import ee # Import Earth Engine API
2. import geemap # Import Geemap package
3. import pandas as pd # Import Pandas as pd
4. import geopandas as gpd # Import GeoPandas as gpd
5. import numpy as np # Import NumPy as np
6. import matplotlib.pyplot as plt # Import Matplotlib as plt
7.
```

After importing the required packages, authenticate and initialize the GEE API in your python environment. This step is only required if you are using the GEE API for the first time in your Python environment. Run the following code and follow the link to complete the authentication in your browser.

```
1. ee.Authenticate()
2. ee.Initialize(project="my-project") # enter the name of your google earth
   project
3.
```

Now you are ready to use GEE in your Python environment. The next step will create a map object that can be used to visualize the GEE outputs in your Jupyter Notebook (Figure 10.14), like the output you will see if you were using the GEE Code Editor in your browser.

```
1. # Create a map centered at (lat, lon) 40, -100
2. Map = geemap.Map(center=[40, -100], zoom=4)
3. Map
4.
```

Now you are set to run the GEE functions in this tutorial in your Jupyter Notebook.

JavaScript Only: Create a new script and open an existing script in the GEE Code Editor

Use the following steps to create and open a script in the GEE Code Editor (Figure 10.15):

1. Navigate to the Script Manager by clicking on the "Scripts" tab in the left pane of the GEE Code Editor.
2. To create a new script, click the "New" dropdown button.
3. Select "File" to open the "Create file" dialogue.
4. Specify the folder and enter the file name.
5. Click the "OK" button to create the new script.
6. To open an existing script, browse through existing scripts and select one to open it in the central pane.

Figure 10.14 GEEmap Map object for visualizing GEE outputs in a Jupyter Notebook. [Credit: Google Earth Engine]

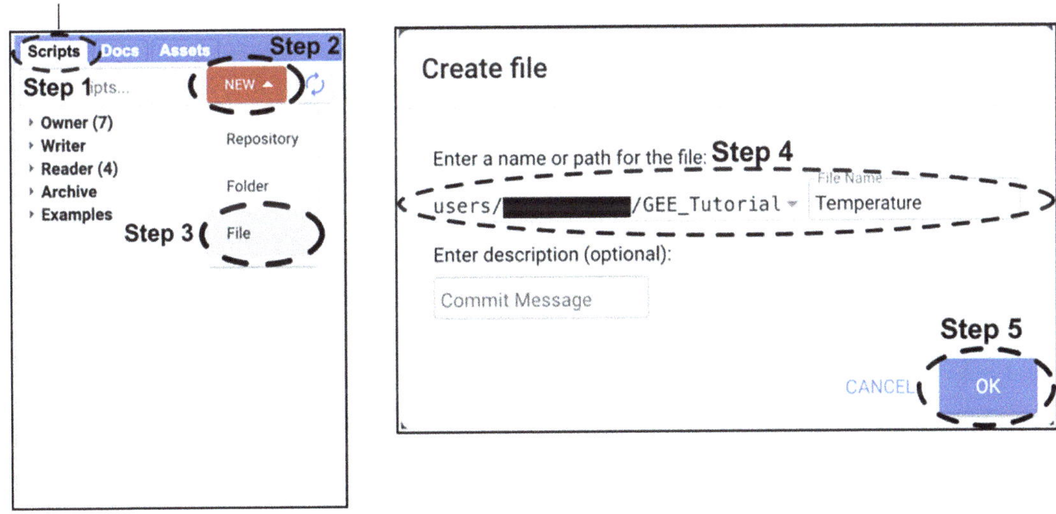

Figure 10.15 Steps for creating a new script in the GEE Code Editor. [Credit: Google Earth Engine]

Tutorial

[**Note:** All necessary supporting files and codes for reproducing this tutorial are available at this link: www.cambridge.org/hossain

To make it easier for the reader, we have provided text versions of the code snippets for completing this tutorial in Python or GEE (Javascript) that can be easily copied and pasted. Although GEE code (Javascript) is provided, readers can access the full GEE script for this tutorial at this link: https://code.earthengine.google.com/dbe1aa76778d6e009b02986be 184d24a?accept_repo=users%2Fgdarkwah-uw%2FGEE_Tutorial

Load and Filter Data

The Landsat 8 dataset on GEE comprises scenes that cover the entire globe, available since the onset of its operations in 2013. However, we are only interested in using data in our ROI, the reservoir and its downstream reach, from 2019 through 2022.

1. Load the Landsat 8 data, reservoir shapefile, and downstream shapefile. Use the "Table ID" of the uploaded shapefiles to specify the location of the dataset. Use the data-set ID of the Landsat 8 data ("LANDSAT/LC08/C02/T1_L2"). Load the Landsat 8 data as an `ee.ImageCollection` object and load the reservoir and downstream shapefiles as an `ee.FeatureCollection` object.

JavaScript:

```
1. var landsat8 = ee.ImageCollection("LANDSAT/LC08/C02/T1_L2");
2. var downstream = ee.FeatureCollection(
3.   "users/john-doe/GEE_Tutorial/FDRooseveltDownstream"
4. );
5. var reservoir = ee.FeatureCollection(
6.   "users/john-doe/GEE_Tutorial/FDRooseveltReservoir"
7. );
8.
```

Python:

```
1. landsat8 = ee.ImageCollection("LANDSAT/LC08/C02/T1_L2")
2. downstream = ee.FeatureCollection(
3.   "users/john-doe/GEE_Tutorial/FDRooseveltDownstream"
4. )
5. reservoir = ee.FeatureCollection(
6.   "users/john-doe/GEE_Tutorial/FDRooseveltReservoir"
7. )
8.
```

2. Filter the dataset to select the specific features in the shapefile. You can skip this step if the shapefiles you uploaded contain only one feature. Also, the filtering condition may differ depending on the specific attributes of your shapefiles. For the shapefiles in this tutorial, filter out the downstream feature with the "reach_id" equal to "FDR_Downstream_2" and the reservoir feature with the "RES_NAME" equal to "Franklin D. Roosevelt".

JavaScript/Python:

```
1. downstream = downstream.filter(ee.Filter.eq("reach_id", "FDR_Downstream_2"))
2. reservoir = reservoir.filter(ee.Filter.eq("RES_NAME", "Franklin D. Roosevelt"))
3.
```

3. Specify the date range of interest by defining the start date ("2019-01-01" for January 1, 2019) and end date ("2022-12-31" for December 31, 2022). The date is defined as an ee.Date object.

<table>
<tr><th>JavaScript:</th><th>Python:</th></tr>
<tr><td>

```
1. var startDate = ee.Date("2022-01-02");
2. var endDate = ee.Date("2022-06-30");
3.
```

</td><td>

```
1. startDate = ee.Date("2022-01-02")
2. endDate = ee.Date("2022-06-30")
3.
```

</td></tr>
</table>

4. Filter out the Landsat 8 dataset within the ROI and the date range. Because we are interested in data from two different shapefiles, merge the reservoir and downstream shapefiles using the merge function to merge two `ee.FeatureCollections`. The .filterBounds function is used to select only scenes from the Landsat dataset that intersect with the ROI. We use `.filterDate` to select scenes that were taken within the specified date range.

JavaScript:

```
1. var landsat8Filtered= landsat8
2.    .filterBounds(reservoir.merge(downstream))
3.    .filterDate(startDate, endDate);
4.
```

Python:

```
1. landsat8Filtered= landsat8.filterBounds(reservoir.merge(downstream)).filterDate(
2.       startDate, endDate
3. )
4.
```

Prepare Landsat Data

Before using the Landsat 8 Collection Level 2, you need to prepare the data. The data is stored as 16-bit unsigned values. Therefore, these are not directly representative of the surface reflectance (unitless) or temperature (K) values unless they are scaled using the scaling factors. Table 10.2 provides a summary of the data types in the surface reflectance and temperature products. Use the `.addBand` function to add the scaled bands to the image.

Thereafter, mask out the unwanted pixels (fill, dilated cloud, cirrus, cloud, cloud shadow and snow) using the QA_PIXEL band. Also, filter out saturated pixels in every image using the

Table 10.2 Landsat Collection 2 Level 2 surface reflectance and surface temperature data types and scaling factors

	Surface reflectance	Surface temperature
Data type	Unsigned 16-bit integer	Unsigned 16-bit integer
Scaling factor	0.0000275 + -0.2	0.00341802 + 149.0
Valid range	1–65,455	1–65,535

QA_RADSAT band. Refer to section 1 of the Appendix in this chapter for extra detail on how to use the QA_PIXEL band. The QA_RADSAT is a mask where saturated pixels are assigned values of 0 and unsaturated pixels are assigned a value of 1. Use the `.updateMask` function to update mask out the unwanted pixels and saturated pixels.

All the data preparation will be bundled into a single function that takes an ee.image object, scales it using the appropriate scaling factors, and masks out unwanted pixels using the QA_PIXEL and QA_RADSAT bands.

JavaScript:

```javascript
1.  // function to prepare Landsat 8 images
2.  function prepareLandsat8(image) {
3.    // mask for unwanted pixels (fill, dilated cloud, cirrus, cloud, cloud
       shadow, snow)
4.    var qa_mask = image
5.      .select("QA_PIXEL")
6.      .bitwiseAnd(parseInt("111111", 2))
7.      .eq(0);
8.    // mask for saturated pixels
9.    var saturation_mask = image.select("QA_RADSAT").eq(0);
10.
11.   // Apply scaling factors to the surface reflectance and thermal bands
12.   var opticalBands = image.select("SR_B.").multiply(0.0000275).add(-0.2);
13.   var thermalBands = image.select("ST_B.*").multiply(0.00341802).add(149.0);
14.
15.   // Add bands to the image and mask unwanted pixels
16.   return image
17.     .addBands(opticalBands, null, true)
18.     .addBands(thermalBands, null, true)
19.     .updateMask(qa_mask)
20.     .updateMask(saturation_mask);
21. }
22.
```

Python:

```python
1.  # function to prepare Landsat 8 images
2.  def prepareLandsat8(image):
```

```
3.    # mask for unwanted pixels (fill, dilated cloud, cirrus, cloud, cloud
      shadow, snow)
4.    qa_mask = image.select("QA_PIXEL").bitwiseAnd(int("111111", 2)).eq(0)
5.    # mask for saturated pixels
6.    saturation_mask = image.select("QA_RADSAT").eq(0)
7.
8.    # Apply scaling factors to the surface reflectance and thermal bands
9.    opticalBands = image.select("SR_B.").multiply(0.0000275).add(-0.2)
10.   thermalBands = image.select("ST_B.*").multiply(0.00341802).add(149.0)
11.
12.   # Add bands to the image and mask unwanted pixels
13.   return (
14.       image.addBands(opticalBands, overwrite=True)
15.       .addBands(thermalBands, overwrite=True)
16.       .updateMask(qa_mask)
17.       .updateMask(saturation_mask)
18.   )
19.
```

Map the `prepareLandsat8` function to all the images in the filtered Landsat 8 collection. Mapping in GEE means applying a function to all the items in a particular collection of objects. Such objects include ImageCollections, FeatureCollections, and Lists. This can be done by:

JavaScript/Python:

```
1. landsat8Prep= landsat8.map(prepareLandsat8)
2.
```

Calculate NDWI

There are two ways to calculate the NDWI. The green and NIR bands are required to calculate NDWI. In the Landsat 8 surface reflectance data, the green band is "SR_B3" and the NIR band is "SR_B5".

The first method is the `.normalizedDifference` function that can be applied on an `ee.Image`. The arguments for the `.normalizedDifference` function are the first and second bands used in calculating the NDWI, i.e. green and NIR bands respectively. One caveat with the `.normalizedDifference` function is that if there is any negative pixel in either of the input bands, that pixel will be masked out in the output.

<table>
<tr><td>

JavaScript:

```
1. var greenBand = "SR_B3";
2. var nirBand = "SR_B5";
3. var ndvi = image.normalizedDifference(
4.    [greenBand, nirBand]
5. );
6.
```

</td><td>

Python:

```
1. greenBand = 'SR_B3'
2. nirBand = 'SR_B5'
3. ndvi = image.normalizedDifference(
4.     [greenBand, nirBand]
5. )
6.
```

</td></tr>
</table>

The second method is to use the `.expression` function where you define the expression for NDWI and provide a dictionary that maps the bands to the variables in the expression. The limitation of this method is that when there is a negative pixel value in either of the bands, the calculated NDWI for that pixel will be greater or less than the −1 and +1 limits of NDWI.

<table>
<tr><td>

JavaScript:

```
1. var ndwi = image.expression(
2.    "NDWI = (green - NIR)/(green + NIR)", {
3.    green: image.select("SR_B3"),
4.    NIR: image.select("SR_B5"),
5. });
6.
```

</td><td>

Python:

```
1. ndwi = image.expression(
2.    "NDWI = (green - NIR)/(green + NIR)",
3.    {"green": image.select("SR_B3"),
4.    "NIR": image.select("SR_B5")},
5. )
6.
```

</td></tr>
</table>

In this tutorial, create a function that can be mapped to images in the Landsat 8 collection to calculate NDWI using the `.expression` method. Rename the resulting band as "NDWI" and add it to the image's bands.

<table>
<tr><td>

JavaScript:

```
1.  // function to calculate and add
    NDWI band
2.  function addNDWI(image) {
3.     var ndwi = image
4.        .expression(
5.           "NDWI = (green - NIR)/(green +
           NIR)",
6.           {
7.              green: image.select("SR_B3"),
8.              NIR: image.select("SR_B5"),
9.           }
10.       )
11.       .rename("NDWI");
12.
13.       return image.addBands(ndwi);
14. }
15.
```

</td><td>

Python:

```
1.  # function to calculate and add
    NDWI band
2.  def addNDWI(image):
3.     ndwi = image.expression(
4.        "NDWI = (green - NIR)/
          (green + NIR)",
5.        {
6.           "green": image.select
             ("SR_B3"),
7.           "NIR": image.select
             ("SR_B5")
8.        },
9.     ).rename("NDWI")
10.
11.    return image.addBands(ndwi)
12.
```

</td></tr>
</table>

Now map the `addNDWI` function to the images in the Landsat 8 collection.

JavaScript/Python:

```
1. landsat8Ndwi = landsat8.map(addNDWI)
2.
```

Add Celsius Band

Band 10 ("ST_B10") of the scaled Landsat 8 images represents the surface temperature values in kelvin (K). To convert to degrees Celsius (°C), subtract 273.15 from Band 10. Rename the output band as "Celsius" and add it to the existing bands in the image. Once again, create a function to calculate the Celsius band and map it to the images in the Landsat 8 collection.

JavaScript:

```javascript
1.  // function to calculate and add Celsius band
2.  function addCelsius(image) {
3.    var celsius = image.select("ST_B10").subtract(273.15).rename("Celsius");
4.
5.    return image.addBands(celsius);
6.  }
7.
8.  // map the addCelsius function over the ImageCollection
9.  landsat8Celsius = landsat8.map(addCelsius);
10.
```

Python:

```python
1.  # function to calculate and add Celsius band
2.  def addCelsius(image):
3.      celsius = image.select("ST_B10").subtract(273.15).rename("Celsius")
4.
5.      return image.addBands(celsius)
6.
7.  # map the addCelsius function over the ImageCollection
8.  landsat8Celsius = landsat8.map(addCelsius)
9.
```

Display Image on the Map

Now, display the Landsat 8 imagery in the map window. For Python users, the map will be displayed in the Map object created previously. This step is just to check and see whether everything worked so far. You will also be introduced to the `.addLayer` function of the Map object. The code below displays the RGB form of the first image in the filtered Landsat 8 `ImageCollection`. In the Landsat 8 surface reflectance product, "SR_B4", "SR_B3", and "SR_B2" represent the red, green, and blue bands respectively. We will call this map layer "Landsat 8 – First Image".

<table>
<tr><td>

JavaScript:

```javascript
1.  Map.addLayer(
2.    landsat8.first(),
3.    { bands: ["SR_B4", "SR_B3", "SR_B2"] },
4.    "Landsat 8 - First Image"
5.  );
6.
```

</td><td>

Python:

```python
1.  Map.addLayer(
2.      landsat8.first(),
3.      {"bands": ["SR_B4",
         "SR_B3", "SR_B2"]},
4.      "Landsat 8 - First Image"
5.  )
6.
```

</td></tr>
</table>

Extract Water Temperature

There are certain steps to follow to extract water temperature from the Landsat 8 collection.

The first step is to prepare the Landsat 8 dataset, as already covered. The next step is to mosaic the images to produce a spatially continuous image for the specified frequency. For example, a monthly mosaic will combine the image collections monthly. In this tutorial, you will explore both bi-weekly and monthly frequencies. Afterward, calculate the average water temperature at the ROI.

Before mosaicking, specify the date ranges for the mosaics. This will also be the basis for extracting the temperature time series. Since we specified the overall date range at the beginning of this tutorial, you can use those dates to generate a list of start and end dates for the mosaicking depending on the frequency. We explore both the bi-weekly and monthly frequencies. The start and end dates are presented as dictionary objects.

JavaScript:

```javascript
1.  // sequence mosaic dates monthly
2.  var n_monthly = endDate.difference(startDate, "month");
3.  var monthlyDates = ee.List.sequence(0, n_monthly.ceil()).map(function (n) {
4.    return ee.DateRange(
5.      startDate.advance(ee.Number(n), "month"),
6.      startDate.advance(ee.Number(n).add(1), "month")
7.    );
8.  });
9.
10. // sequence mosaic dates bi-weekly
11. var n_biWeekly = ee.Number(endDate.difference(startDate, "week")).divide(2);
12. var biWeeklyDates = ee.List.sequence(0, n_biWeekly.ceil()).map(function (n) {
13.   return ee.DateRange(
14.     startDate.advance(ee.Number(n).multiply(2), "week"),
15.     startDate.advance(ee.Number(n).add(1).multiply(2), "week")
16.   );
17. });
18.
```

Python:

```python
1.  # sequence mosaic dates monthly
2.  n_monthly = endDate.difference(startDate, "month")
3.  monthlyDates = ee.List.sequence(0, n_monthly.ceil()).map(
4.      lambda n: ee.DateRange(
5.          startDate.advance(ee.Number(n), "month"),
6.          startDate.advance(ee.Number(n).add(1), "month"),
7.      )
8.  )
9.
10. # sequence mosaic dates bi-weekly
11. n_biWeekly = ee.Number(endDate.difference(startDate, "week")).divide(2)
12. biWeeklyDates = ee.List.sequence(0, n_biWeekly.ceil()).map(
```

```
13.        lambda n: ee.DateRange(
14.            startDate.advance(ee.Number(n).multiply(2), "week"),
15.            startDate.advance(ee.Number(n).multiply(2).add(2), "week"),
16.        )
17. )
18.
```

All the processes involved in extracting water temperature will be combined into a single function that can be mapped to the date ranges above. But first, let us look at each of the steps in this function individually.

Mosaicking Images   Given a specific date range for the mosaic, you can filter out the prepared Landsat 8 collection within the date range. Prepare the Landsat 8 data, add the NDWI and Celsius bands and then use the `.mosaic` function to combine all the images within the date range – in this example, mosaic images in the month of May 2022.

JavaScript:

```javascript
1. // Specify mosaic date range
2. // var mosaicStartDate = ee.DateRange(monthlyDates.get(0)).start();
3. // var mosaicEndDate = ee.DateRange(monthlyDates.get(0)).end();
4. var mosaicStartDate = ee.Date("2022-05-01");
5. var mosaicEndDate = ee.Date("2022-05-30");
6.
7. // Create a mosaic of the Landsat 8 images
8. var mosaic = landsat8
9.    .filterBounds(reservoir.merge(downstream))
10.    .filterDate(mosaicStartDate, mosaicEndDate)
11.    .map(prepareLandsat8)
12.    .map(addNDWI)
13.    .map(addCelsius)
14.    .mosaic();
15.
```

Python:

```python
1. # Specify mosaic date range
2. # mosaicStartDate = ee.DateRange(dateRange).start()
3. # mosaicEndDate = ee.DateRange(dateRange).end()
4. mosaicStartDate = ee.Date("2022-05-01")
5. mosaicEndDate = ee.Date("2022-05-30")
6.
7. # Create a mosaic of the Landsat 8 images
8. mosaic = (
9.        landsat8.filterBounds(reservoir.merge(downstream))
10.        .filterDate(mosaicStartDate, mosaicEndDate)
11.        .map(prepareLandsat8)
12.        .map(addNDWI)
13.        .map(addCelsius)
14.        .mosaic()
```

```
15. )
16.
```

Select Water Pixels   After mosaicking, the next step is to extract the water pixels. You can use the NDWI band to extract the water pixel. Using the NDWI band requires a threshold that requires careful tuning. As a rule of thumb, you can start with a threshold of 0.2. Another way of extracting the water pixels is to use the QA_PIXEL band. Using the QA_PIXEL band is more reproducible. Experiment with the two methods to see the difference. In this tutorial, we will stick with the QA_PIXEL band for consistency.

JavaScript:

```
1. 1. // Create a mask for the water pixels using the QA_PIXEL band
2. 2. var waterMask = mosaic
3. 3.    .select("QA_PIXEL")
4. 4.    .bitwiseAnd(parseInt("10000000", 2)) // bit value for water
5. 5.    .neq(0);
6. 6. mosaic = mosaic.updateMask(waterMask);
7. 7.
8.
```

Python:

```
1. # Create a mask for the water pixels using the NDWI band
2. waterMaskNdwi = mosaic.select("NDWI").gt(0.2)
3.
4. # Create a mask for the water pixels using the QA_PIXEL band
5. waterMask = (
6.     mosaic.select("QA_PIXEL").bitwiseAnd(int("10000000", 2)).neq(0)
7. ) # bit value for water
8. mosaic = mosaic.updateMask(waterMask)
9.
```

Now, display the temperature band of the mosaicked and masked image to see the spatial variation of temperature in the water pixels (Figure 10.16).

JavaScript:

```
1. // Add the mosaicked and masked image to the map
2. Map.addLayer(
3.    mosaic,
4.    { bands: ["Celsius"], min: 5, max: 35, palette: ["blue", "green", "red"] },
5.    "Temperature (C)"
6. );
7.
```

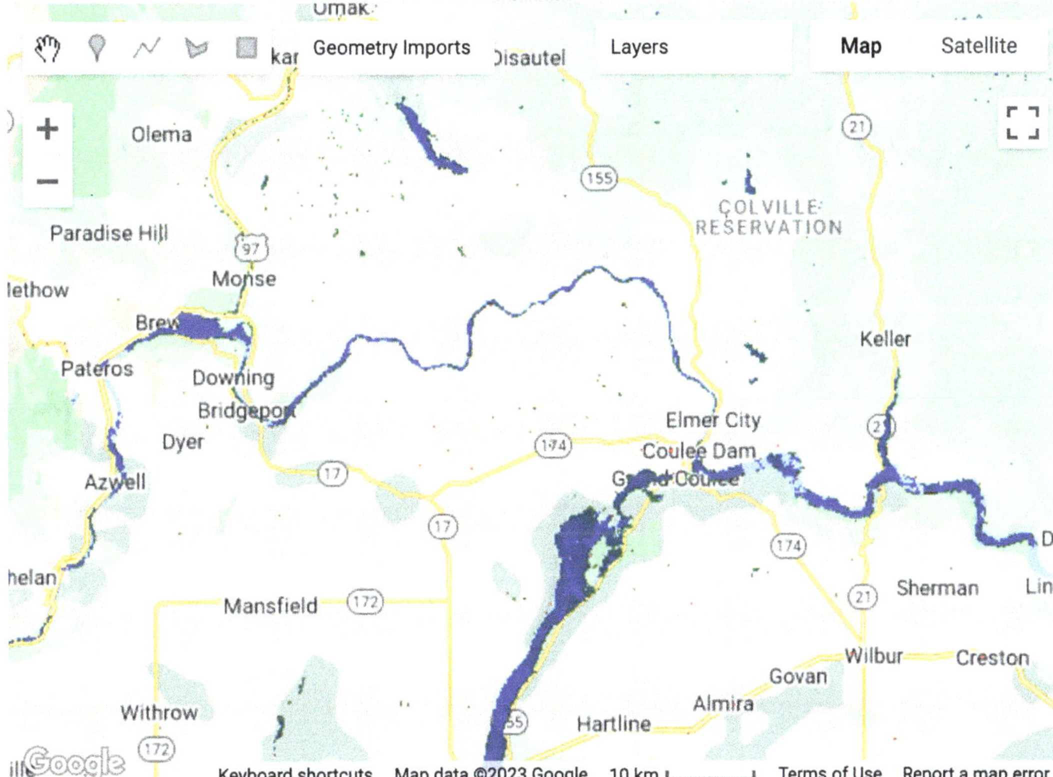

Figure 10.16 Temperature map over water pixels. [Credit: Google Earth Engine]

Python:

```python
1. # Add the mosaicked and masked image to the map
2. Map.addLayer(
3.     mosaic,
4.     {"bands": ["Celsius"], "min": 5, "max": 35, "palette": ["blue", "green",
       "red"]},
5.     "Temperature (C)",
6. )
7.
```

Use the inspector tool to assess the water temperature at different points. Also, change the date range to a different month and visually compare the temperature in different months.

Calculate the Mean Water Temperature   Use the `.reduceRegion` function to calculate the mean water temperature over the reservoir and the downstream reach. For this step, we will use the mean reducer (`ee.Reducer.mean`) to calculate the mean. For the scale, we will use the resolution of the Landsat 8 data.

JavaScript:

```
1. // Find the mean values of the Celsius over the reservoir
2. var meanReservoirTemperature = mosaic
3.     .select("Celsius")
4.     .reduceRegion(ee.Reducer.mean(), reservoir.geometry(), 30);
5. // Find the mean values of the Celsius downstream
6. var meanDownstreamTemperature = mosaic
7.     .select("Celsius")
8.     .reduceRegion(ee.Reducer.mean(), downstream.geometry(), 30);
9.
```

Python:

```
1. # Find the mean values of the Celsius over the reservoir
2. meanReservoirTemperature = mosaic.select("Celsius").reduceRegion(
3.     ee.Reducer.mean(), reservoir.geometry(), 30
4. )
5. # # Find the mean values of the Celsius downstream
6. meanDownstreamTemperature = mosaic.select("Celsius").reduceRegion(
7.     ee.Reducer.mean(), downstream.geometry(), 30
8. )
9.
```

Now, combine all these steps into a single function that returns a feature that comprises the date, reservoir temperature, and downstream temperature.

JavaScript:

```
1.  // Extract water temperature
2.  function extractWaterTemperature(dateRange) {
3.      // Specify mosaic date range
4.      var mosaicStartDate = ee.DateRange(dateRange).start();
5.      var mosaicEndDate = ee.DateRange(dateRange).end();
6.
7.      // Create a mosaic of the Landsat8 images
8.      var mosaic = landsat8
9.         .filterBounds(reservoir.merge(downstream))
10.        .filterDate(mosaicStartDate, mosaicEndDate)
11.        .map(prepareLandsat8)
12.        .map(addNDWI)
13.        .map(addCelsius)
14.        .mosaic();
15.
16.     // Create a mask for the water pixels using the NDWI band
17.     var waterMaskNdwi = mosaic.select("NDWI").gt(0.2);
18.
19.     // Create a mask for the water pixels using the QA_PIXEL band
```

```javascript
20.    var waterMask = mosaic
21.      .select("QA_PIXEL")
22.      .bitwiseAnd(parseInt("10000000", 2)) // bit value for water
23.      .neq(0);
24.    mosaic = mosaic.updateMask(waterMask);
25.
26.    // Find the mean values of the Celsius over the reservoir
27.    var meanReservoirTemperature = mosaic
28.      .select("Celsius")
29.      .reduceRegion(ee.Reducer.mean(), reservoir.geometry(), 30);
30.    // Find the mean values of the Celsius downstream
31.    var meanDownstreamTemperature = mosaic
32.      .select("Celsius")
33.      .reduceRegion(ee.Reducer.mean(), downstream.geometry(), 30);
34.
35.    return ee.Feature(null, {
36.      date: mosaicStartDate.format("YYYY-MM-dd"),
37.      reservoirTemp: ee.Number(meanReservoirTemperature.get("Celsius")),
38.      downstreamTemp: ee.Number(meanReservoirTemperature.get("Celsius"))
39.    });
40. }
41.
```

Python:

```python
1.  # Extract water temperature
2.  def extractWaterTemperature(dateRange):
3.      # Specify mosaic date range
4.      mosaicStartDate = ee.DateRange(dateRange).start()
5.      mosaicEndDate = ee.DateRange(dateRange).end()
6.
7.      # Create a mosaic of the Landsat 8 images
8.      mosaic = (
9.          landsat8.filterBounds(reservoir.merge(downstream))
10.         .filterDate(mosaicStartDate, mosaicEndDate)
11.         .map(prepareLandsat8)
12.         .map(addNDWI)
13.         .map(addCelsius)
14.         .mosaic()
15.     )
16.
17.     # Create a mask for the water pixels using the NDWI band
18.     waterMaskNdwi = mosaic.select("NDWI").gt(0.2)
19.
20.     # Create a mask for the water pixels using the QA_PIXEL band
21.     waterMask = (
22.         mosaic.select("QA_PIXEL").bitwiseAnd(int("10000000", 2)).neq(0)
23.     ) # bit value for water
24.     mosaic = mosaic.updateMask(waterMask)
25.
26.     # Find the mean values of the Celsius over the reservoir
27.     meanReservoirTemperature = mosaic.select("Celsius").reduceRegion(
28.         ee.Reducer.mean(), reservoir.geometry(), 30
29.     )
30.     # # Find the mean values of the Celsius downstream
```

```
31.        meanDownstreamTemperature = mosaic.select("Celsius").reduceRegion(
32.            ee.Reducer.mean(), downstream.geometry(), 30
33.        )
34.
35.        return ee.Feature(
36.            None,
37.            {
38.                "date": mosaicStartDate.format("YYYY-MM-dd"),
39.                "reservoirTemp": ee.Number(meanReservoirTemperature.
                   get("Celsius")),,
40.                "downstreamTemp": ee.Number(meanDownstreamTemperature.
                   get("Celsius")),,
41.            },
42.        )
43.
```

Now, map this to the sequence of dates to generate the temperature time series. Cast the results as a feature collection combining all the features output by the function. Use the monthly sequence of dates in this example.

JavaScript:

```javascript
1. // Map the extractWaterTemperature function over date sequence
2. var temperatureSeries = ee.FeatureCollection(
3.    monthlyDates.map(extractWaterTemperature)
4. );
5.
```

Python:

```python
1. # Map the extractWaterTemperature function over date sequence
2. temperatureSeries = ee.FeatureCollection(monthlyDates.map
   (extractWaterTemperature))
3.
```

Plot Temperature Time Series

Plotting temperature time series in the GEE Code Editor is different from the Python environment. The GEE Code Editor has inbuilt `ui.Chart` options that can be used to plot the data in `temperatureSeries`. In Python, other packages are used to generate the temperature time series plot.

GEE Code Editor Use the `ui.Chart.feature.byFeature` function to plot the time series for the feature collection that consists of the dates, reservoir temperature, and downstream temperature. This function requires the input FeatureCollection, and specifications for the x- and y-axes. The function does not handle null values, so the `ee.Filter.notNull` function is used to filter out the null values before plotting the data. Additional modifications and styling can be added to specify the axis labels, title, etc. If you follow the code in this tutorial, the resulting chart should look like Figure 10.17.

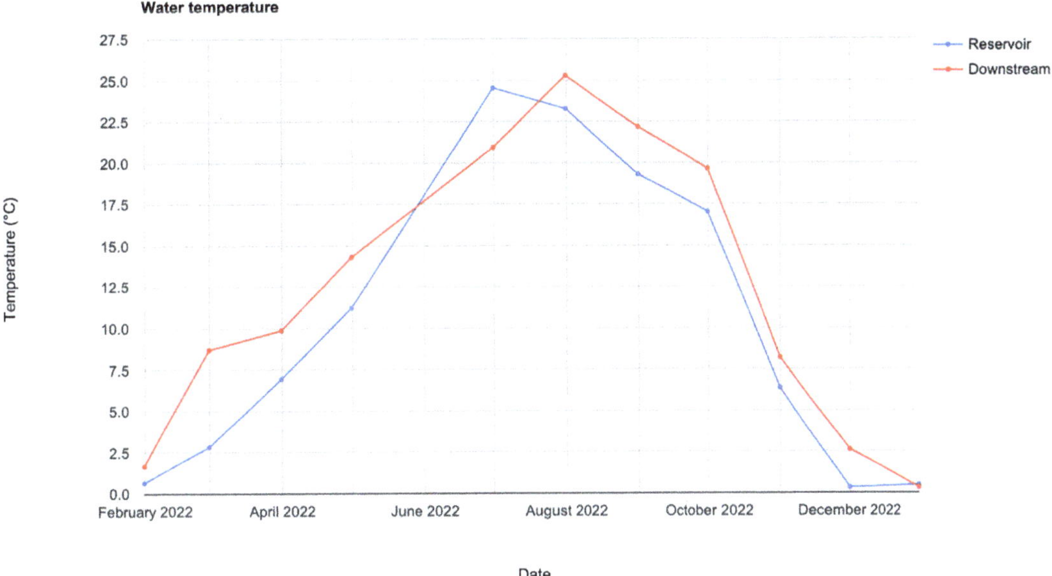

Figure 10.17 Water temperature time series in the GEE Code Editor environment.

JavaScript:

```javascript
1.  // plot temperature time series
2.  var chart = ui.Chart.feature
3.    .byFeature(
4.      temperatureSeries.filter(
5.        ee.Filter.notNull(["reservoirTemp", "downstreamTemp"])
6.      ),
7.      "date",
8.      ["reservoirTemp", "downstreamTemp"]
9.    )
10.   .setSeriesNames(["Reservoir", "Downstream"])
11.   .setChartType("LineChart")
12.   .setOptions({
13.     title: "Water Temperature",
14.     hAxis: {
15.       title: "Date",
16.     },
17.     vAxes: {
18.       0: {
19.         title: "Temperature (C)",
20.       },
21.     },
22.   lineSize: 2,
23.   pointSize: 4,
24.   });
25.
26. print(chart);
27.
```

Python Environment   In the Python environment, use the `geemap.ee_to_pandas` to convert the feature collection into a Pandas DataFrame. Then, cast the dates column as a "datetime" object so that it can be plotted appropriately.

Use `matplotlib.plt` to create a plot of the temperature time series. Like the GEE Code Editor, you can modify the elements of the plot to add the necessary details. If you followed the steps in this tutorial, your final plot should look like Figure 10.18. Note the gap in the time series. This represents a period when there was no quality temperature measurement for reasons such as high cloud cover.

```
 1. # convert temperature series to pandas dataframe
 2. temperatureSeriesDf = geemap.ee_to_df(temperatureSeries)
 3. temperatureSeriesDf["date"] = pd.to_datetime(temperatureSeriesDf["date"])
 4.
 5. # plot temperature series
 6. fig, ax = plt.subplots(figsize=(10, 5))
 7. ax.plot(
 8.     temperatureSeriesDf["date"], temperatureSeriesDf["reservoirTemp"],
        label="Reservoir"
 9. )
10. ax.plot(
11.     temperatureSeriesDf["date"],
12.     temperatureSeriesDf["downstreamTemp"],
13.     label="Downstream",
14. )
15. ax.set_xlabel("Date")
16. ax.set_ylabel("Temperature (C)")
17. ax.set_title("Water Temperature")
18. ax.legend()
19. plt.show()
20.
```

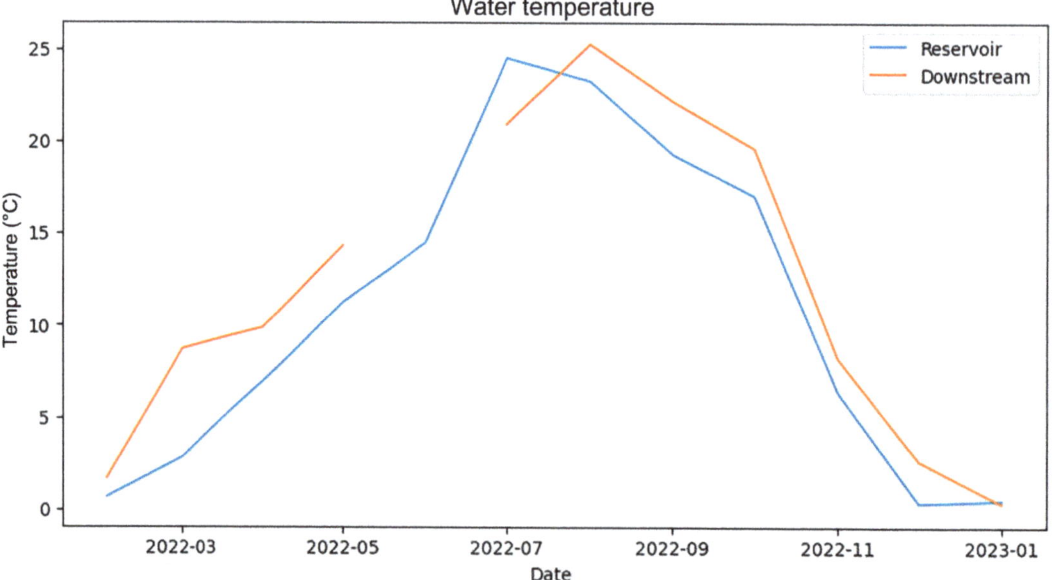

Figure 10.18 Water temperature time series in the Python environment.

APPENDIX

A.1   Understanding the Quality Assessment Band Values

Landsat and most other satellite data come with quality bands that provide additional information about the quality of each pixel in a particular scene. The quality band is a convenient way of masking out certain pixels in the data. The QA Band can be used to identify and extract water pixels in the satellite images. The quality information of a pixel is designated using a binary assignment in bits. For example, the QA_PIXEL band of Landsat 8, Collection 2 products is encoded in a 16-bit binary assignment (Table 10.3). The value of the fifth bit indicates whether there is a high confidence of snow/ice in the pixel or not, while the seventh bit indicates whether the pixel is water or land. The pixel value in the quality band is the decimal integer equivalent of the binary assignment. For instance, the pixel value that represents a dilated cloud over water with low confidence has a value of 21,890 (Figure 10.19).

A.2   Deriving Surface Temperature Using the Single-Channel Algorithm

The single-channel method, developed by Jiménez-Muñoz et al. (2009), is a method used to obtain land surface temperature (T_s) from raw Landsat data by applying atmospheric corrections to the registered radiance at the sensor (L_{sen}) by:

$$T_s = \gamma \left[\frac{1}{\varepsilon} (\varphi_1 L_{sen} + \varphi_2) + \varphi_3 \right] + \delta \tag{10.3}$$

where ε is the surface emissivity; φ_1, φ_2, and φ_3 are atmospheric factors that are dependent on water vapor; and γ and δ are parameters that are dependent on the Planck equation (Jiménez-Muñoz et al., 2009).

The atmospheric factors φ_1, φ_2, and φ_3 can be calculated by:

$$\left[\varphi_1 \, \varphi_2 \, \varphi_3 \right] = \left[c_{11} \, c_{12} \, c_{13} \, c_{21} \, c_{22} \, c_{23} \, c_{31} \, c_{32} \, c_{33} \right] \left[w^2 \, w \, 1 \right] \tag{10.4}$$

Table 10.3 **Bit assignment for the Landsat 8 QA_PIXEL band. The value for each bit position represents a unique description of the pixel. Some bit positions such as 8–9, 10–11, 12–13, and 14–15 are combined to describe a single piece of information respectively.**

Bit position	Flag description	Values
0	Fill	0 for image data 1 for fill data
1	Dilated Cloud	0 for cloud is not dilated or no cloud 1 for cloud dilation
2	Cirrus	0 for cirrus confidence: no confidence level set or low confidence 1 for high confidence cirrus
3	Cloud	0 for cloud confidence is not high 1 for high-confidence cloud

Table 10.3 **(cont.)**

Bit position	Flag description	Values
4	Cloud Shadow	0 for cloud shadow confidence is not high
		1 for high-confidence cloud shadow
5	Snow	0 for snow/ice confidence is not high
		1 for high confidence snow cover
6	Clear	0 if cloud or dilated cloud bits are set
		1 if cloud and dilated cloud bits are not set
7	Water	0 for land or cloud
		1 for water
8–9	Cloud Confidence	00 for no confidence level set
		01 low confidence
		10 medium confidence
		11 high confidence
10–11	Cloud Shadow Confidence	00 for no confidence level set
		01 low confidence
		10 reserved
		11 high confidence
12–13	Snow/Ice Confidence	00 for no confidence level set
		01 low confidence
		10 reserved
		11 high confidence
14–15	Cirrus Confidence	00 for no confidence level set
		01 low confidence
		10 reserved
		11 high confidence

(Source: Landsat Level-2 Surface Reflectance Science Product courtesy of the US Geological Survey).

Bit position	15	14	13	12	11	10	9	8	7	6	5	4	3	2	1	0
Flag description		Cirrus Confidence		Snow/Ice Confidence		Cloud Shadow Confidence		Cloud Confidence	Water	Clear	Snow	Cloud Shadow	Cloud	Cirrus	Dilated Clouds	Fill

Example Illustration

Pixel description: Dilated cloud over water with low confidence values set

0	1	0	1	0	1	0	1	1	0	0	0	0	0	1	0

16-bit binary value: 0101010110000010
Pixel value (decimal integer): 21890

Figure 10.19 An illustration of the pixel value information of a 16-bit quality band of the Landsat 8 Collection 2 data products. The pixel values of the quality band are the decimal integer equivalents of the binary value. [Credit: US Geological Survey and NASA]

where c_{ij} are coefficients that are obtained by simulation; w is the water vapor that can be obtained from the National Centers for Environmental Prediction (NCEP) and the National Center for Atmospheric Research (NCAR) reanalysis dataset. For the thermal band (Band 10) of Landsat 8, the atmospheric factors can be calculated using the following coefficients (García-Santos et al., 2018):

$$[\varphi_1 \; \varphi_2 \; \varphi_3] = [0.04019 \, 0.02916 \, 1.01523 - 0.38333 - 1.50294 \, 0.20324 \, 0.00918 \, 1.36072 \; - 0.27514][w^2 \; w \; 1] \tag{10.5}$$

The parameters γ and δ are given by:

$$\gamma = \frac{T_{\text{sen}}^2}{bL_{\text{sen}}} \tag{10.6}$$

$$\delta = T_{\text{sen}} - \frac{T_{\text{sen}}^2}{b} \tag{10.7}$$

where b is a constant and T_{sen} is the brightness temperature observed at the sensor. For Landsat 8, Band 10, $b = 1{,}324\,\text{K}$ (Sharaf et al., 2019). T_{sen} is calculated from:

$$T_{\text{sen}} = \frac{K_2}{\ln\left(\dfrac{K_1}{L_{\text{sen}}} + 1\right)} \tag{10.8}$$

where K_1 and K_2 are calibration constants for temperature conversion that can be obtained from the Landsat dataset properties. For Landsat 8 Band 10 (thermal band), K_1 and K_2 can be obtained from the image properties "K1_CONSTANT_BAND_10" and "K2_CONSTANT_BAND_10" respectively.

The single-channel algorithm can be lumped up into a single function that can be mapped to the images in the raw Landsat 8 image collection to obtain the temperature values in kelvin (or degrees Celsius by subtracting 273.15). The function can be written as:

JavaScript:

(https://code.earthengine.google.com/8656295440741853d7434bf4ea3b8b2d)

```
1.  // load raw landsat 8 data
2.  var landsat8Raw = ee.ImageCollection("LANDSAT/LC08/C02/T1"); // raw landsat 8
    data
3.
4.  // load NCEP Reanalysis water vapor data
5.  var water_vapor = ee.ImageCollection("NCEP_RE/surface_wv"); // NCEP
    Reanalysis water vapor data
6.
7.  // single channel algorithm
8.  function singleChannel(image) {
9.    // # Constants
10.   var b = ee.Number(1324),
11.     // K1 = ee.Number(774.89), // K1 constant for band 10
12.     K1 = ee.Number(image.get("K1_CONSTANT_BAND_10")), // K1 constant for band 10
```

```
13.       // K2 = ee.Number(1321.08), // K2 constant for band 10
14.       K2 = ee.Number(image.get("K2_CONSTANT_BAND_10")), // K2 constant for band 10
15.       e = ee.Number(0.99); // surface emissivity
16.
17.    var B10 = ee.Image(image.select("B10"));
18.
19.    // // scaling factors for band 10 radiance
20.    // var multiplicationFactor = ee.Number(0.0003342);
21.    // var additionFactor = ee.Number(0.1);
22.
23.    // Alternative scaling factors for band 10 radiance
24.    var multiplicationFactor = ee.Number(image.get("RADIANCE_MULT_BAND_10"));
25.    var additionFactor = ee.Number(image.get("RADIANCE_ADD_BAND_10"));
26.
27.    // registered radiance at the sensor
28.    var Lsen = B10.multiply(multiplicationFactor).add(additionFactor);
29.
30.    // brightness temperature observed at the sensor
31.    var Tsen = Lsen.pow(-1).multiply(K1).add(1).log().pow(-1).multiply(K2);
32.
33.    // gamma and sigma
34.    var gamma = Tsen.pow(2).divide(b).divide(Lsen);
35.    var sigma = Tsen.subtract(Tsen.pow(2).divide(b));
36.
37.    // water filter and convert water vapor
38.    var w = water_vapor
39.      .filterDate(
40.        ee.Date(image.get("system:time_start")),
41.        ee.Date(image.get("system:time_start")).advance(1, "day")
42.      )
43.      .first()
44.      .divide(10); // convert from km/m^2 to g/cm2
45.
46.    // atmospheric factors
47.    var psi1 = w.pow(2).multiply(0.04019).add(w.multiply(0.02916)).add(1.01523);
48.    var psi2 = w.pow(2).multiply(-0.38333).add(w.multiply(-
       1.50294)).add(0.20324);
49.    var psi3 = w.pow(2).multiply(0.00918).add(w.multiply(1.36072)).add(-0.27514);
50.
51.    // Surface Temperature from the single channel algoritm
52.    var Ts = gamma
53.      .multiply(Lsen.multiply(psi1).add(psi2).multiply(e.pow(-1)).add(psi3))
54.      .add(sigma);
55.    // .subtract(273.15); // convert from K to C
56.
57.    return image.addBands(Ts.rename("Temperature")); // add temperature band to
       image
58. }
59.
60. var startDate = ee.Date("2022-01-01");
61. var endDate = ee.Date("2022-12-31");
62.
63. // filter the landsat 8 raw data and map the single-channel algorithm
64. var landsat8RawFiltered = landsat8Raw
65.   .filterDate(startDate, endDate)
66.   // .filterBounds(geometry) // filter by geometry
```

```
67.    .map(singleChannel); // map the function to add the derived temperature band
68.
69.
```

Python:

```python
1. # load raw landsat 8 data
2. landsat8Raw = ee.ImageCollection("LANDSAT/LC08/C02/T1") # raw landsat 8 data
3.
4. # load NCEP Reanalysis water vapor data
5. water_vapor = ee.ImageCollection(
6.     "NCEP_RE/surface_wv"
7. ) # NCEP Reanalysis water vapor data
8.
9. # single channel algorithm
10. def singleChannel(image):
11.     # # Constants
12.     b = ee.Number(1324)
13.     # K1 = ee.Number(774.89) # K1 constant for band 10
14.     K1 = ee.Number(image.get("K1_CONSTANT_BAND_10")) # K1 constant for band 10
15.     # K2 = ee.Number(1321.08) # K2 constant for band 10
16.     K2 = ee.Number(image.get("K2_CONSTANT_BAND_10")) # K2 constant for band 10
17.     e = ee.Number(0.99) # surface emissivity
18.
19.     B10 = ee.Image(image.select("B10"))
20.
21.     # # scaling factors for band 10 radiance
22.     # multiplicationFactor = ee.Number(0.0003342)
23.     # additionFactor = ee.Number(0.1)
24.
25.     # Alternative scaling factors for band 10 radiance
26.     multiplicationFactor = ee.Number(image.get("RADIANCE_MULT_BAND_10"))
27.     additionFactor = ee.Number(image.get("RADIANCE_ADD_BAND_10"))
28.
29.     # registered radiance at the sensor
30.     Lsen = B10.multiply(multiplicationFactor).add(additionFactor)
31.
32.     # brightness temperature observed at the sensor
33.     Tsen = Lsen.pow(-1).multiply(K1).add(1).log().pow(-1).multiply(K2)
34.
35.     # gamma and sigma
36.     gamma = Tsen.pow(2).divide(b).divide(Lsen)
37.     sigma = Tsen.subtract(Tsen.pow(2).divide(b))
38.
39.     # water filter and convert water vapor
40.     w = (
41.         water_vapor.filterDate(
42.             ee.Date(image.get("system:time_start")),
43.             ee.Date(image.get("system:time_start")).advance(1, "day"),
44.         )
45.         .first()
46.         .divide(10)
47.     ) # convert from km/m^2 to g/cm2
48.
```

```
49.        # atmospheric factors
50.        psi1 = w.pow(2).multiply(0.04019).add(w.multiply(0.02916)).add(1.01523)
51.        psi2 = w.pow(2).multiply(-0.38333).add(w.multiply(-1.50294)).add(0.20324)
52.        psi3 = w.pow(2).multiply(0.00918).add(w.multiply(1.36072)).add(-0.27514)
53.
54.        # Surface Temperature from the single channel algoritm
55.        Ts = (
56.            gamma.multiply(Lsen.multiply(psi1).add(psi2).multiply(e.pow(-1))
                .add(psi3))
57.            .add(sigma)
58.            .subtract(273.15)
59.        ) # convert from K to C
60.
61.        return image.addBands(Ts.rename("Temperature")) # add temperature band to
           image
62.
63. startDate = ee.Date("2022-01-01")
64. endDate = ee.Date("2022-12-31")
65.
66. # filter the landsat 8 raw data and map the single-channel algorithm
67. landsat8RawFiltered = (
68.     landsat8Raw.filterDate(startDate, endDate)
69.     .filterBounds(geometry) # filter by geometry
70.     .map(singleChannel)
71. ) # map the function to add the derived temperature band
72.
```

EXERCISES: CHAPTER 10

Q10.1 What are the wavelengths typically used for remote sensing of surface temperature?

Q10.2 What is the basic premise of thermal infrared (TIR)-based remote sensing for surface water temperature?

Q10.3 What are the limitations of satellite TIR-based remote sensing of surface water temperature?

Q10.4 Can microwave-based temperature estimation resolve some of these limitations?

Q10.5 What is thermal stratification of reservoirs, and what role does hydropower play in thermal alteration of water downstream?

Q10.6 Despite the limitations of satellite TIR for water temperature sensing, how far back can we track the surface water temperature of many years using the Landsat mission?

Q10.7 What do you think are some of the potential uses of multi-decadal tracking of river's surface temperature of rivers to improve water management in regulated rivers?

REFERENCES

Ahmad, S. K., F. Hossain, G. W. Holtgrieve, T. Pavelsky, and S. Galelli (2021). Predicting the likely thermal impact of current and future dams around the world. *Earth's Future*, 9, e2020EF001916. https://doi.org/10.1029/2020EF001916

Allen, G. H. and T. M. Pavelsky (2018). Global river widths from Landsat (GRWL) database (Version V01.01) [Dataset]. *Zenodo*. http://doi.org/10.5281/zenodo.1297434

Bonnema, M., F. Hossain, B. Nijssen, and G. Holt (2020). Hydropower's hidden transformation of rivers in the Mekong, *Environmental Research Letters*, vol. 15. https://doi.org/10.1088/1748-9326/ab763d.

García-Santos, V., J. Cuxart, D. Martínez-Villagrasa, M. Jiménez, and G. Simó (2018). Comparison of three methods for estimating land surface temperature from Landsat 8-TIRS sensor data. *Remote Sensing*, vol. 10, 1450. https://doi.org/10.3390/rs10091450

Jiménez-Muñoz, J. C., J. Cristóbal, J. A. Sobrino, et al. (2009). Revision of the single-channel algorithm for land surface temperature retrieval from Landsat thermal-infrared data. *IEEE Transactions on Geoscience and Remote Sensing*, vol. 47, 339–349. https://doi.org/10.1109/TGRS.2008.2007125

Landsat Collection 2 (2021). Fact Sheet 2021–3002. US Geological Survey. https://doi.org/10.3133/fs20213002

Lehner, B., C. R. Liermann, C. Revenga, et al. (2011). High-resolution mapping of the world's reservoirs and dams for sustainable river-flow management, *Frontiers in Ecology & Environment*, vol. 9, 494–502. https://doi.org/10.1890/100125

McFeeters, S. K. (1996). The use of the Normalized Difference Water Index (NDWI) in the delineation of open water features. *International Journal of Remote Sensing*, vol. 17, 1425–1432. https://doi.org/10.1080/01431169608948714

Sharaf, N., A. Fadel, M. Bresciani, et al. (2019). Lake surface temperature retrieval from Landsat-8 and retrospective analysis in Karaoun Reservoir, Lebanon. *Journal of Applied Remote Sensing*, vol. 13, 1. https://doi.org/10.1117/1.JRS.13.044505

SUGGESTED READING

Dugdale, S. J. (2016). A practitioner's guide to thermal infrared remote sensing of rivers and streams: recent advances, precautions and considerations. *WIREs Water*, vol. 3, 251–268. https://doi.org/10.1002/wat2.1135

11 Social Justice in Water Management

11.1 Chapter Overview

In this chapter and the next, we will be switching gears to transition to less-technical but more social and governance-related issues in which satellite remote sensing of water can play a positive role. So far, what we have learned up to Chapter 10 are technical aspects of satellite remote sensing of water and their applications in water management. In this chapter, we will explore the potential of satellite remote sensing for social justice in water management. As mentioned in the preface of this book, this chapter is intentionally short and lacks the rigor and depth that a course on social justice or citizen science may demand. This is because these topics are not an active area of research by the author, and therefore expertise is not claimed. However, the community at large has begun to address these issues in water management. In fact, there is now a new topic called *critical remote sensing* (Bennett et al., 2022) that looks at these nontechnical issues. Thus, we believe that readers trying to gain basic literacy in satellite remote sensing for water management need to start thinking about these social justice and equity issues in the context of the "potential" of satellite remote sensing data due to increasing availability and access.

In this chapter, we will first define social justice. The chapter will cover three potential examples in which satellite remote sensing could be used for disadvantaged communities or nations to mitigate the negative impacts of past water management. The case studies and descriptions in this and the next chapter are mainly for building awareness of the social justice and governance issues in mind when exploring how equitable are the satellite-based water management solutions.

11.2 Introduction

Social justice at the intersection of water management and satellites is still a fledgling field. The goal here will be to understand the potential of satellite remote sensing of water to empower communities who are vulnerable or disproportionately affected by water management challenges. Using some case studies and examples, we will explore how the plethora of satellite water data and their easy availability for accessing and processing data can level the playing field, mitigating many water management challenges (Figure 11.1).

In this chapter, the goal will be to understand how remote sensing can be used as a tool to identify, quantify, and explore solutions for social justice in water management. We will do this by sharing case studies and examples. We will first address the definition of social justice at high level with some conceptual examples. Next, we will overview two social justice issues related to

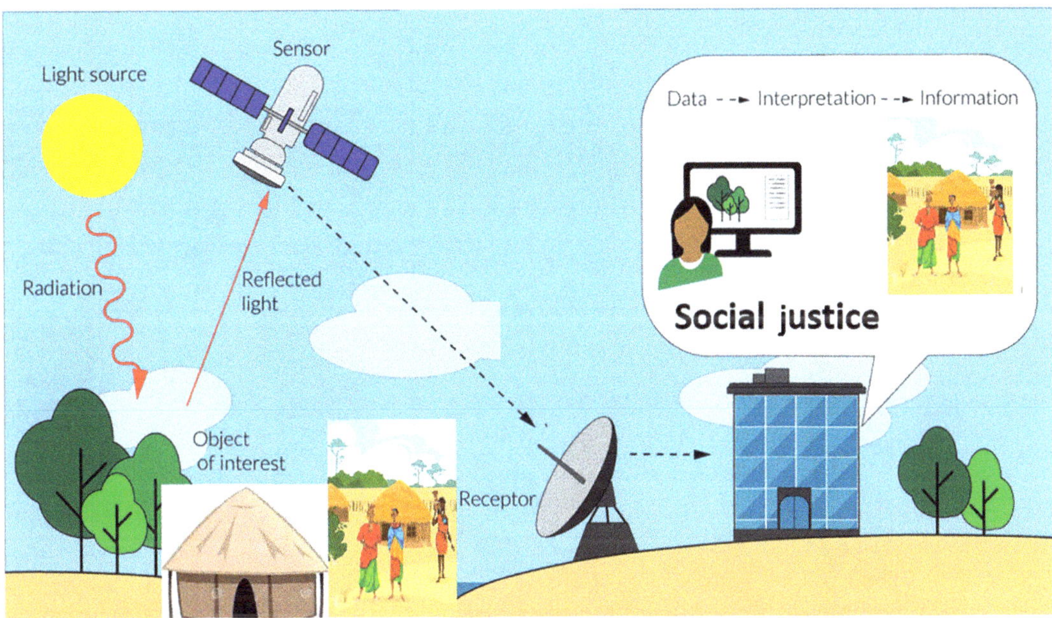

Figure 11.1 Using satellite remote sensing to understand social justice issues of a distant region by converting the data on that region to information through interpretation.

surface water in the Amazon and the Mesopotamian region where the community is deprived of sufficient access to water. Using these two examples, our goal will be to show how satellite remote sensing of the water fluxes can potentially help to empower vulnerable communities to develop strategies of their own that can potentially be a pathfinder to more equitable access to water.

11.3 What Is Social Justice?

The definition for social justice can vary, but there are three common and underlying themes: equal rights, equal opportunity, and equal treatment. We will stick to the definition the United Nations used in 2006 for social justice as follows:

Social justice may be broadly understood as the fair and compassionate distribution of the fruits of economic growth. (UN, 2006)

In water management, this can be defined as follows:

Every human and every nation have a fundamental right to clean and usable water for their basic needs.

Since we are talking about satellite remote sensing and therefore dealing with data, a more remote-sensing-style definition for this chapter is:

Every human and every nation or community has a fundamental right to information on water resources that affects their survival and livelihood.

Figure 11.2 The concept of equity, and its difference from equality. [Image credit: Craig Froehle from https://medium.com/@CRA1G/the-evolution-of-an-accidental-meme-ddc4e139e0e4#.pqiclk8pl)]

When we think of social justice in water, the three underlying themes of "equal rights", "equal opportunity", and "equal treatment" boil down to the concept of equity. Figure 11.2 summarizes best what equity means at the individual human level. The key issue is the recognition that no two communities or stakeholders or users are exactly the same in terms of their capacity to withstand or overcome barriers.

In the water management context, we can understand equity using the example of water distribution among different nations or communities that make up a river basin. Imagine there is a nation or community located upstream at the source of the water, another located midstream, and a third one located downstream. If we divide the available river water into three equal parts, that would be an example of equality. Even if we distribute the water based on relative population or area of each group, that would be a population or area-weighted equal distribution. This will not be an equitable distribution for the following reasons. The dependence on water resources and the ability to cope with shortages or excesses may not be the same for each nation or community. The downstream community or nation is obviously at the mercy of the two upstream nations for supply of water. Yet this nation perhaps also faces the threat of sea level rise, and saltwater intrusion into its freshwater rivers. The rivers in the downstream country probably sustain fertile agricultural land and a diverse ecosystem. So, this nation would need a minimum freshwater flow to be maintained in its rivers to flush out salt water from creeping upstream – a problem that midstream and upstream nations do not have to face much. A gradually more saline waterfront not only will damage ecosystems but could also lower soil fertility and agricultural productivity. Consequently, the downstream nation may have to spend a disproportionately greater amount of money to treat its water for consumption but may not be able to afford the high costs. So, a more equitable distribution of water for all three nations would be one that also accounts for the unique vulnerabilities of each nation. For the case of the downstream nation, it is the saltwater intrusion from the oceans. Just sharing water among the nations based on the amount of water needed for each population will not be an equitable solution for water management.

Today, there are numerous such examples of upstream–midstream–downstream country dynamics for social justice, such as Egypt (downstream), Sudan (midstream), and Ethiopia (upstream) for the Blue Nile; China (upstream), Laos and Thailand (midstream), and Cambodia for the Mekong River; and India (upstream) and Bangladesh (downstream) for the Ganges River.

11.4 Author's Personal Story on Social Justice

My father was born (in 1946) into a farming family in Bangladesh where he was the youngest of six. In the 1950s, the Green Revolution hadn't quite arrived in South Asia, and farm water practices were human-labor intensive. It was expected that all siblings would continue farming as the day job to keep the family financially afloat.

Fortunately, being the youngest and having excelled in math, my father was allowed to complete high school, after which he expressed an interest in becoming an architect or an engineer. Unfortunately, my father could not take the entrance exam for the only university that provided such an education. This is because in those days (in the 1960s), there used to be a mandatory "drawing test". All applicants were required to bring their own drawing tools (a board and drafter) to the exam. My father could not afford the drawing kit, and reluctantly he had to give up his dream of becoming an architect or an engineer. Later, he ended up with a training in physics from the University of Durham.

I always think of this as a social justice and equity issue. My father would have probably done well in the entrance exam if someone had loaned him the drawing board and drafter, or if there had been a system for underprivileged applicants to access the equipment to prepare for the entrance exam at very low or no cost.

Many decades later, I was able to enroll into an engineering program because my father, who, by the mid-1990s, was financially solvent, was able to buy me the drawing equipment I needed to compete in the exams and courses. It had never dawned on me until very recently that what my father faced could have been easily addressed with some simple tweaks in the "education system" to level the playing field and give everyone equal opportunity in the spirit of equity.

11.5 Examples of Social Justice in Water

11.5.1 The Ogallala Aquifer

The Ogallala is a major groundwater aquifer of the United States extending from South Dakota to North Texas (Figure 11.3). The aquifer has stored fossil water that was formed many thousands of years ago and thus is not recharged at annual cycles. Usage of the Ogallala aquifer's water for food production began in the 1930s. It is estimated that about 20 percent of total agricultural harvest in the US currently relies on this aquifer. Some estimates claim that at current rates, approximately 60–70 percent of the aquifer will be depleted in 50 years (Figure 11.4).

The Ogallala aquifer today presents a social justice issue for water management. This is because the groundwater table has declined over many decades of use. However, this decline is not uniform, as can be seen from Figure 11.4. The level of decline is a good proxy for farmers on how much additional money to spend for fuel to pump the water to the surface for irrigation. Not all farmers, however, have a similarly high income level to manage rising costs for fuel. Naturally, this gradual decline in the groundwater table negatively affects the low-income farming communities more than others. One way to verify the disproportionately greater negative impact is by looking at median household income, which is not the same or proportional to the extent of groundwater decline within the aquifer. So, to be equitable and grounded in social justice, any solution regarding the water management of the Ogallala to mitigate challenges faced by farmers must take into

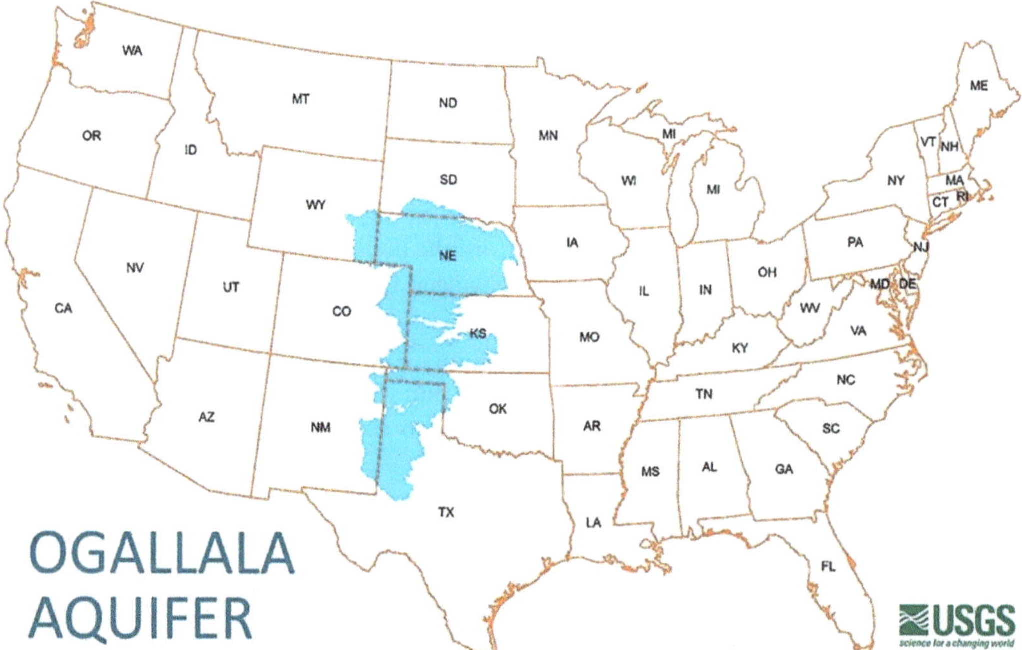

Figure 11.3 Extent of the Ogallala aquifer in the United States, shown in blue. [Image credit: US Geological Survey]

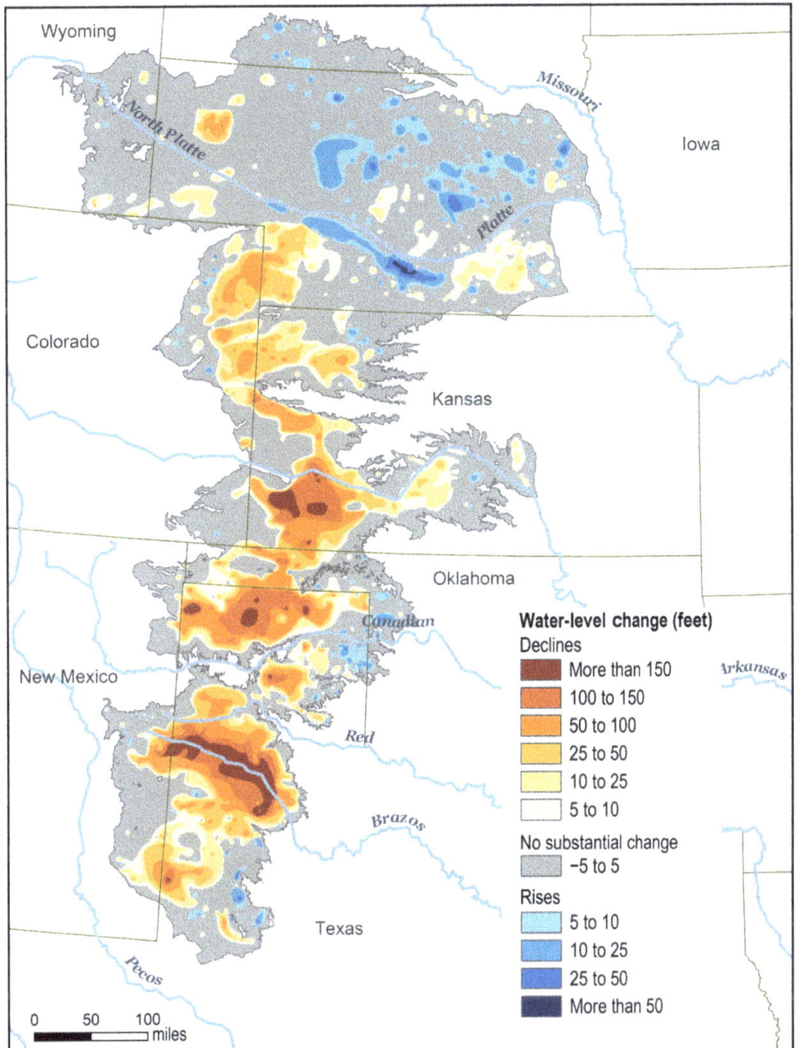

Figure 11.4 Changes in Ogallala water levels from the period before the aquifer was tapped to 2015. [Source: National Climate Assessment figure 10.3, by National Oceanic and Atmospheric Administration (NOAA)]

account the underlying factors of income capacity. If we derive a solution based entirely and only on quantitative decline of groundwater without any regard for the socio-economic background of farmers, the solution may not be as societally impactful as it should be.

11.5.2 Wildfires in the Western United States

Wildfires need water for their control, and hence this can be considered a social justice issue for water management. Recently, in a study by Masri et al. (2021), it was reported that wildfires affect a diverse range of populations in the western US but very disproportionately. Figure 11.5

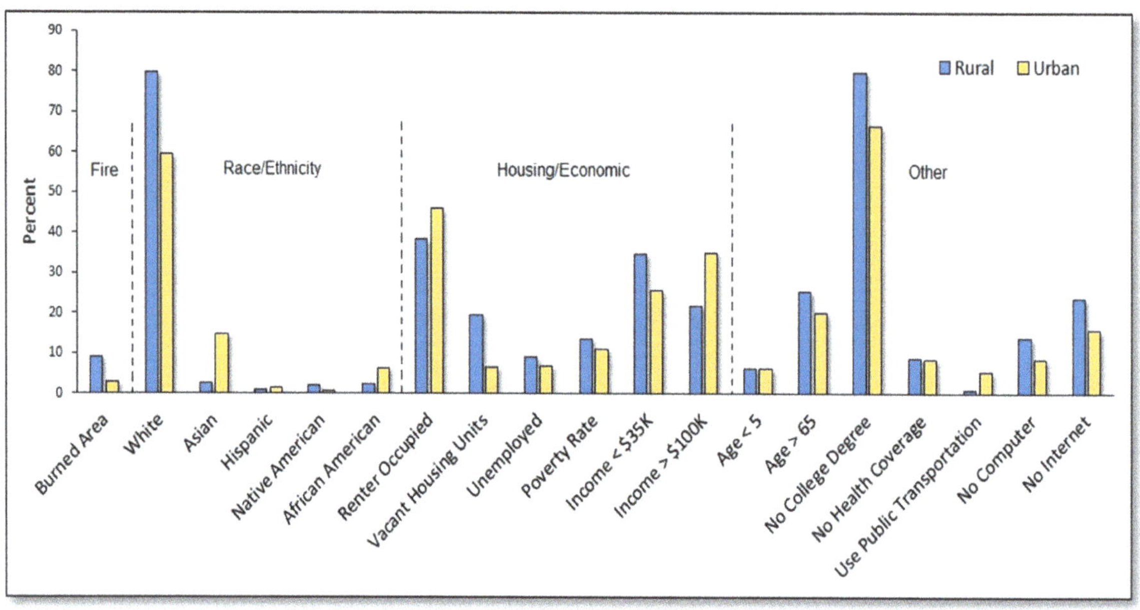

Figure 11.5 Wildfire vulnerability characteristics presented as percentages of population averaged across urban and rural Census tracts. [After Masri et al., 2021]

shows that rural populations there are more negatively affected by wildfires. Within this rural group, which has lower income levels and less access to the internet, because disproportionate numbers of Native Americans and older Americans live rurally, they are likely to be more affected than other groups. So, any solution for managing wildfires through an equity-based water management strategy needs to take these underlying factors of affected communities into account. We need to ask whether our water management solution is working for all.

Case Study 11.1: **The Xingu River of the Amazon**[1]

The Xingu River in the Amazon is revered as the "house of God" by the Indigenous people living along its Volte Grande, or Big Bend, in the Brazilian Amazon (Figure 11.6). The river is essential to their culture and religion, and a crucial source of fish, transportation, and water for trees and plants.

Five years ago, the Big Bend was a broad river valley interwoven with river channels teaming with fish, turtles, and other wildlife. Today, as much as 80 percent of the water flow is gone (Figure 11.7).

That's because in late 2015, the massive Belo Monte Dam project began redirecting water from the Xingu River upstream from the Big Bend, channeling it through a canal to a giant new

---

[1] Reproduced from an article written by the author and others, and originally published in www.TheConversation.com: https://theconversation.com/satellites-over-the-amazon-capture-the-choking-of-the-house-of-god-by-the-belo-monte-dam-they-can-help-find-solutions-too-182012

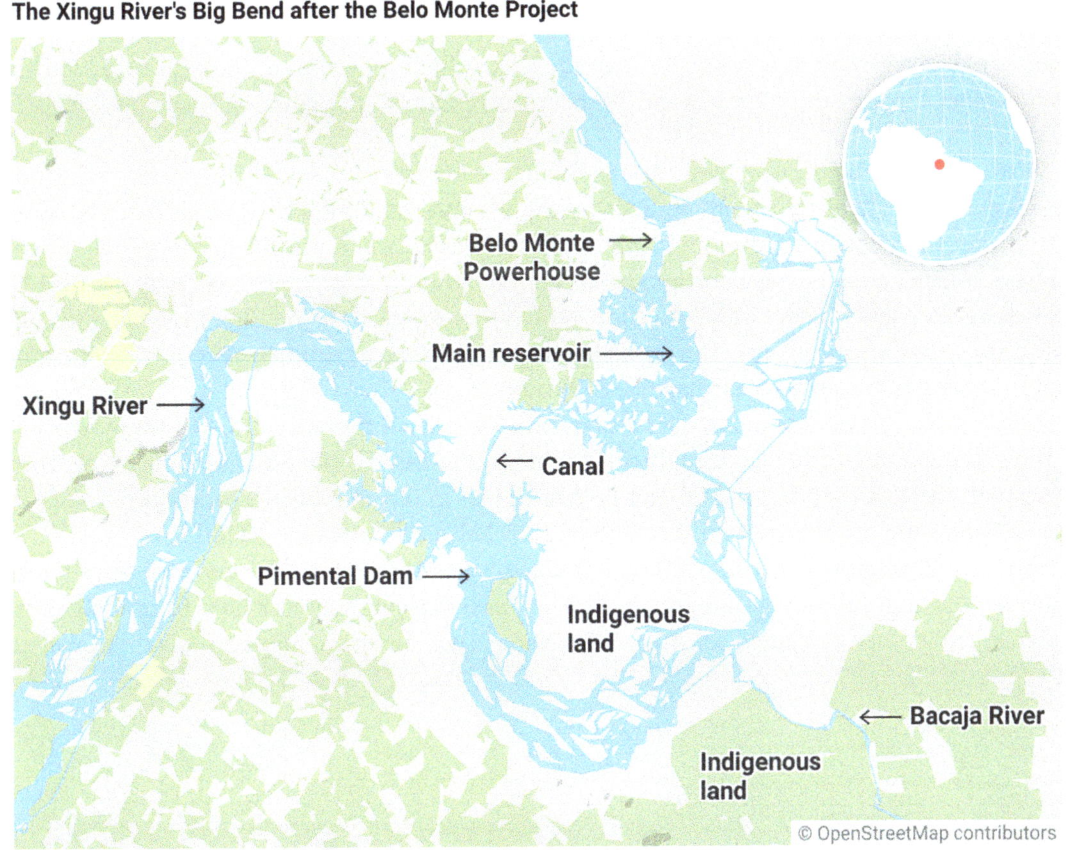

Figure 11.6 Locations of the dams of the Belo Monte hydropower project, showing the drying up of the Big Bend, and the Indigenous communities of the Xingu River. [© Openstreet Maps]

Figure 11.7 Satellite imagery (visible) showing the evolution of the landscape near Xingu River before, during, and after the construction of the Belo Monte Dam on the Xingu River. [Image credit: US Geological Survey]

Case Study 11.1 (cont.)

reservoir. The reservoir now powers one of the largest hydropower dams in the world, designed with enough capacity to power around 20 million households, though it has been producing far less.

Most of the river's flow now bypasses the Big Bend (Figures 11.6 and 11.7), and the Indigenous peoples who live there are watching their livelihoods and way of life become endangered. Some of the most devastating effects are during the rainy season, when wildlife and trees rely heavily on having high water. The consortium of utilities and mining companies that runs the dam has pushed back on government orders to allow more water to reach the Big Bend, claiming it would cut their generation and profits. The group has argued in the past that there was no scientific proof that the change in water flow harmed fish or turtles.

As scientists who work with remote sensing, we believe satellite observations can empower populations around the world who face threats to their resources. The fact that satellite observations of surface water of the Xingu River can be clearly tied to the construction and operation of the Belo Monte Dam offers hope that this kind of knowledge can no longer be hidden.

Satellites have been monitoring changes in Earth's landscapes for 50 years, ever since the US launched the first Landsat satellite on July 23, 1972. By piecing together data from the Landsat program and other satellites, scientists can reconstruct historical patterns of change in the

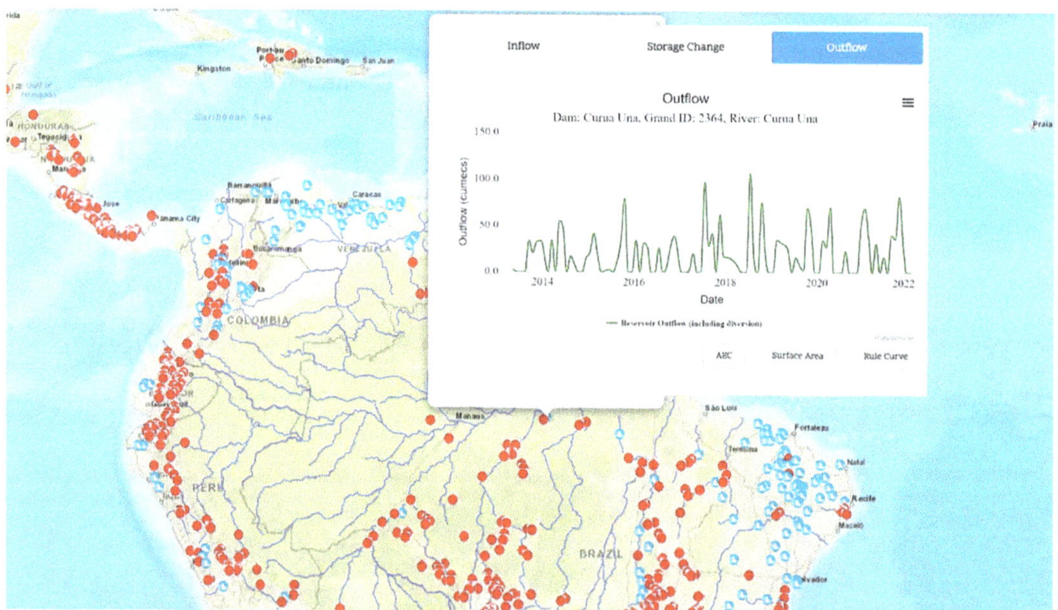

Figure 11.8 Screenshot of the satellite-based Reservoir Assessment Tool (RAT) tracking reservoir operations (as an example; see www.satellitedams.net). Note the RAT tool currently monitors dams built before 2000, but it can be modified to incorporate newer dams, such as Belo Monte on the Xingu River.

landscape and predict current and future trends. They can monitor forest cover, drought, wildfire damage, and desert expansion, as well as river flows and reservoir operations around the world.

An example of how that data can be used to help threatened communities is the global Reservoir Assessment Tool (see Chapter 8, "Reservoir Management from Space"). Figure 11.8 is a screenshot of the tool showing a map of Brazil and an example dam's chart of water outflow (see www.satellitedams.net). This Reservoir Assessment Tool allows communities to track river flow changes caused by nearby dams and understand the potential impact of proposed dams.

Dam operators already collect thorough on-site data about water flow, but their datasets are rarely shared with the public. Remote sensing doesn't face the same restrictions. Making that data public can help hold operators to account, and protect local communities and their rivers.

Satellite Data as a Tool for Social Justice in Water Management of the Xingu River

Satellite monitoring can provide unprecedented insight into the operations of dams like the Belo Monte and their impact on downstream populations. Existing satellite data can be used to monitor recent historical behavior of a dam's operations, track the state of the river and patterns of inflow and outflow at the dam, and even forecast the likely state of the reservoir. Much of that data is easily accessible and free. For example, a tool created for the regional governing body of the Mekong River Commission is empowering communities along the river in Southeast Asia by giving them access to satellite data about water flow at each dam – data that cannot be hidden or modified by those in power (see http://depts.washington.edu/saswe/mekong).

While estimates based on remote sensing have higher uncertainty than on-site measurements, unfettered access to such information can provide local populations with evidence to argue, in court if necessary, for more water releases.

Members of Indigenous groups living in the Big Bend region talk about changes they have seen since the dam was built. Long-term observations of dams and hydroclimate records show that it is possible to revise the standard operating procedures of dams so they allow more water to flow downstream when needed (Eldardiry and Hossain, 2020). A compromise with the Belo Monte Dam could ensure that enough water flows to the Xingu's Big Bend region while also providing hydropower benefits.

By making the impact of the Belo Monte Dam, and others like it, public to the world, agencies and the general public can put pressure on the dam's operators and its investors to release more water. Public pressure will become increasingly important, as water disputes in the Amazon are expected to worsen as the planet warms and deforestation continues. Climate change will affect river flow patterns in the Amazon and likely increase droughts, leaving less water during some periods.

Monitoring dams is a powerful way that satellites can make a difference. Nearly two-thirds of Brazil's electricity comes from more than 200 large and 400-plus small hydropower plants, and more large dams are expected to be built in the Amazon this decade. Many are in areas with Indigenous populations.

Remote sensing may not directly solve the problem of social injustice in water management, but it offers the tools needed to recognize the problems and explore solutions. Being able to monitor changes in near real-time and compare them with historical operations can help maintain the checks and balances required for equitable growth of communities.

Case Study 11.2: Mesopotamian River Revival[2]

Why Restoration of Mesopotamian Rivers Is Important

For several thousand years, the sediment-rich annual flood pulses travelling down the Tigris and Euphrates rivers have sustained the fertile crescent of Mesopotamia that we know today as the home of the earliest human civilization. This is where the world's first irrigated agricultural system was developed. It was perhaps in Mesopotamia that humans began living in organized settlements as they mastered the ability to grow enough food to reduce dependence on hunting and gathering.

In modern times, the annual flood pulse has been systematically chipped away by national water development plans of four nations – Türkiye, Syria, Iran, and Iraq. These plans competed with or reduced the power of the rivers to replenish the soil's fertility and maintain a healthy ecosystem. From the 1970s, dams began to appear in Syria and Iraq on the Tigris and Euphrates rivers to produce hydropower, provide flood control, and ensure a more reliable supply of water for irrigation. More recently, a series of large dams built by Türkiye as part of the Southeastern Anatolia project in the aforementioned rivers upstream have now dramatically diminished the water availability for the downstream nations of Iraq and Syria. In addition, Iran has built dams on Tigris tributaries on the border with Iraq and reduced the tributary flow into the eastern and southeastern parts of Iraq. In many instances, development of such water resources management structures has led to the drying up of the marshland in southern Iraq that is a vital source of livelihood for millions.

It is not an understatement to say that a thriving and healthy river system for the Mesopotamian region means a thriving economy for all, with a productive agriculture and a sustainable manufacturing base that needs reliable water supply and cheap power. When looked at from 30,000 feet above the ground, the Mesopotamian region began experiencing a steady decline in the health of its rivers from the 1980s, with a series of unfortunate man-made events ranging from wars to uncoordinated dam development. While there have been many analyses and historical dissection of the current water situation, one question that we all need to ask now is "How can we restore the Mesopotamian rivers?" This question is a social justice question for water management, given how disproportionately dependent the Iraqi population is on the waters of the Tigris–Euphrates rivers, compared with other nations.

How Do We Start Restoring Mesopotamian Rivers?

At its core, we believe we need to use the latest science and technology with as much historical information and wisdom as is at our disposal to co-develop actionable knowledge, user-ready tools, and more inclusive policy for all Mesopotamian stakeholders. Our quest for an answer to the question of Mesopotamian river restoration needs to begin with seeking an answer to *"How much water does the region have, how much will it have, where and when, and in what quality?"* The answer has fundamental implications for a myriad of critical activities supported by the

---

[2] Reprinted from an article by the author titled "Restoring Mesopotamian rivers for future generations", *Water Resources Research*, 59(5), 2023.

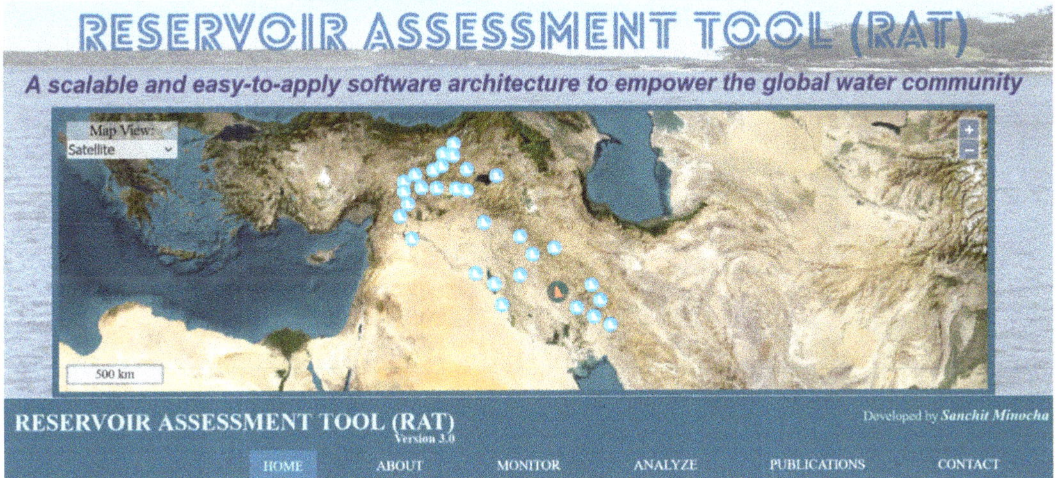

Figure 11.9 Recently developed Reservoir Assessment Tool for the Tigris–Euphrates (RAT-TE) river system (accessible at www.satellitedams.net). Note that the tool is not yet operational and is presented for demonstration purposes only, to showcase what is possible for Mesopotamian river restoration.

Mesopotamian river system, such as biodiversity management, fisheries, irrigation and agriculture, urban water supply, and waste management, among others.

Currently, it is almost impossible to answer the question of water availability using the conventional methods of the twentieth century that rely on ground-based methods and data from gauges. This is due either to the simple fact that countries refuse to share hydrologic data on a common platform for the greater good of the region, or because most areas of the river basin lack a well-maintained and continuously running hydrologic measurement network.

Space can be the answer in this case. Using satellite remote sensing of water from space-based platforms, we can now track most of the water in its various forms (precipitation, surface water, soil moisture). The good news is that we can piece together all the information with physical models and create tools, provided there is some calibration, and can track at a high frequency how the dams are operating in terms of release and hedging of water. Such a granular synopsis can also be created from the early 1980s to present day, to understand how we have altered and choked the Mesopotamian rivers over the past four decades. In fact, such a tool, called the Reservoir Assessment Tool (RAT), was recently developed for the Tigris–Euphrates river system with the explicit goal of triggering a co-developed decision-making system to empower the Mesopotamian river stakeholders (Figure 11.9).

What Can a Satellite-Based Reservoir Monitoring Tool Like RAT Offer for Mesopotamia?

Space-based platforms for reservoir tracking such as RAT-TE (see Figure 11.10) require minimal internet bandwidth as they leverage the advances made in information technology with cloud computing. Furthermore, the satellite data is publicly available from NASA and other space agencies. With such a tool, we can now create a common platform for developing a coherent multi-year

Case Study 11.2 (cont.)

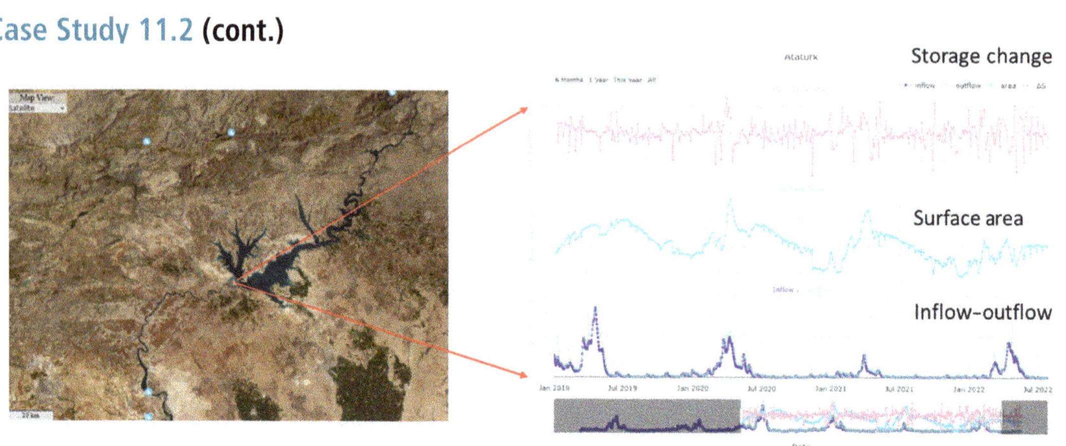

Figure 11.10 An example of what RAT for Tigris–Euphrates can produce in terms of reservoir tracking of a mega dam (Ataturk Dam) in upstream Türkiye for downstream nations such as Iraq (accessible from www.satellitedams.net). Such information on reservoir operations can be compared with downstream dams and used as input for scientific research collaboration, planning, and policy development.

agenda for Mesopotamian river restoration and build a multi-sectoral and multi-disciplinary strategy. That said, RAT-TE might not necessarily be able to address all the issues regarding river revival. However, at its core, such a tool provides a bigger strategic overview to focus on seeking an answer to the fundamental question that has eluded Mesopotamia for many decades – "How much water is or will be available, where and when?" Such an approach using remote sensing and models has already been explored for deriving adaptation plans for Egypt, which faces a similar situation to Mesopotamia (Eldardiry and Hossain, 2020). Almost all processes and decision-making that are important for sustaining a healthy river, such as biodiversity, vegetation health, agricultural productivity, fisheries, and soil health, depend on the answer to the former question.

The vantage of space-based Earth observations within the RAT framework also provides a basin-wide overview for all stakeholders to observe the hydrological and ecological anomalies from the Turkish headwaters to the Persian Gulf outlet, spanning all four countries. Such a synopsis is not afforded by the currently fragmented and opaque in-situ data-sharing scenario on the ground. Using past satellite missions, RAT now makes it possible to create a collection of climate-driven and man-made changes that have plagued the region in terms of water scarcity or excess.

A Vision of Hope

The four countries making up the Mesopotamian rivers need to be in this restoration business together, preferably via some type of cooperative arrangement similar to other multi-national river agencies such as the Mekong River Commission (www.mrcmekong.org). For restoring the Mesopotamian rivers, a diplomatic track will be as essential as the technical one. As noted in past studies reported by Akanda et al. (2007), it is important to consider the historical failures and successes of past mediating efforts to understand how a tool like RAT can best complement diplomatic efforts that are also required for the successful restoration of Mesopotamian rivers.

Second, the water sharing and its management needs to be cast as a win–win for all nations in the region (Akanda et al., 2007). What if Türkiye was allowed to store more of the water in Tigris–Euphrates, given the lower evaporation rates, and in return Iraq received benefits from Türkiye with cheap hydropower? If Türkiye withholds more water than needed by Iraq during drought conditions, can it be convinced to release more in exchange of faster access to the Asian market for Turkish goods via the Shatt Al Arab waterway? In this case, keeping the marshes wet by making sure there is timely release of the flood pulse from the largest dams of eastern Türkiye would be in Türkiye's best economic interest. Investigating such potential win–win scenarios will require complementing SWOT (Surface Water and Ocean Topography) data-driven tools like RAT that can reconstruct and predict reservoir impacts, with water management scenario assessment models such as WEAP (Water Evaluation, Analysis and Planning) or BASINS (Better Assessment Science Integrating Point and Non-point Sources). For example, Chowdhury et al. (2022) have reported leveraging RAT outputs to explore the development of low-carbon electricity supply systems for Southern Africa.

Case Study 11.3: Using Remote Sensing of River Temperature and Reservoirs for Tribal Fishing Rights in Columbia River, USA

The Columbia River is an extensively dammed river system, so dammed that the reservoir and river regulation has negatively affected fish abundance (notably salmon which swims from the ocean to spawn). Consequently, the systematically increasing regulation and management of the Columbia River have resulted in a gradual decline in livelihood for Indigenous communities who have depended on fishing for thousands of years. In 1974, the famous Boldt Decision was issued to protect tribal fishing rights and to legally enforce reservoir operators to do their fair share to mitigate the negative impacts of reservoir operations on fish abundance (for more details on the Boldt Decision, see www.historylink.org/file/21084). It is estimated that, today, the major water management agencies spend about 300 million USD per year just on restoring fish abundance in the Columbia River system that has been negatively impacted by the man-made regulation by dams.

Since 2022, the author and his research group have been engaging with the Columbia River Inter-tribal Fish Commission (CRITFC) (https://critfc.org/) to co-build satellite-based water management tools to empower the tribal government. Recall from Chapter 10 that we can estimate river and reservoir temperature using satellite remote sensing in the thermal infrared wavelengths. If we can combine this capability with satellite-based reservoir tracking that we covered in Chapter 8, then we should be able to see how the surface water temperatures of the Columbia River system are co-varying with reservoir storage change, inflow, and outflow (releases). Because river temperature directly affects fish abundance, among other ecosystem functions, and is often affected by reservoir operations (as seen in Chapter 10), the tracking of this co-variation of river surface temperature with reservoir dynamics can empower CRITFC to better understand where the stress

Case Study 11.3 (cont.)

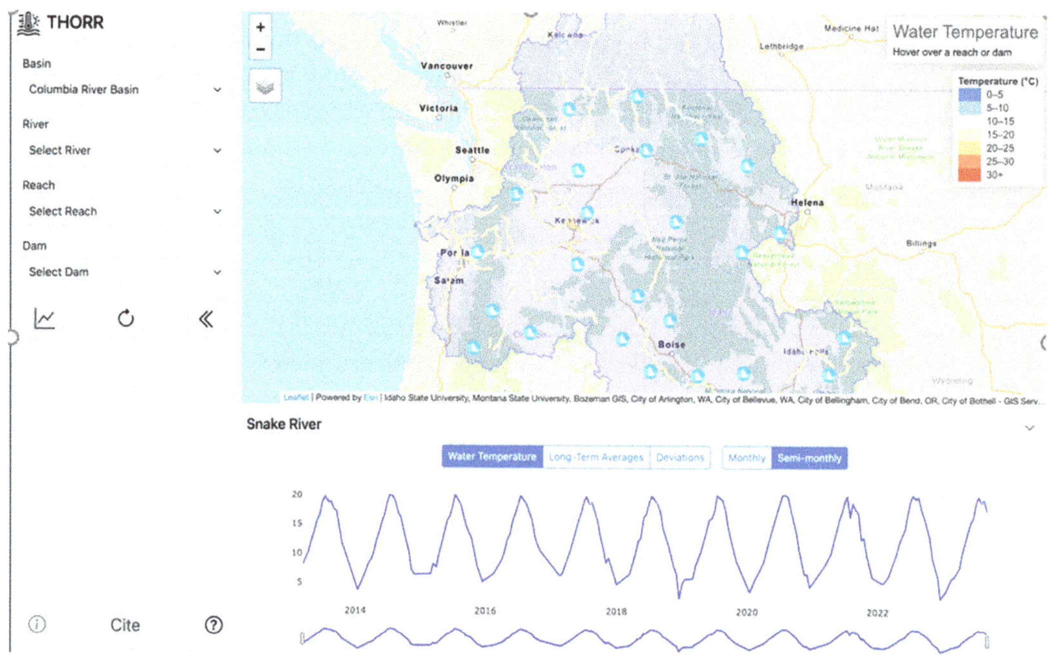

Figure 11.11 Screenshot of the satellite-based Columbia River temperature and reservoir tracking portal called THORR – Thermal History of Regulated Rivers – that has been co-built for the tribal government agency of CRITFC, on the 50th anniversary of the Boldt Decision. Available at https://depts.washington.edu/saswe/thorr

points are for their livelihood. Consequently, CRITFC can be in a better position to engage with reservoir management authorities to explore mitigation action plans. This is not as trivial as it sounds, and readers can refer to Darkwah et al. (2023) for details on how we can combine multi-decadal remote sensing data, machine learning, and ground information to build an operational system for CRITFC which we call THORR – Thermal History of Regulated Rivers. One can see the online portal of THORR which CRITFC can now use to explore regions of abnormal stress for fish here: https://depts.washington.edu/saswe/thorr (Figure 11.11 shows a screenshot).

There is now a case to be made that although the vantage of space to monitor the currently unmonitored vast swaths of tribal fishing spots cannot make the need to spend 300 million USD per year go away, it can certainly shave off a few million USD in annual operations and mainte-nance for mitigation efforts. And with those few millions saved, the entire system could be owned, maintained, and operated from the permanent budget line of reservoir management agencies who have a vested interest in seeing everyone thrive, particularly the indigenous communities, while maintaining a healthy eco-system. Only time will tell how effective THORR becomes in empow-ering a tribal government agency such as CRITFC.

11.6 Conclusion

In this chapter, we have reviewed the topic of social justice, its basic definition, and what it means in water management. We provided two examples of social justice to help us distinguish between equity and equality, and the importance of addressing many underlying factors for users affected by water management challenges. For example, in the case of the Ogallala aquifer, the decline in the groundwater table does not affect all farmers equally. Pumping from a deeper level requires more fuel, and those who have lower income are naturally more negatively affected. Therefore any water management solution to address the issue, perhaps with surface water supplementation, should prioritize the inherent capacity and resources of farmers to withstand rising costs for fuel. Similarly, for wildfires, we have seen from recent research that rural communities that already have lower income and lower access to internet are disproportionately affected. Within that group, the native American populations are found to be most vulnerable. We also showed two case studies illustrating potential ways to mitigate the social justice challenges related to water in the Amazon and in the parched Mesopotamian region. These are examples to help us see the potential of satellite remote sensing as a tool for empowerment when it comes to equitable management and sharing of water resources.

In the next chapter, we will tackle another nontechnical issue, called Citizen Science, for monitoring of surface water resources.

REFERENCES

Akanda, A. S. (2007). The Tigris–Euphrates River basin: mediating a path towards regional water stability. *Al Nakhlah, The Fletcher School Journal for issues related to Southwest Asia and Islamic Civilization*, vol. 31, 1–12 (available at: https://ciaotest.cc.columbia.edu/olj/aln/aln_spring07/aln_spring07g.pdf, last accessed March 28, 2023).

Chowdhury, A. F. M. K., R. Deshmukh, G. Wu, et al. (2022). Enabling a low-carbon electricity system for Southern Africa. *Joule*, vol. 6, 1826–1824. https://doi.org/10.1016/j.joule.2022.06.030

Darkwah, G., K. Hossain, F., Tchervenski, V., Holtgrieve, G., Graves, D., Seaton, C., et al. (2024). Reconstruction of the hydro-thermal behavior of regulated river networks of the Columbia River Basin using satellite remote sensing and data-driven techniques. Earth's Future, 12, e2024EF004815. https://doi.org/10.1029/2024EF004815

Eldardiry, H. and F. Hossain (2020). A blueprint for adapting High Aswan Dam operation in Egypt to challenges of filling and operation of the Grand Ethiopian Renaissance Dam. *Journal of Hydrology*, vol. 598. https://doi.org/10.1016/j.jhydrol.2020.125708

Hossain, F. (2023). Restoring Mesopotamian rivers for future generations. *Water Resources Research*, 59, e2023WR034514.

Masri, S., E. Scaduto, Y. Jin, and J. Wu (2021). Disproportionate impacts of wildfires among elderly and low-income communities in California from 2000–2020. *International Journal of Environmental Research and Public Health*, vol. 18, 3921. https://doi.org/10.3390/ijerph18083921

United Nations (2006). *Social Justice in an Open World: The Role of the United Nations*. United Nations (available at www.un.org/esa/socdev/documents/ifsd/SocialJustice.pdf)

SUGGESTED READING

Bennett, M. M., J. K. Chen, L. F. Alvarez León, and C. J. Gleason (2022). The politics of pixels: a review and agenda for critical remote sensing. *Progress in Human Geography*, vol. 46, 729–752. https://doi.org/10.1177/03091325221074691

Weigand, M., M. Wurm, S. Dech, and H. Taubenböck (2019). Remote sensing in environmental justice research – a review. *ISPRS International Journal of Geo-Information* 8, 20. https://doi.org/10.3390/ijgi8010020

12   Citizen Science and Water Management

12.1 Chapter Overview

This chapter will explore the topic of citizen science in the context of water management using satellite remote sensing. This is a broad field, and the goal here is to expose readers to an important social issue of using citizens to carry out science for building more robust management solutions. As mentioned in Chapter 11, this chapter is in no way comprehensive. The objective here is to encourage readers to start thinking about the idea of citizen science and the positive role it can play in building more equitable satellite-based water management solutions.

12.2 Taking Stock in the Final Chapter

By now, we have covered several topics on the application of satellite remote sensing for water management. First we overviewed the why, where (Chapter 1), and later (Chapters 2 and 3) the how (the fundamentals of remote sensing). In Chapter 4, we covered cloud computing, which is here to stay, to help water managers deal with datasets such as those from satellite remote sensing related to water. Today's modern water managers around the world need such data literacy skill to reap the full benefits of the vantage of space and innovate new types of water management decision making that were not possible in the past. We also covered precipitation estimation in Chapter 5 and surface water in Chapters 6 and 7 (water extent, storage change, and river discharge). We dedicated a full chapter to piecing all this together to monitor a reservoir's dynamic state in terms of inflow, outflow, storage change, and losses. Being able to monitor reservoir operations almost anywhere for medium to large dams from the mid-1980s and also having an ability to forecast impact is quite an exciting capability for water management today. We therefore introduced our readers to a freely available framework for such satellite-based reservoir monitoring, the Reservoir Assessment Tool (RAT). In Chapter 9, we saw how satellite data can be used for crop area monitoring and crop water management. In Chapter 10, we introduced readers to the satellite remote sensing of temperature, and in particular water surface temperature. Temperature is an important water quality parameter controlling many biological and biochemical processes that water management needs to consider. At a minimum, it is a proxy for aquatic habitat or for unusual thermal anomalies of managed water caused by anthropogenic activities such as dam operations and irrigation. Water managers of today need to find ways to use such water quality information with water quantity, both remotely sensed from space, to improve decision making that respects the ecosystem.

In the previous chapter (Chapter 11), we introduced ourselves to the topic of social justice in water management. So, can we claim we have come full circle now on the topic of satellite remote sensing for water management? There are still a few important water targets we had to leave out, such as soil moisture, groundwater, and snow. Hopefully, these will be covered in a future edition; or readers can find relevant books to supplement the knowledge gained here (such as Seidel and Martinec, 2004; Adams et al., 2022). Overall, we have covered enough to get the water managers and water practitioners started on the path to leveraging the full power of satellite remote sensing.

12.3 Citizen Science

In this chapter, we will close the book with an overview of a topic that is gaining ground in the field of satellite remote sensing. That topic is "citizen science", using the power of connectivity to the information highway that most people (citizens) have, in order to be a citizen scientist for satellite remote sensing. Because of the connected world we all share in information space, each one of us has the potential to be a "gauge" or a "sensor" to generate additional data on the target that is being remotely sensed. Consequently, such data, if harvested well, can be applied to improve satellite remote sensing products for water management in a very cost-effective manner.

12.3.1 Understanding Citizen Science Through the Lens of Surface Water Tracking

The active participation of the public in the design of research, data collection, and improvement of understanding of processes together with scientists is often called citizen science. Citizen science has existed for a long time; however, the developments in satellite remote sensing technology have created a wide range of new opportunities for public participation in scientific research. Here, we will try to understand citizen science and the role that citizens can play in improving satellite-based monitoring of surface water storage change (which we covered in Chapter 6).

12.4 Lake Observations by Citizen Scientists and Satellites (LOCSS)[1]

The LOCSS Project, as the name implies, is designed to combine measurements of water level by citizen scientists with satellite observations to address a range of science questions and to provide useful information to communities near lakes. The reader can check www.locss.org for more information. The principle behind LOCSS is quite simple. Citizen scientists representing the general public from around the world have partnered with LOCSS to install a simple staff gauge (ruler) in a lake (Figure 12.1). These citizen scientists then observe the water level on the gauge and transmit that observation back to the LOCSS team. Transmission method depends on the particular location. The water level is then archived and made available to the public as rapidly as possible on the LOCSS Project website (https://locss.org).

---

[1] Credit to www.locss.org, LOCSS Director Dr. Tamlin Pavelsky, and NASA Earth Systems Program.

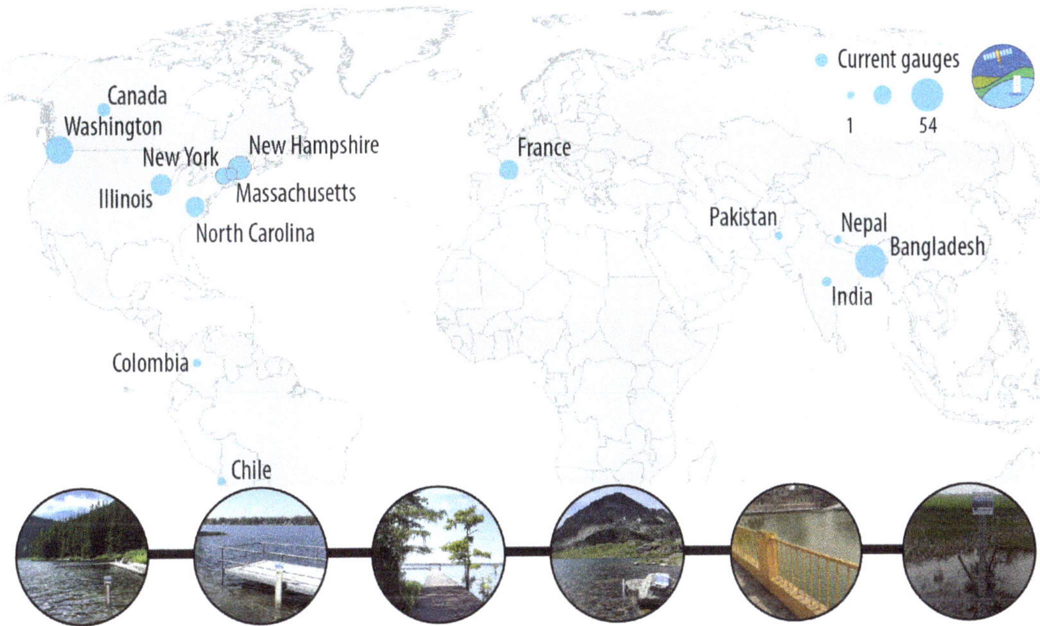

Figure 12.1 Global outreach of citizen science project called LOCSS – Lake Observations by Citizen Scientists and Satellites (www.locss.org). [Image credit: Megan Lane, University of North Carolina]

The LOCSS Project began in April 2017 with the installation of gauges in several lakes in eastern North Carolina as a pilot project. It was found that citizen scientists generally provided very accurate data, with a mean absolute error of 1.6 cm relative to an automated sensor (Little et al., 2021), and thus did not generally require training prior to making measurements. Currently, there are more than 160 installed gauges across seven countries. As of 2023, citizen scientists have contributed more than 38,000 water level readings to the LOCSS Project. These citizen-contributed water levels have been used to validate satellite altimeter measurements of water level (Ahmad et al., 2020) and estimate storage change of lakes around the world using satellite data on lake surface area (Khan et al., 2023). Recall from Chapter 6 on surface water that the estimation of storage change of a water body requires observations of surface area and water elevation.

By harvesting in-situ data on lake levels practically for free via citizen engagement through the presence of general public who are connected to the internet, the LOCSS Project is an example of how satellite remote sensing of water levels from altimeters, satellite data on water extent, and their use in storage change estimation can be improved (Figure 12.2). For example, Khan et al. (2023) studied variations in the volume of water stored in small lakes and wetlands using satellite remote sensing and data on lake water height contributed by citizen scientists from the LOCSS Project. A total of 94 water bodies across the globe were studied using satellite data in the optical and microwave wavelengths from Landsat 8, Sentinel-2, and Sentinel-1. The uncertainty in volume estimation as a function of geography and geophysical factors such as cloud cover, precipitation, and water surface temperature was studied. The key finding that emerged from this global citizen science study is that uncertainty in estimation of volume change with satellite data is highest in regions with a distinct precipitation season, such as in

Figure 12.2 An example of a citizen-science-based lake level gauge in Bangladesh, for the LOCSS Project. [Image credit: Faisal Hossain; location: Ram Sagar, Dinajpur, Bangladesh]

monsoon-dominated South Asia or the Pacific Northwest region of the United States. This uncertainty is further compounded when small lakes and wetlands are seasonal, with alternating land use as a water body and agricultural land, such as the wetlands of northeastern Bangladesh. By complementing satellite remote sensing of surface water extent with ground-based data on water levels from citizen scientists, we can now build a better understanding of surface water availability. This point will be further elaborated using a real-world case study to help readers understand the value of citizen science in satellite-based water management.

Case Study 12.1: Citizen Scientists and NASA Satellites Join Forces to Track Water Supplies for Bangladesh[2]

Bangladesh is a small country basin with a huge amount of water that moves over its surface. The Ganges, Brahmaputra, and Meghna (GBM) rivers form the world's largest river delta as they flow south to the Bay of Bengal and ultimately into the Indian Ocean (Figure 12.3). Bangladesh also has seasonal monsoon rains that create ecologically and agriculturally important wetlands

[2] Source: NASA Applied Science Program.

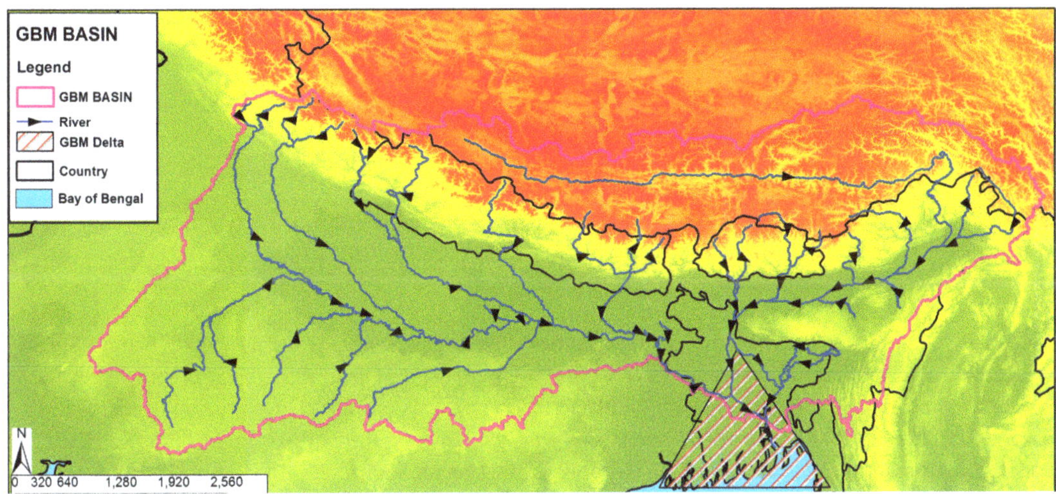

Figure 12.3 Bangladesh, as a delta in the Ganges, Brahmaputra, and Meghna river basins, receives surface water from a drainage area far larger than its size.

known as haors. Just how much water fills these vast floodplains? That has been a tricky question, and one that was only recently possible to answer thanks to citizen science, LOCSS, and satellite remote sensing.

Through the LOCSS Project, combined data from satellites on surface water extent and water level observations from citizen scientists on the ground has now been used to quantify the total volume of water stored in Bangladesh's large wetland complexes for the first time.

This information has proved useful to the government water agency of Bangladesh (the Bangladesh Water Development Board, BWDB), which is navigating new challenges caused by climate change. Knowing how much water is stored on the country's surface can help the government make critical water management and food security decisions. This information on how much surface water is stored in terms of volume is coming just in time, because sea level rise, cyclones, the erosion of coastlines and riverbanks, and other climate-related disasters threaten Bangladesh's population of more than 160 million. As part of an adaptation strategy called the Delta Plan 2100, the government found that improving its water storage capability in the dry season would help make it more resilient. Thanks to this work involving citizen science, the BWDB now knows that, on average, the northeastern flood-prone region stores 30 cubic kilometers of water (Figure 12.4), about the same amount of water as is in Lake Mead at full capacity. This amount changes owing to weather conditions.

Using the same methodology, it was also possible to estimate that the 2022 historic floods that occurred in the northeastern region of Bangladesh brought that number up to 42 cubic kilometers.

Citizen science projects like this can also be meaningful for the recently launched NASA SWOT mission, which is expected to be a vanguard for tracking the world's surface water. As mentioned in Chapter 6, the SWOT mission is the first satellite to provide high-resolution surface water

Case Study 12.1 (cont.)

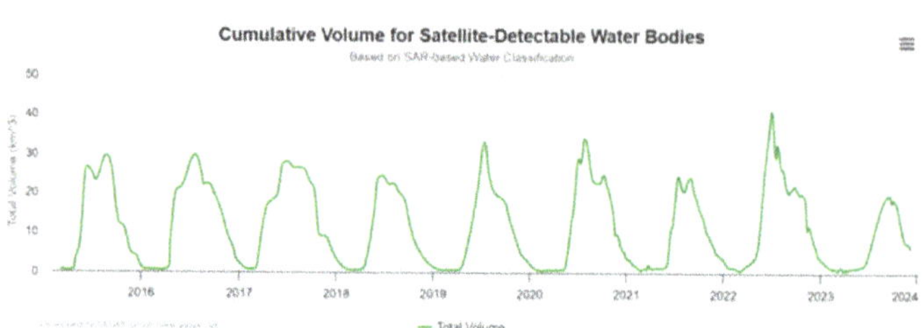

Figure 12.4 The surface water monitoring system based on citizen science gauging of lakes, from LOCSS Project and satellite data. This screenshot from http://depts.washington.edu/saswe/haors shows the time series of total surface water (called "haors") stored in the northeastern Bangladesh region from 2016 to present. The unusually wet year of 2022 and drier year of 2023 are quite distinguishable in terms of the volumetric storage that was derived using citizen science together with satellite data.

elevation maps, allowing scientists to monitor lakes, rivers, and oceans with unprecedented accuracy. The SWOT mission does not, however, erase the need for citizen scientists; in fact, networks of observers will be crucial for ground-truthing the new satellite.

The ability to track volumetric change for surface water using satellites, citizen scientists, and in-situ gauges in a cost-effective manner has also prompted BWDB and the Bangladesh Government to expand the lake gauging network country-wide to build a more "national" monitoring capability for tracking surface water volume change. Such an estimate of surface water volume change is desired because of the potential to drive policy planning on water security at the national level. For example, based on discussions spurred by the tool for the northeastern region in 2020, the BWDB pitched an innovative idea to the Water Ministry of Bangladesh Government on using a small fraction of this "excess" flooded freshwater (about 1 km^3) in an ecologically safe manner to build a drinking water bottling plant. The motivation here is to address drinking water scarcity in Dhaka city, a megacity of more than 20 million, and generate revenue for BWDB to carry out more water research and development.

Lessons Learned in Establishing LOCSS Citizen Science Surface Water Gauging Network in Developing Countries

For the water manager and practitioner eager to explore how best to leverage citizen scientist power of connectivity in the application of satellite data, it is worth keeping in mind that one size does not fit all when it comes to participation by the public and the establishment of gauges for citizen scientists. Because satellite remote sensing is often most needed in ungauged regions of the developing world, we share below some experiences we have gathered in setting up the LOCSS network.

South Asia, as a region comprising Pakistan, India, Bangladesh, and Nepal, is not as diverse in governance and underlying challenges for fieldwork as it is in geography and climate. Throughout this region, adhering to government bureaucracy appears a requirement for any fieldwork. Yet, within this pervasive bureaucratic environment, some nuances emerged during the implementation of the LOCSS network, as we elaborate below.

The first key lesson we learned when implementing the LOCSS program in South Asia is that citizen science programs are effective only if there already exists prior institutional trust with a local partnering agency that understands the governance and bureaucratic structure. These partners were found to be essential in helping our LOCSS team navigate the challenges of getting gauges set up and become operational at selected lake sites. Unlike in developed nations, citizen science is still a relatively new concept in South Asia, and the idea of the general public contributing to science by acting as scientific observers is quite new. In general, each country needed local engagement and support from the partnering agencies for LOCSS to become successful. Bangladesh and Pakistan, being administratively quite similar, required similar levels of strong government support to initiate the LOCSS program. India, on the other hand, being a country where water is a state compact, required a more "provincial" approach. Nepal, with its limited technological and infrastructural resources and a vastly remote terrain, was found to be one where complete government agency support was essential for initiating LOCSS cost-effectively.

12.5 Conclusion

Early in Chapter 1 we covered at length why satellite remote sensing for water management is critical in this century for managing our water resources. Our world faces many challenges when it comes to measurement and sharing of data on water using traditional and in-situ networks. The vantage of space using satellites and the increasingly easy access to data via advanced information technology are becoming a mainstay in addressing new challenges to water management. We have already seen many examples of how satellite remote sensing of water and its application can improve decision making in water management. However, no matter how advanced our remote sensing techniques are, they are always indirect and based on electromagnetic energy interactions, as we have learned in Chapters 2 and 3. So, what does citizen science have to offer to keep satellite remote sensing honest and impactful as a continuously improving field? As we close this chapter and this book, we believe that citizen science will be and should be an integral part of satellite remote sensing for water management, because of the highly connected world we now live in. By engaging with the general public near our water targets (lakes, reservoirs, rivers, precipitation, soil moisture) we can relatively easily generate a plethora of quasi in-situ data for comparison and validation against satellite data. Given the right infrastructure, we can "harvest" such data for water by taking advantage of the millions of cellphones with internet connectivity, as is already done in other fields by tech companies, for example in real-time traffic monitoring. By building the right synergy between citizen scientists monitoring/gauging our water targets and the satellites estimating those same targets, water managers can build formidable management systems more cost-effectively and accurately for improved stewardship of our precious water resources.

REFERENCES

Adams, K. H., J. T. Reager, P. Rosen, et al. (2022). Remote sensing of groundwater: current capabilities and future directions. *Water Resources Research*, vol. 58, e2022WR032219. https://doi.org/10.1029/2022WR032219

Ahmad, S., F. Hossain, T. Pavelsky, et al. (2020). Understanding volumetric water storage in monsoonal wetlands of northeastern Bangladesh. *Water Resources Research*, vol. 56, e2020WR027989. https://doi.org/10.1029/2020WR027989

Khan, S., F. Hossain, T. Pavelsky, and G. M. Parkins (2023). Understanding volume estimation uncertainty of lakes and wetlands using satellites and citizen science. *IEEE Journal of Selected Topics in Applied Earth Observations and Remote Sensing*, vol. 16, 2386–2401. https://doi:10.1109/JSTARS.2023.3250354

Little, S. B., T. M. Pavelsky, F. Hossain, et al. (2021). Monitoring variations in lake water storage with satellite imagery and citizen science. *Water*, vol. 13, 949. https://doi.org/10.3390/w13070949.

Seidel, K. and J. Martinec (2004). *Remote Sensing in Snow Hydrology: Runoff Modelling, Effect of Climate Change*. Springer.

SUGGESTED READING

Buytaert, W., Z. Zulkafli, S. Grainger, et al. (2014). Citizen science in hydrology and water resources: opportunities for knowledge generation, ecosystem service management, and sustainable development. *Frontiers in Earth Science* 2, 26. https://doi.org/10.3389/feart.2014.00026

Strasser, B. J., J. Baudry, D. Mahr, G. Sanchez, and E. Tancoigne (2019). "Citizen Science"? Rethinking science and public participation. *Science & Technology Studies*, vol. 32, 52–76. https://doi.org/10.23987/sts.60425

Index